Selected Titles in This Series

(*Continued in the back of this publication*)

Dynamical, Spectral, and Arithmetic Zeta Functions

CONTEMPORARY MATHEMATICS

290

Dynamical, Spectral, and Arithmetic Zeta Functions

AMS Special Session on
Dynamical, Spectral, and Arithmetic Zeta Functions
January 15–16, 1999
San Antonio, Texas

Michel L. Lapidus
Machiel van Frankenhuysen
Editors

American Mathematical Society
Providence, Rhode Island

This volume contains the proceedings of the AMS Special Session on Dynamical, Spectral, and Arithmetic Zeta Functions held at the Annual Meeting of the American Mathematical Society in San Antonio, Texas on January 15–16, 1999.

2000 *Mathematics Subject Classification.* Primary 11F67, 11Mxx, 11Y35, 11N05, 28A80, 30F40, 37Axx, 58J35.

Library of Congress Cataloging-in-Publication Data
AMS Special Session on Dynamical, Spectral, and Arithmetic Zeta Functions (1999: San Antonio, Texas)
    Dynamical, spectral, and arithmetic zeta functions: AMS Special Session on Dynamical, Spectral, and Arithmetic Zeta Functions, January 15–16, 1999, San Antonio, Texas / Michel L. Lapidus, Machiel van Frankenhuysen, editors.
    p. cm. — (Contemporary mathematics, ISSN 0271-4132 ; 290)
    Includes bibliographical references.
    ISBN 0-8218-2079-6 (alk. paper)
    1. Functions, Zeta—Congresses.   I. Lapidus, Michel L. (Michel Laurent), 1956–   II. van Frankenhuysen, Machiel, 1967–   III. Title.   IV. Contemporary mathematics (American Mathematical Society) ; v. 290.

QA351.A73   1999
515′.56—dc21
                                                                          2001053944

Contents

Preface

These are the proceedings of the Special Session on 'Dynamical, Spectral and Arithmetic Zeta Functions', held at the Annual Meeting of the American Mathematical Society in San Antonio, Texas, January 13–16, 1999.

The purpose of the session was to bring together leading researchers working in these fields, thereby helping to find links, analogies and explore new methods. The volume consists of eleven papers, seven of which were presented at the meeting. The subjects of the papers range from dynamical systems or spectral geometry on hyperbolic manifolds (Chang-Mayer, Fan-Jorgenson) and dynamical zeta functions (Fel'shtyn, Lapidus-van Frankenhuysen, Stark-Terras), to trace formulas in geometry (Deninger-Singhof, Perry) and in arithmetic (Haran, Soulé) as well as to computational work on the Riemann zeta function (Galway, Odlyzko).

We hope that the reader will benefit from the different perspectives of the various authors.

July, 2001
Michel L. Lapidus
Machiel van Frankenhuysen

Alphabetical list of contributors with affiliation

Cheng-Hung Chang
Institut für Theoretische Physik
Technische Universität Clausthal
Arnold-Sommerfeld-Straße 6
38678 Clausthal-Zellerfeld, Germany
ch.chang@tu-clausthal.de

Carol E. Fan
Department of Mathematics MC-8130
Loyola Marymount University
Los Angeles, CA 90045-8130
cfan@lmumail.lmu.edu

Christopher Deninger
Mathematisches Institut
Westfälische Wilhelms-Universität
Einsteinstraße 62
48149 Münster, Germany
deninge@math.uni-muenster.de

Alexander Fel'shtyn
Institut für Mathematik
Ernst-Moritz-Arndt-Universität
Jahn-Straße 15a
D-17489 Greifswald, Germany
felshtyn@mail.uni-greifswald.de

William F. Galway
Department of Mathematics
University of Illinois
1409 W. Green Street
Urbana, IL 61801
galway@math.uiuc.edu

Jay Jorgenson
Department of Mathematics
City College of New York
New York, NY 10031
jjorgen@math0.sci.ccny.cuny.edu

Dieter H. Mayer
Institut für Theoretische Physik
Technische Universität Clausthal
Arnold-Sommerfeld-Straße 6
38678 Clausthal-Zellerfeld, Germany
dieter.mayer@tu-clausthal.de

Peter Perry
Department of Mathematics
University of Kentucky
Lexington, Kentucky 40506–0027
perry@ms.uky.edu

Christophe Soulé
Institut des Hautes Études Scientifiques
35, Route de Chartres
91440, Bures-sur-Yvette, France
soule@ihes.fr

Audrey A. Terras
Department of Mathematics
9500 Gilman Drive
University of California, San Diego
La Jolla, CA 92093-0112
aterras@ucsd.edu

Shai Haran
Department of Mathematics
Technion
Haifa 32000, Israel
haran@tx.technion.ac.il

Michel L. Lapidus
Department of Mathematics
University of California
Riverside, CA 92521-0135
lapidus@math.ucr.edu

Andrew M. Odlyzko
AT&T Labs - Research
Florham Park, NJ 07932
amo@research.att.com

Wilhelm Singhof
Mathematisches Institut
Universität Düsseldorf
Universitätsstraße 1
40225 Düsseldorf, Germany
singhof@cs.uni-duesseldorf.de

Harold M. Stark
Department of Mathematics
9500 Gilman Drive
University of California, San Diego
La Jolla, CA 92093-0112
stark@math.ucsd.edu

Machiel van Frankenhuysen
Rutgers University
Department of Mathematics
110 Frelinghuysen Road
Piscataway, NJ 08854-8019
machiel@math.rutgers.edu

Contemporary Mathematics
Volume **290**, 2001

Eigenfunctions of the transfer operators and the period functions for modular groups

Cheng-Hung Chang and Dieter H. Mayer

ABSTRACT. We extend the transfer operator approach to the period functions of Lewis and Zagier for the group $PSL(2, \mathbb{Z})$ [**LZ97**] to general subgroups of $PSL(2, \mathbb{Z})$ with finite index. Thereby we derive functional equations for the eigenfunctions of the transfer operators with eigenvalue $\lambda = 1$ which generalize the one derived by J. Lewis for $PSL(2, \mathbb{Z})$. For special congruence subgroups we find that Lehner and Atkins theory of old and new forms [**AL70**] is realized also on the level of the eigenfunctions of the transfer operators for these groups. It turns out that the old eigenfunctions can be related through Lewis' transform for $PSL(2, \mathbb{Z})$ to the old forms for these subgroups, whereas a similar relation for new eigenfunctions and the corresponding new forms is not yet known.

1. Introduction

In the Eichler-Manin-Shimura theory of periods [**Eic57**] to every holomorphic modular cusp form one can associate a period polynomial with certain cocycle properties under the action of the group elements. This theory was extended to general holomorphic forms for $PSL(2, \mathbb{Z})$ by Zagier in [**Zag91**]. Regularizing the corresponding integrals of Eichler in the case of the holomorphic Eisenstein series for $PSL(2, \mathbb{Z})$ leads to certain rational functions. Recently, J. Lewis found a further extension of this theory in which to every Maaß wave form for $PSL(2, \mathbb{Z})$ there is related a certain holomorphic function in the cut z-plane which fulfills a functional equation depending on a complex parameter β. This equation combines the two cocycle relations for the polynomials of the two generators of $PSL(2, \mathbb{Z})$ [**Lew97**], [**LZ97**], [**LZ**]. The polynomial solutions of this equation are just the period polynomials and its rational solutions are the rational functions of Zagier. Zagier found another discrete series of solutions of the Lewis-Zagier functional equation which are transforms of the nonholomorphic Eisenstein series of $PSL(2, \mathbb{Z})$ for parameter values β corresponding to the resonances of the Laplacian on the modular surface [**CM99**] given by the nontrivial zeros of Riemann's zeta function $\zeta_R(2\beta)$.

1991 *Mathematics Subject Classification.* (MSC **2000**) Primary 37C30, 81Q50, 11F67; Secondary 11F66, 37D40, 11M36.

This work is supported by DFG Schwerpunktprogramm 'Ergodentheorie, Analysis und effiziente Simulation dynamischer Systeme'.

It was found [**CM99**], [**LZ97**] that all the above mentioned solutions of the Lewis functional equation are eigenfunctions to the eigenvalues $\lambda = +1$ or $\lambda = -1$ of the so called transfer operator $\mathcal{L}_\beta$ of the geodesic flow on the modular surface, that means, they are closely related to the zeros of the function $Z_S(\beta) = \det(1 - \mathcal{L}_\beta)\det(1 + \mathcal{L}_\beta)$. This function on the other hand is nothing else than Selberg's zeta function for the group $PSL(2, \mathbb{Z})$ [**CM99**], [**May91b**]. Selberg's zeta function relates in a highly non trivial way classical mechanics to quantum mechanics: the geodesic flow describes a classical point particle moving freely on the corresponding surface with constant velocity, the spectrum of the Laplace-Beltrami operator on the other hand describes the same free particle in quantum mechanics. This relation between classical and quantum mechanics is the central problem within the theory of quantum chaos in physics. Selberg's theory, based on his trace formula, relates the energy spectrum of the free quantum particle to the length spectrum of the classical particle. The new approach through the transfer operator on the other hand gives also a connection between the bound and scattering states of the quantum system and the eigenfunctions of the transfer operator which obviously themselves are purely classical objects defined by the geodesic flow and its dynamical properties. In this sense in this special case the classical system completely determines the quantum system.

In a recent paper [**CM00**] we showed how the transfer operator approach to the geodesic flow can be extended to subgroups Γ of the modular group $PSL(2, \mathbb{Z})$ of finite index. It was shown that for such groups Selberg's zeta function can be expressed, as in the case $PSL(2, \mathbb{Z})$, as the Fredholm determinant of the corresponding transfer operator $\tilde{\mathcal{L}}_\beta$ such that its zeros and poles are related to those β values where 1 is an eigenvalue of $\tilde{\mathcal{L}}_\beta$. It was found that the operator $\tilde{\mathcal{L}}_\beta$ can be interpreted also as the transfer operator of $PSL(2, \mathbb{Z})$ together with the representation of this group induced from the trivial representation of the subgroup Γ. This gives a new proof of the well known fact that Selberg's zeta function for $\Gamma \subseteq PSL(2, \mathbb{Z})$ is identical to the generalized zeta function for $PSL(2, \mathbb{Z})$ with the above mentioned induced representation.

In the present paper we continue our investigations of the transfer operators for subgroups of the group $PSL(2, \mathbb{Z})$. In chapter 2 we briefly recall the explicit form of the transfer operator $\tilde{\mathcal{L}}_\beta$ for an arbitrary subgroup $\Gamma \subseteq PSL(2, \mathbb{Z})$ of finite index, the theory of Atkin and Lehner of old and new forms for congruence subgroups and Eichler's definition of period polynomials for cusp forms. In chapter 3 we derive, starting from the transfer operator for Γ, a functional equation for its eigenfunctions and discuss which solutions of this equation define eigenfunctions of the transfer operator with eigenvalue 1 for the special groups $\Gamma(1)$, Γ_2, $\Gamma_0(2)$, $\Gamma^0(2)$, Γ_ϑ and $\Gamma(2)$. It turns out that in these cases the eigenfunctions of the transfer operator fall into two classes, old ones and new ones, reflecting in a nice way the theory of old and new forms for these groups: The old eigenfunctions of $\tilde{\mathcal{L}}_\beta$ are related to the old forms of the subgroup Γ through the transformations of Lewis for $PSL(2, \mathbb{Z})$. A similar relation between the new eigenfunctions and the new forms however is not yet known. Also the general relation between the polynomial eigenfunctions and Eichler's period polynomials for general subgroups Γ is not yet clear.

2. The transfer operators for modular groups

2.1. Modular groups and modular surfaces. The hyperbolic plane $\mathbb{H}$ is the complex upper half plane $\mathbb{H} = \{x + iy \,|\, x, y \in \mathbb{R}, y > 0\}$ equipped with the Poincaré metric $d s^2 = y^{-2} (\, d x^2 + d y^2\,)$. The group

$$PSL(2, \mathbb{R}) = \{\begin{pmatrix} a & b \\ c & d \end{pmatrix} \,|\, a, b, c, d, \in \mathbb{R}, ad - bc = 1\}/\{\pm id\}$$

is the isometry group of this space acting on $\mathbb{H}$ as $g\,z = \frac{az+b}{cz+d}$. The (full) modular group

$$\Gamma(1) := PSL(2, \mathbb{Z}) = \{\begin{pmatrix} a & b \\ c & d \end{pmatrix} \,|\, a, b, c, d, \in \mathbb{Z}, ad - bc = 1\}/\{\pm id\}$$

is a discrete subgroup of $PSL(2, \mathbb{R})$ with two generators:

(2.1) $$Q := \begin{pmatrix} 0 & 1 \\ -1 & 0 \end{pmatrix} \quad \text{and} \quad T := \begin{pmatrix} 1 & 1 \\ 0 & 1 \end{pmatrix},$$

which fulfill the relations

(2.2) $$Q^2 = id \quad \text{and} \quad (QT)^3 = id.$$

The subgroups $\Gamma \subseteq \Gamma(1)$ with finite index $\mu = [\Gamma(1) : \Gamma]$ are called modular groups. Examples are the principal congruence subgroup of order N, $N \in \mathbb{N}$,

$$\Gamma(N) := \{\begin{pmatrix} a & b \\ c & d \end{pmatrix} \in \Gamma(1) \,|\, \begin{pmatrix} a & b \\ c & d \end{pmatrix} \equiv \begin{pmatrix} 1 & 0 \\ 0 & 1 \end{pmatrix} \, mod \, N \}$$

which is a normal subgoup of the modular group $\Gamma(1)$ or the congruence subgroups

$$\Gamma_0(N) := \{\begin{pmatrix} a & b \\ c & d \end{pmatrix} \in \Gamma(1) \,|\, c = 0 \, mod \, N, \, N \in \mathbb{N}\},$$

$$\Gamma^0(N) := \{\begin{pmatrix} a & b \\ c & d \end{pmatrix} \in \Gamma(1) \,|\, b = 0 \, mod \, N, \, N \in \mathbb{N}\}$$

which are not normal. Of interest for the following discussion is also the theta group

$$\Gamma_\vartheta := \Gamma(2) \cup \Gamma(2)Q$$

which has index three in $\Gamma(1)$ and is conjugated to $\Gamma_0(2)$ and $\Gamma^0(2)$. The group $\Gamma_2 := \Gamma(2) \cup \Gamma(2)QT \cup \Gamma(2)(QT)^2$ is another normal group of $\Gamma(1)$ with index two. The modular surface defined by such a modular group Γ is the quotient space $M_\Gamma = \Gamma \backslash \mathbb{H}$ [**Ter85**].

2.2. The geodesic flow on M_Γ.

2.2.1. *The Poincaré map of the geodesic flow on M_Γ.* The physical phase space of a free particle on M_Γ with unit velocity is the unit tangent bundle $T_1 M_\Gamma$. The dynamics on $T_1 M_\Gamma$ can be described by the geodesic flow

(2.3) $$\begin{aligned} \hat{\phi}_t \; &: \; T_1 M_\Gamma \to T_1 M_\Gamma \\ &(\hat{\gamma}_{M_\Gamma}(0), \dot{\hat{\gamma}}_{M_\Gamma}(0)) \mapsto (\hat{\gamma}_{M_\Gamma}(t), \dot{\hat{\gamma}}_{M_\Gamma}(t)) \end{aligned}$$

where $\hat{\gamma}_{M_\Gamma} : \mathbb{R} \to M_\Gamma$ denotes the geodesic in M_Γ, parameterized by the time t through the initial position $\hat{\gamma}_{M_\Gamma}(0)$ with the initial tangent vector $\dot{\hat{\gamma}}_{M_\Gamma}(0)$. When choosing an appropriate Poincaré section X_Γ for this flow [**CM00**] its Poincaré return map can be described as

$$P_\Gamma \quad : \quad [0,1] \times [0,1] \times \mathbb{Z}_2 \times \Gamma\backslash\Gamma(1) \to [0,1] \times [0,1] \times \mathbb{Z}_2 \times \Gamma\backslash\Gamma(1)$$

with

$$
\begin{aligned}
P_\Gamma(x_1, x_2, \varepsilon, \{g\}) &= (T^{-n\varepsilon}Qx_1, QT^{n\varepsilon}x_2, -\varepsilon, \{gT^{n\varepsilon}Q\}) \\
(2.4) \qquad &= (T_G x_1, \frac{1}{[\frac{1}{x_1}] + x_2}, -\varepsilon, \{gT^{n\varepsilon}Q\}),
\end{aligned}
$$

where $T_G : [0,1] \to [0,1]$ denotes the Gauss map $T_G z = \frac{1}{z} - [\frac{1}{z}]$ and $n = n(x) = [\frac{1}{x_1}]$ denotes the integer part for $\frac{1}{x_1}$ with $x := (x_1, x_2, \varepsilon, \{g\})$. The ergodic properties of this map are determined by its expanding directions, namely

$$(2.5) \qquad P_\Gamma|_{ex}(z, \varepsilon, \{g\}) = (T_G z, -\varepsilon, \{gT^{n\varepsilon}Q\}) = (\frac{1}{z} - [\frac{1}{z}], -\varepsilon, \{gT^{n\varepsilon}Q\}),$$

with $n = n(x) = [\frac{1}{z}]$ and $x = (z, \varepsilon, \{g\})$.

2.2.2. *The transfer operator for Γ.* The transfer operator for the map (2.5) has the explicit form [**CM00**]

$$(2.6) \qquad \tilde{\mathcal{L}}_\beta^{\Gamma(1), \chi^\Gamma} \underline{f}(z, \varepsilon) = \sum_{n=1}^\infty (\frac{1}{z+n})^{2\beta} \chi^\Gamma(QT^{n\varepsilon}) \underline{f}(\frac{1}{z+n}, -\varepsilon),$$

where χ^Γ denotes the representation of $\Gamma(1)$ induced from the trivial representation of the subgroup Γ. Let $B(D)$ be the Banach space of all holomorphic functions on the disk

$$(2.7) \qquad D := \{z | z \in \mathbb{C}, |z - 1| < \frac{3}{2}\}$$

which are continuous on the closure $\bar{D}$ of the disk. Then the transfer operator (2.6) is well defined on $\oplus_{i=1}^{2\mu} B(D)$, where $\mu = [\Gamma(1) : \Gamma]$:

$$\tilde{\mathcal{L}}_\beta^{\Gamma(1), \chi^\Gamma} : \oplus_{i=1}^{2\mu} B(D) \to \oplus_{i=1}^{2\mu} B(D).$$

Denoting the operator $\tilde{\mathcal{L}}_\beta^{\Gamma(1), \chi^\Gamma}$ simply as $\tilde{\mathcal{L}}_\beta$, the decomposition

$$(2.8) \qquad \tilde{\mathcal{L}}_\beta = \tilde{\mathcal{A}}_\beta^{(\kappa)} + \tilde{\mathcal{L}}_\beta^{(\kappa)} \qquad \text{for } \kappa \in \mathbb{N}_0 = \mathbb{N} \cup \{0\}$$

with

$$(2.9) \qquad \tilde{\mathcal{A}}_\beta^{(\kappa)} \underline{f}_\beta(z, \varepsilon) = \sum_{l=0}^\kappa (\frac{1}{r})^{2\beta+l} \sum_{m=1}^r \chi^\Gamma(QT^{m\varepsilon}) \frac{f_\beta^{(l)}(0, -\varepsilon)}{l!} \zeta(2\beta+l, \frac{z+m}{r})$$

and

$$(2.10) \qquad \tilde{\mathcal{L}}_\beta^{(\kappa)} \underline{f}_\beta(z, \varepsilon) = \sum_{n=1}^\infty \sum_{m=1}^r \chi^\Gamma(QT^{m\varepsilon}) \left(\frac{1}{z+m+r(n-1)}\right)^{2\beta}$$

$$\times \left[\underline{f}\left(\frac{1}{z+m+r(n-1)}, -\varepsilon\right) - \sum_{l=0}^\kappa \frac{f_\beta^{(l)}(0, -\varepsilon)}{l!} \left(\frac{1}{z+m+r(n-1)}\right)^l \right],$$

shows that $\tilde{\mathcal{L}}_\beta$ is meromorphic in the half plane $\Re\beta > -\frac{\kappa}{2}$ with possible singularities at $\beta = \beta_l := \frac{1-l}{2}$, $l = 0, 1, \cdots, \kappa$ [**CM00**]. The number r is the minimal natural number with $\chi^\Gamma(T^r) = 1$. Since κ is an arbitrary natural number, the operator $\tilde{\mathcal{L}}_\beta$ is meromorphic in the entire complex β-plane.

The operator $\tilde{\mathcal{L}}_\beta$ can be written as [**CM00**]

$$(2.11) \qquad \tilde{\mathcal{L}}_\beta = \begin{pmatrix} 0 & \mathcal{L}_{\beta,-} \\ \mathcal{L}_{\beta,+} & 0 \end{pmatrix}$$

with

$$\mathcal{L}_{\beta,\pm} : \oplus_{i=1}^\mu B(D) \to \oplus_{i=1}^\mu B(D)\,.$$

where

$$(2.12) \qquad (\mathcal{L}_{\beta,\pm}\underline{f}_{\beta,\pm})(z) = \sum_{n=1}^\infty (\frac{1}{z+n})^{2\beta} \chi^\Gamma(QT^{\pm n})\underline{f}_{\beta,\mp}(\frac{1}{z+n})\,,$$

and

$$\underline{f}_\beta(z) := \begin{pmatrix} \underline{f}_\beta(z,+1) \\ \underline{f}_\beta(z,-1) \end{pmatrix} = \begin{pmatrix} \underline{f}_{\beta,+}(z) \\ \underline{f}_{\beta,-}(z) \end{pmatrix}\,.$$

If the induced representation χ^Γ has the property $\chi^\Gamma(T^2) = 1$, then $\mathcal{L}_{\beta+} = \mathcal{L}_{\beta-} := \mathcal{L}_\beta$. In this case the eigenvalues and the eigenfunctions of $\mathcal{L}_\beta$ determine the eigenvalues and the eigenfunctions of $\tilde{\mathcal{L}}_\beta$ [**CM00**].

In [**CM00**] we showed, that Selberg's zeta function $Z_S(\beta;\Gamma)$ can be written as

$$(2.13) \qquad Z_S(\beta;\Gamma) = Z_S(\beta;\Gamma(1);\chi^\Gamma) = \det(1 - \tilde{\mathcal{L}}_\beta^{\Gamma(1),\chi^\Gamma})\,.$$

Due to this relation, the β values where the operator $\tilde{\mathcal{L}}_\beta^{\Gamma(1),\chi^\Gamma}$ has eigenvalue $+1$ play a very special role, since these β's are related to the zeros of Selberg's zeta function. The zeros of Selberg's zeta function fall into three classes, namely the trivial zeros at $\beta = \beta_l = \frac{1-l}{2}$, $l \in \mathbb{N}_0$, the Riemann zeros on the line $\Re\beta = \frac{1}{4}$ (if Riemann's conjecture is true) and the spectral zeros on the line $\Re\beta = \frac{1}{2}$, which are related to the quantum energies $\beta(1 - \beta)$ of a free particle on M_Γ [**Hej76**]. All these zeros are closely related to the automorphic forms and the Maaß wave forms for the group Γ, whose definition we briefly recall for the readers' convenience.

2.3. Automorphic forms and Maaß wave forms for Γ.

2.3.1. *Automorphic forms.* Let k be a nonnegative even integer. A complex function f on $\mathbb{H}$ is called an automorphic form of weight k (or degree $-k$) for Γ if f is holomorphic on $\mathbb{H}$ and at every cusp of Γ [**Ter85**] and $f(z)|_k\,\sigma = f(z)$ for all $\sigma \in \Gamma$, where

$$(2.14) \qquad f(z)\,|_k\,\sigma := (cz + d)^{-k}\,f(\sigma z)\,.$$

For $\Gamma \subseteq \Gamma(1)$ the automorphic forms are also called modular forms.

REMARK 2.1. Some authors define the automorphic forms and the modular forms as meromorphic functions [**Miy89**], [**Shi71**]. However, we follow to the definition in [**Ter85**] and [**Zag92a**].

Denote by $A_k(\Gamma)$ the space of all automorphic forms of weight k for the group Γ. A function in $A_k(\Gamma)$ is called a holomorphic cusp form if $f(z)$ vanishes at every cusp of Γ [**Lan76**]. The set of cusp forms is denoted by $S_k(\Gamma)$. Formulas for the dimensions of the spaces $S_k(\Gamma)$ for different Γ can be found for instance in [**Shi71**], [**Miy89**]. For $\Gamma = \Gamma(1)$ one has

$$(2.15) \qquad dim\, S_k(\Gamma(1)) = \begin{cases} 0 & (k = 2)\,, \\ [k/12]\text{-}1 & (k > 2,\, k \equiv 2\; mod\; 12)\,, \\ [k/12] & (k \not\equiv 2\; mod\; 12)\,, \end{cases}$$

where $[x]$ is the integer part of x.

The holomorphic Eisenstein series are modular forms which do not vanish at the cusps. The Eisenstein series of weight k for the full modular group $\Gamma(1)$ reads

$$(2.16) \qquad E_k(z) = \sum_{m,n\in\mathbb{Z},\,(m,n)\neq(0,0)} (mz + n)^{-k}\,.$$

For examples of Eisenstein series for subgroups of $\Gamma(1)$ see [**Kob93**] p.131, [**Ran77**] p.174 and [**Sar90**] p.16. The number of Eisenstein series for $\Gamma \subseteq \Gamma(1)$ is equal to the number of Γ-inequivalent cusps in $\Gamma\backslash\mathbb{H}$ [**Ran77**].

2.3.2. *The Maaß wave forms.* Let z_i, $i = 1, \cdots, n$ be the Γ-inequivalent cusps of the hyperbolic surface M_Γ and s_i be a generator of the stabilizer S_i of z_i, i.e., the subgroup $S_i \subseteq \Gamma$ such that $s_i\, z_i = z_i$ for $s_i \in S_i$. The Laplace-Beltrami operator

$$(2.17) \qquad -\triangle = -y^2 \left(\frac{\partial^2}{\partial x^2} + \frac{\partial^2}{\partial y^2} \right)\,, \qquad z = x + i\,y \in \mathbb{H}$$

for $\Gamma \subseteq \Gamma(1)$ can be defined on the Hilbert space

$$\mathcal{H}(\Gamma) \;:=\; \Big\{ \underline{f} : \mathbb{H} \to V \mid \underline{f}(\sigma\, z) = \chi(\sigma)\, \underline{f}(z),\, \forall \sigma \in \Gamma,\, \forall z \in \mathbb{H},$$

$$\|\underline{f}\|^2 = \int_{\mathcal{F}_\Gamma} |\underline{f}(z)|^2\, d\,m(z) < \infty \Big\}\,,$$

where χ is a representation of Γ in the finite dimensional Hilbert space V and $d\,m = \frac{dx\,dy}{y^2}$ is the Γ-invariant hyperbolic measure. The representation χ is called singular respectively regular at the cusp z_i if $\chi(s_i) = 1$ respectively $\chi(s_i) \neq 1$ for $s_i \in S_i$ [**BV97**].

The Maaß wave forms (or non-holomorphic modular forms) $u(z)$ for the group Γ are the eigenfunctions of the Laplace-Beltrami operator $-\triangle$ on $\mathbb{H}$ fulfilling $u(\sigma z) = u(z)$ for all $\sigma \in \Gamma$ and $z \in \mathbb{H}$ which increase at every cusp at most like a polynomial. The Maaß wave forms which vanish at infinity are called Maaß cusp forms. Besides the constant function $u(z) = c$ the only explicitly known Maaß wave forms are the non-holomorphic Eisenstein series which do not vanish at infinity. These Eisenstein series can be defined as follows: Let z_i, $i = 1, \cdots, n$ denote the Γ-inequivalent cusps of $\Gamma\backslash\mathbb{H}$ with $z_1 = \infty$ and S_1 be the stabilizer of z_1, i.e., $gz_1 = z_1$ for all $g \in S_1$. Suppose $g_i \in PSL(2,\mathbb{R})$ is chosen such that $g_i\infty = z_i$. Then $S_i := g_i S_1 g_i^{-1}$ is the stabilizer of z_i. The Eisenstein series $E_s^{(i)}(z)$ for the cusp z_i of Γ is then defined as [**Kub73**]

$$(2.18) \qquad E_s^{(i)}(z) = \sum_{\sigma \in S_i\backslash\Gamma} y^s\, (g_i^{-1}\sigma z)\,, \qquad \Re s > 1\,,$$

with $y(z) := \Im z$. For $\Gamma(1)$ the point $z = \infty$ is the only cusp. The Eisenstein series in this case reads

$$E_s(z) = \sum_{\sigma \in \Gamma_\infty \backslash \Gamma} y^s(\sigma z) \quad \text{with} \quad \Gamma_\infty = \left\{ \begin{pmatrix} \pm 1 & n \\ 0 & \pm 1 \end{pmatrix} \middle| n \in \mathbb{Z} \right\}.$$

Another definition for the Eisenstein series of $\Gamma(1)$ is

$$(2.19) \qquad \tilde{E}_s(z) = \frac{1}{2} y^s \sum_{m,n \in \mathbb{Z},\, (m,n) \neq (0,0)} |mz + n|^{-2s}, \quad \Re s > 1.$$

It differs from (2.18) only by the factor $\zeta(2s)$ [**Ter85**] p.207. A more detailed discussion of the non-holomorphic Eisenstein series can be found in [**Kub73**] and [**Ter85**].

2.3.3. *The theory of old and new forms.* The theory of old and new modular forms goes back to A. Atkin and J. Lehner [**AL70**]. It is a classification of modular forms of Γ depending on whether they are related to modular forms for some group Γ' with $\Gamma \subset \Gamma'$ or not. In the literature this theory is usually developed for congruence subgroups [**AL70**], [**Zag92a**] or the principal congruence subgroups [**Lan76**]. We restrict ourselves to the congruence subgroups:

Let $\Gamma_0(N)$ be a subgroup of $\Gamma_0(N')$ with $N, N' \in \mathbb{N}$ and $N = kN'$, $k > 1$. Then obviously $\begin{pmatrix} a & bm \\ c/m & d \end{pmatrix} \in \Gamma_0(N')$ for $\begin{pmatrix} a & b \\ c & d \end{pmatrix} \in \Gamma_0(N)$, when m is a divisor of k. Suppose $f(z)$ is a modular form of $\Gamma_0(N')$. Then it follows that $c/m = l\,N/m = l\,k\,N'/m = n\,N'$ with $n, l \in \mathbb{N}$ and one gets

$$f(z) = (\frac{c}{m} z + d)^{-k} f\left(\begin{pmatrix} a & bm \\ c/m & d \end{pmatrix} z \right).$$

However this implies

$$f(mz) = (\frac{c}{m} mz + d)^{-k} f\left(\frac{amz + bm}{\frac{c}{m} mz + d} \right) = (cz + d)^{-k} f\left(m \begin{pmatrix} a & b \\ c & d \end{pmatrix} z \right).$$

That is, $f(mz)$ is a modular form of $\Gamma_0(N)$ if $f(z)$ was a modular form of $\Gamma_0(N')$. In this sense the function $f(mz)$ is called an old form of $\Gamma_0(N)$. We define $A_k(\Gamma)^{old}$ as the space of old forms and $A_k(\Gamma)^{new}$ as the space of new forms which is the complement of $A_k(\Gamma)^{old}$, i.e., $A_k(\Gamma) = A_k(\Gamma)^{old} \oplus A_k(\Gamma)^{new}$ with respect to the Petersson scalar product [**Lan76**]. The concept of old and new modular forms can be generalized to Maaß wave forms [**Iwa95**] and to arbitrary subgroups Γ of $\Gamma(1)$.

2.4. Period polynomials and period functions associated to automorphic forms. In the classical theory of Eichler-Shimura and Manin and its extension by Zagier [**Zag91**] to every holomorphic form for a Fuchsian group Γ one associates period polynomials or rational period functions which fulfill certain cocycle relations under the action of the group Γ. This theory was recently generalized in the case of the modular group $\Gamma(1)$ to nonholomorphic automorphic forms by J. Lewis and D. Zagier [**LZ97**]. The resulting period functions are holomorphic in the complex plane cut along the negative real axis and fulfill a certain functional equation depending on a complex parameter β. For negative integer β the solutions of this functional equation are just the aforementioned classical period polynomials of Eichler respectively rational period functions of Zagier for $\Gamma(1)$. For preparing our discussion of this theory for subgroups $\Gamma \subseteq \Gamma(1)$ we briefly recall the results of Zagier and Lewis for $\Gamma(1)$.

2.4.1. *Period polynomials in the Eichler-Shimura-Manin theory.* M. Eichler studied in [**Eic57**] the $(k-1)$-th indefinite integral of cusp forms of weight k ($k \geq 2$ even number) leading to so called 'period polynomials' for $u(z)$ a holomorphic cusp form for a Fuchsian group $\Gamma \subseteq PSL(2, \mathbb{R})$ of weight k. He considered the integral

$$(2.20) \qquad \Phi(\tau) = \frac{1}{(k-2)!} \int_{\tau_0}^{\tau} (\tau - z)^{k-2} u(z) \, dz \,,$$

for $\tau_0, \tau \in \mathbb{H}$. This integral is obviously independent of the path in $\mathbb{H}$ connecting τ_0 and τ. It can be easily verified that

$$(2.21) \qquad \frac{d^{k-1}}{d\tau^{k-1}} \Phi(\tau) = u(\tau) \,.$$

Obviously an arbitrary polynomial $\Theta(\tau)$ of degree $k-2$ can be added to $\Phi(\tau)$ in (2.21) as an 'integration constant' without destroying relation (2.21). That is, instead of (2.20) one can also consider the function

$$(2.22) \qquad \Phi(\tau) = \frac{1}{(k-2)!} \int_{\tau_0}^{\tau} (\tau - z)^{k-2} u(z) \, dz + \Theta(\tau) \,.$$

Changing the integration variable z to σz with $\sigma = \begin{pmatrix} a & b \\ c & d \end{pmatrix} \in \Gamma \subseteq PSL(2, \mathbb{R})$ one finds

$$(2.23) \qquad \Phi(\tau)\,|_{2-k}\,\sigma = \Phi(\tau) + \Omega_\sigma(\tau) \,,$$

with

$$(2.24) \qquad \begin{aligned} \Omega_\sigma(\tau) \;=\; & \Theta(\tau)\,(\,|_{2-k}\,\sigma - |_{2-k}\,1\,) \\ & + \frac{1}{(k-2)!} \int_{\tau_0}^{\sigma \tau_0} (a\tau + b - (c\tau + d)z)^{k-2} u(z) \, dz \,, \end{aligned}$$

where we used the notation $|_k\,\sigma$ as in (2.14). The function $\Omega_\sigma(\tau)$ is obviously a polynomial of degree $k-2$ which is called a period polynomial of $\sigma \in \Gamma$ (originally called 'period' by Eichler). A function $\Phi(\tau)$ obeying relation (2.23) with a polynomial $\Omega_\sigma(\tau)$ for all $\sigma \in \Gamma$ is called an automorphic integral. Obviously the modular forms of weight $k-2$ are also automorphic integrals [**Kno78**].

Now we rewrite (2.23) as

$$\Omega_\sigma = \Phi\,|_{2-k}\,(\,\sigma - 1\,)$$

where for simplicity the dependence on τ is suppressed. Suppose σ_1, $\sigma_2 \in \Gamma$. Then the equality $\Phi|_{2-k}\,(\sigma_1 \sigma_2) = (\Phi|_{2-k}\,\sigma_1)\,|_{2-k}\,\sigma_2$ induces the cocycle relation

$$(2.25) \qquad \Omega_{\sigma_1 \sigma_2} = \Omega_{\sigma_1}\,|_{2-k}\,\sigma_2 + \Omega_{\sigma_2} \,.$$

Hence the period polynomials for the cusp forms of weight k are solutions of these equations and vice versa every family of polynomial solutions of (2.25) defines period polynomials for a cusp form [**Eic57**].

2.4.2. *The period polynomials of* $\Gamma(1)$. Let us briefly recall the special case of the full modular group $\Gamma(1)$ with the two generators Q and T in (2.1). Let $u(z)$ be a holomorphic cusp form. Setting $\Theta = 0$, $\sigma = Q$, $\tau_0 = 0$ in (2.24) and using the condition $u|_k(z) = u(z)$ for the automorphic form $u(z)$ of weight k one gets up to a constant factor the polynomial

$$(2.26) \qquad \Omega_Q(\tau) = \int_0^{i\infty} (\tau - z)^{k-2} u(z)\, dz.$$

Relations (2.23) for the group $\Gamma(1)$ can be reduced to the two equations for the generators Q and T

$$(2.27) \qquad \Omega_Q = \Phi\,|_{2-k}\, Q - \Phi \quad \text{and}$$
$$(2.28) \qquad \Omega_T = \Phi\,|_{2-k}\, T - \Phi.$$

Since there exists a polynomial P_T of degree smaller than $k-2$ with $\Omega_T = P_T|T-P_T$, one can assume that $\Omega_T = 0$ [**Kno78**], since the polynomials Ω_σ, $\sigma \in \Gamma(1)$ are determined only up to an arbitrary coboundary $P|_\sigma - P$. On the other hand, every element $\sigma \in \Gamma(1)$ can be generated by Q and T, hence an arbitrary period polynomial Ω_σ can be expressed by means of the polynomial $\Omega = \Omega_Q$ by repeated application of the cocycle relation (2.25). The relations for the generators Q and T of $\Gamma(1)$.

$$Q^2 = S^3 = id \quad \text{with} \quad S := QT = \begin{pmatrix} 0 & 1 \\ -1 & -1 \end{pmatrix}$$

induce the two cocycle relations

$$(2.29) \qquad \Omega\,|_{2-k}\,(1+Q) = 0 \quad \text{and} \quad \Omega\,|_{2-k}\,(1+S+S^2) = 0$$

for the generating period polynomial $\Omega_Q = \Omega$.

2.4.3. *The period polynomials for the holomorphic cusp forms of* $\Gamma(1)$. Let $u(z)$ be a cusp form of weight k. Then the constant term $a_u(0)$ in the Fourier expansion $u(z) = \sum_{n=0}^{\infty} a_u(n)e^{2\pi i n z}$ vanishes. The period polynomial in (2.26) for $u(z)$ then reads [**Zag91**]:

$$(2.30) \qquad r_u(z) \;=\; \int_0^{i\infty} (t-z)^{k-2} u(t)\, dt$$

$$(2.31) \qquad\qquad =\; -\sum_{n=0}^{k-2} \binom{k-2}{n} \frac{L(u, n+1)}{(2\pi i)^{n+1}} n!\, z^{k-2-n},$$

where $L(u, n+1)$ is the analytic continuation of the L function defined through the series $\sum_{n=1}^{\infty} a_u(n)n^{-s}$. This polynomial is of degree $\leq k-2$.

A simple calculation shows that the function $r_u|\sigma := (r_u|_{2-k}\,\sigma)(z)$ can be expressed through the same integral in (2.30) with the path of integration from $\sigma^{-1}(0)$ to $\sigma^{-1}(i\infty)$. Then the two relations (2.29) can be verified easily [**Lan76**]:

$$(2.32) \qquad r_u\,|\,(1+Q) = \int_0^{i\infty} + \int_{i\infty}^{0} = 0,$$

$$(2.33) \qquad r_u\,|\,(1+S+S^2) = \int_0^{i\infty} + \int_{-1}^{0} + \int_{i\infty}^{-1} = 0.$$

Since the functions $r_u(z)$ are polynomials of degree at most $k-2$ and fulfill the two relations (2.32) and (2.33), the period polynomials of the cusp forms of weight k

lie in the space

$$W_{S_k(\Gamma(1))} := \left\{ p(z) \in \oplus_{n=0}^{k-2} \mathbb{C} z^n \mid p \,|\, (1+Q) = p \,|\, (1+S+S^2) = 0 \right\}.$$

Every polynomial p can be decomposed into an even and an odd polynomial p_+ respectively p_-. The Eichler-Shimura-Manin theory [**Lan76**] then tells us that the map $u \mapsto r_u^-$ induces an isomorphism of the space $S_k(\Gamma(1))$ of cusp forms of weight k to the space of odd polynomials in $W_{S_k(\Gamma(1))}$. On the other hand, the map $u \to r_u^+$ gives an isomorphism of the space $S_k(\Gamma(1))$ to the space of even polynomials in $W_{S_k(\Gamma(1))}$ modulo the polynomial $z^{k-2} - 1$.

2.4.4. *The period functions for the holomorphic Eisenstein series of* $\Gamma(1)$. The theory of period polynomials for the holomorphic cusp forms was generalized by Zagier in [**Zag91**] to the holomorphic Eisenstein series $E_k(z) = \sum_{n=0}^{\infty} a_{E_k}(n) \, e^{2\pi i n z}$ of weight k. Instead of (2.30) Zagier considered

$$(2.34) \qquad r_{E_k}(z) = \sum_{n \in \mathbb{Z}} i^{1-n} \binom{k-2}{n} L^*(E_k, n+1) z^{k-2-n},$$

where L^* for $\Re s \gg 0$ is defined as

$$(2.35) \qquad L^*(u, s) = \int_0^\infty \left(u(i\,y) - a_u(0) \right) y^{s-1} d\,y = (2\pi)^{-s} \Gamma(s) \, L(u, s).$$

The function $L^*(u, s)$ has only two singularities at $s = 0$ and $s = k$ with residua $-a_{E_k}(0)$ and $(-1)^{k/2} a_{E_k}(0)$ [**Zag91**]. The binomial coefficients $\binom{k-2}{n}$ which are given by $\frac{\Gamma(k-1)}{\Gamma(n+1)\,\Gamma(k-1-n)}$ vanish for all $n \in \mathbb{Z}$ besides the numbers $n = 0, \cdots, k-2$. For $n = -1$ and $n = k-1$ the singularities of $L^*(u, s)$ for $s = n+1$ are cancelled by the zeros of $\binom{k-2}{n}$. Thus comparing with $r_u(z)$ in (2.31) shows that $r_{E_k}(x)$ has two additional terms proportional to $\frac{1}{z}$ and z^{k-1}:

PROPOSITION 2.2. [**Zag91**] *For $k > 2$ let p_k^+ respectively p_k^- be the following even respectively odd rational functions:*

$$(2.36) \qquad p_k^+(z) \quad := \quad z^{k-2} - 1,$$

$$(2.37) \qquad p_k^-(z) \quad := \sum_{-1 \le n \le k-1, \, n \text{ odd}} \frac{B_{n+1}}{(n+1)!} \frac{B_{k-n-1}}{(k-n-1)!} z^n,$$

where B_i denotes the Bernoulli numbers. Then the period function r_{E_k} of the holomorphic Eisenstein series E_k is

$$(2.38) \qquad r_{E_k}(z) = \omega^- \, p_k^-(z) + \omega^+ p_k^+(z),$$

with $\omega^- = -\frac{(k-2)!}{2}$ and $\omega^+ = \frac{\zeta(k-1)}{(2\pi i)^{k-1}} \omega^-$.

The period functions respectively period polynomials of the modular forms of weight k of the group $\Gamma(1)$ hence belong to the following space [**Zag91**]

$$(2.39) \quad W_{M_k(\Gamma(1))} := \left\{ \phi(z) \in \oplus_{n=-1}^{k-1} \mathbb{C} z^n \mid \phi \,|\, (1+Q) = \phi \,|\, (1+S+S^2) = 0 \right\}.$$

In [**Zag91**] and [**CM99**] it is shown that the even respectively odd functions in $W_{M_k(\Gamma(1))}$ are exactly the solutions of the equation

$$(2.40) \qquad \lambda \left(\phi(z) - \phi(z+1) \right) = z^{k-2} \phi(1 + \frac{1}{z}),$$

with $\lambda = 1$ respectively -1 in the space $\oplus_{n=-1}^{k-1} \mathbb{C} z^n$.

2.4.5. *The period functions for the Maaß wave forms of* $\Gamma(1)$. The theory of period polynomials of modular forms of even positive weight for the group $\Gamma(1)$ was generalized by D. Zagier and J. Lewis to nonholomorphic automorphic forms like the Maaß wave forms [**LZ97**]. Let $u_s(z)$ be a Maaß wave form of the Laplace-Beltrami operator $-\triangle$ with eigenvalue $s(1-s)$, i.e., $-\triangle u_s(z) = s(1-s)u_s(z)$ and assume $u_s(z)$ obeys the conditions in section 2.3.2. Then the function $u_s(z)$ can be decomposed into an even and odd part under the reflection $z \mapsto -\bar{z}$. This follows from the invariance of the Laplace-Beltrami operator under this reflection with respect to the imaginary axis. Writing $z = x+iy$, the Taylor expansion of the even respectively the odd Maaß wave form $u_s(z)$ has the form [**LZ97**]

$$u_s^+(x+iy) = \sqrt{y} \sum_{n=1}^{\infty} a_n K_{s-\frac{1}{2}}(2\pi ny)\, cos(2\pi nx) \,\cdot$$

respectively

$$u_s^-(x+iy) = \sqrt{y} \sum_{n=1}^{\infty} a_n K_{s-\frac{1}{2}}(2\pi ny)\, sin(2\pi nx)\,,$$

where K_ν denotes the modified Bessel function of order ν. None of these functions, besides the constant function $u_s^+(z) = c$ is explicitly known. In his work on harmonic analysis for the modular group $\Gamma(1)$ J. Lewis established an interesting new relation between the Maaß wave forms and certain holomorphic functions [**Lew97**]:

THEOREM 2.3. *Let* Γ *be the modular group* $\Gamma(1)$ *and* $0 < \Re s \le 1$. *For every even Maaß wave form with eigenvalue* $s(1-s)$ *there exists a holomorphic function* ψ_s *on* $\mathbb{C} \setminus (-\infty, 0]$ *with* $\psi_s(z) = O(1/z)$ *for* $z \to \infty$ *which fulfills the functional equation*

$$(2.41) \qquad \psi_s(z) - \psi_s(z+1) = z^{-2s}\psi_s(1+z^{-1})\,.$$

On the other hand, for any solution ψ_s *of (2.41) which is holomorphic on* $\mathbb{C}\setminus(-\infty,0]$ *and* $\psi_s(z) = O(1/z)$ *for* $z \to \infty$, *there exists an even Maaß wave form with eigenvalue* $s(1-s)$.

An analogous result holds for the odd Maaß wave forms when equation (2.41) is replaced by

$$(2.42) \qquad -\psi_s(z) + \psi_s(z+1) = z^{-2s}\psi_s(1+z^{-1})\,.$$

The even Maaß wave forms $u_s(z)$ are related to the solutions $\psi_s(z)$ of (2.41) by [**LZ97**]:

$$(2.43) \qquad \psi_s(z) = z \int_0^{\infty} y^s\, u_s(iy)\,(z^2+y^2)^{-s-1}\,dy\,, \quad \Re(z) > 0$$

and the odd Maaß wave forms $u_s(z)$ are related to the solutions $\psi_s(z)$ of (2.42) by

$$(2.44) \qquad \psi_s(z) = \int_0^{\infty} \frac{\partial}{\partial x}\, u_s(iy)\, y^s\,(z^2+y^2)^{-s}\,dy\,, \quad \Re(z) > 0\,.$$

Comparing with (2.41) respectively (2.42) shows that (2.40) is just a special case of (2.41) respectively (2.42) with $s = \frac{2-k}{2}$. For this reason Lewis and Zagier called the functions $\psi_s(z)$ the period functions of the Maaß wave forms in analogy to the period polynomials for the holomorphic modular forms. Zagier found also a special family of solution of equation (2.41) for arbitrary $s \neq 1$ [**Zag92b**] given for $\Re s > 1$ as

$$(2.45) \qquad \psi_s(z) = \sum_{m,n \geq 1} (\frac{1}{mz+n})^{2s} + \frac{1}{2}\zeta(2s)(1+(\frac{1}{z})^{2s}),$$

where ζ denotes the Riemann zeta function. This function can be analytically continued to the entire complex s-plane and defines for $s \in \mathbb{C}\backslash\{1\}$ a holomorphic family of solutions of equation (2.41).

Up to a factor the analytic extension of $\psi_s(z)$ in (2.45) is just the transform (2.43) of the non-holomorphic Eisenstein series (2.19) [**CM98**] for $\Gamma(1)$ and coincides for $s = -N$, $N \in \mathbb{N}$ with the odd part of the period functions (2.37) of the holomorphic Eisenstein series [**CM99**] for this group.

2.4.6. *The period functions and the transfer operator of* $\Gamma(1)$. Now let $f_\beta(z)$ for $\Re\beta > \frac{1}{2}$ be an eigenfunction of the transfer operator [**CM99**]

$$\mathcal{L}_\beta f_\beta(z) = \sum_{n=1}^{\infty} (\frac{1}{z+n})^{2\beta} f_\beta(\frac{1}{z+n}).$$

for $\Gamma(1)$ with eigenvalue λ_β

$$(2.46) \qquad \left[\lambda_\beta - \mathcal{L}_\beta \right] f_\beta(z) = 0.$$

A simple calculation shows that $\mathcal{L}_\beta\, f_\beta(z) - \mathcal{L}_\beta\, f_\beta(z+1) = (\frac{1}{z+1})^{2\beta} f_\beta(\frac{1}{z+1})$ and therefore f_β fulfills the functional equation

$$(2.47) \qquad \lambda_\beta \left[f_\beta(z) - f_\beta(z+1) \right] = (z+1)^{-2\beta} f_\beta(\frac{1}{1+z}).$$

Hence every eigenfunction $f_\beta(z)$ of $\mathcal{L}_\beta$ is a solution of equation (2.47). On the other hand, let $f_\beta(z)$ be a solution of (2.47), i.e.,

$$\lambda_\beta f_\beta(z) = \lambda_\beta f_\beta(z+1) + (z+1)^{-2\beta} f_\beta(\frac{1}{1+z}),$$

then after N iterations one gets

$$(2.48) \qquad \lambda_\beta f_\beta(z) = \lambda_\beta f_\beta(z+N) + \sum_{n=1}^{N}(z+n)^{-2\beta} f_\beta(\frac{1}{n+z}).$$

Assuming $\lim_{N\to\infty} f(z+N) = 0$ this yields

$$\lambda_\beta f_\beta(z) = \sum_{n=1}^{\infty}(z+n)^{-2\beta} f_\beta(\frac{1}{n+z}) = \mathcal{L}_\beta\, f_\beta(z).$$

That is, every solution f_β of (2.47) for $\Re\beta > \frac{1}{2}$ with $\lim_{z\to\infty} f_\beta(z) = 0$ is an eigenfunction of $\mathcal{L}_\beta$ with eigenvalue λ_β.

Inserting $f_\beta(z) = \psi_\beta(z+1)$ into (2.47), the functional equation (2.47) for the eigenfunction f_β of $\mathcal{L}_\beta$ turns out to be

$$(2.49) \qquad \lambda_\beta\,(\psi_\beta(z) - \psi_\beta(z+1)) = z^{-2\beta}\psi_\beta(1+z^{-1}),$$

which for $\lambda_\beta = \pm 1$ is nothing but Lewis functional equation of the period polynomials respectively period functions in (2.41). An analogous result holds true for arbitrary β [**CM99**] for the analytic extension of the transfer operator.

In the present paper this relation between the eigenfunctions of $\mathcal{L}_\beta$ and the solutions of the Lewis equation (2.49) and hence the period functions of modular forms shall be generalized to an arbitrary subgroup $\Gamma \subseteq \Gamma(1)$ with finite index.

2.4.7. *The period polynomials of Eichler for* $\Gamma_0(2)$. Let us briefly consider Eichler's theory for an arbitrary subgroup $\Gamma \subseteq \Gamma(1)$. For simplicity we restrict ourselves to the group $\Gamma_0(2)$. This group has three generators, namely for example

$$e_1 := T, \quad e_2 := QT^{-2}Q = \begin{pmatrix} -1 & 0 \\ -2 & -1 \end{pmatrix} \quad \text{and} \quad e_3 := T^{-1}QT^{-2}Q = \begin{pmatrix} 1 & 1 \\ -2 & -1 \end{pmatrix},$$

which fulfill the relations [**BV97**]

$$(2.50) \qquad\qquad e_1 e_2 e_3 = id \quad \text{and} \quad e_3^2 = id.$$

An argument analogous to the case $\Gamma(1)$ shows that one can choose $\Omega_{e_1} = \Omega_T = 0$. We write Ω for Ω_{e_3}. Then Ω_{e_2} can obviously be expressed in terms of Ω.

Let $u(z)$ be a cusp form of weight k for the group $\Gamma_0(2)$. Setting $\Theta = 0$ in (2.24) one gets for any $\sigma = \begin{pmatrix} a & b \\ c & d \end{pmatrix} \in \Gamma_0(2)$

$$(2.51) \qquad\qquad \Omega_\sigma(\tau) = \int_{\tau_0}^{\sigma \tau_0} (a\tau + b - (c\tau + d)z)^{k-2} u(z)\, dz,$$

The relation $e_3^2 = 1$ leads to the condition

$$(2.52) \qquad\qquad \Omega \,|_{2-k} e_3 + \Omega = 0$$

for $\Omega = \Omega_{e_3}$, which has the explicit form

$$(2.53) \qquad\qquad \Omega\left(\frac{z+1}{-2z-1}\right)(-2z-1)^{k-2} + \Omega(z) = 0.$$

From the work of Eichler [**Eic57**] and Knopp [**Kno78**] it follows that the polynomial solutions of this equation generate all period polynomials of $\Gamma_0(2)$. Trivial solutions of this equation are given by the period functions $\Omega_Q(z)$ for the group $\Gamma(1)$: one just expresses $\Omega_{e_3}(z)$ with the cocycle relations (2.25) through Ω_Q which gives

$$(2.54) \qquad\qquad \Omega_{e_3}(z) = z^{k-2}\Omega_Q(-\frac{2z+1}{z}) + \Omega_Q(z).$$

This Ω_{e_3} obviously fulfills equation (2.53). The solutions of (2.53) shall be compared with the eigenfunctions of the transfer operator for $\Gamma_0(2)$ in section 4.5.3.

3. The functional equations for the eigenfunctions of the transfer operator

3.1. Lewis equations and master equations. Let Γ be a subgroup of $\Gamma(1)$ with finite index and λ_β be an eigenvalue of the analytically continued transfer operator (2.8) with eigenfunction $\underline{f}_{-\beta}(z, \varepsilon)$, i.e.,

$$(3.1) \qquad\qquad \lambda_\beta \underline{f}_{-\beta}(z, \varepsilon) - \tilde{\mathcal{L}}_\beta \underline{f}_{-\beta}(z, \varepsilon) = 0.$$

This eigenequation implies immediately that the eigenfunction $\underline{f}_{-\beta}(z, \varepsilon)$ can be analytically continued from D to $\mathbb{C}^*$, where $\mathbb{C}^*$ denotes the complex plane $\mathbb{C}\backslash(-\infty, -1]$

cut along the negative real axis $(-\infty, -1]$. If $\underline{f}_\beta(z, \varepsilon)$ is a solution of (3.1), the function $\underline{f}_\beta(z, \varepsilon)$ is automatically also a solution of the equation:

$$\lambda_\beta \underline{f}_\beta(z, \varepsilon) - \tilde{\mathcal{L}}_\beta \underline{f}_\beta(z, \varepsilon)$$
$$(3.2) \qquad = \chi^\Gamma(QT^\varepsilon Q) \left[\lambda_\beta \underline{f}_\beta(z+1, \varepsilon) - \tilde{\mathcal{L}}_\beta \underline{f}_\beta(z+1, \varepsilon) \right].$$

The factor $\chi^\Gamma(QT^\varepsilon Q)$ has been introduced for convenience as will be seen later. Conversely, a solution of (3.2) which is holomorphic in D is a solution of (3.1) only if after N iterations of relation (3.2), i.e.,

$$\lambda_\beta \underline{f}_\beta(z, \varepsilon) - \tilde{\mathcal{L}}_\beta \underline{f}_\beta(z, \varepsilon)$$
$$(3.3) \qquad = \chi^\Gamma(QT^{N\varepsilon} Q) \left[\lambda_\beta \underline{f}_\beta(z+N, \varepsilon) - \tilde{\mathcal{L}}_\beta \underline{f}_\beta(z+N, \varepsilon) \right],$$

the right-hand side of (3.3) vanishes in the limit $N \to \infty$ uniformly for $z \in D$. That is, a solution of equation (3.2) fulfills

$$(3.4) \qquad \lim_{N \to \infty} \chi^\Gamma(QT^\varepsilon Q) \left[\lambda_\beta \underline{f}_\beta(z+N, \varepsilon) - \tilde{\mathcal{L}}_\beta \underline{f}_\beta(z+N, \varepsilon) \right] = 0$$

uniformly for $z \in D$ iff $\underline{f}_\beta(z, \varepsilon)$ is an eigenfunction of $\tilde{\mathcal{L}}_\beta$. However, in the limit $N \to \infty$ one gets for $\Re\beta > -\frac{\kappa}{2}$ from (2.8)

$$\tilde{\mathcal{L}}_\beta \underline{f}_\beta(z+N, \varepsilon) \sim \tilde{A}_\beta^{(\kappa)} \underline{f}_\beta(z+N, \varepsilon)$$
$$(3.5) \qquad = \sum_{l=0}^{\kappa} \sum_{m=1}^{r} (\frac{1}{r})^{2\beta+l} \chi^\Gamma(QT^{m\varepsilon}) \frac{f_\beta^{(l)}(0, -\varepsilon)}{l!} \zeta(2\beta+l, \frac{z+m+N}{r}).$$

Hence in the region $\Re\beta > -\frac{\kappa}{2}$ the condition

$$\lim_{z \to \infty} \left[\lambda_\beta \underline{f}_\beta(z, \varepsilon) - \right.$$
$$(3.6) \qquad \left. \sum_{l=0}^{\kappa} \sum_{m=1}^{r} (\frac{1}{r})^{2\beta+l} \chi^\Gamma(QT^{m\varepsilon}) \frac{f_\beta^{(l)}(0, -\varepsilon)}{l!} \zeta(2\beta+l, \frac{z+m}{r}) \right] = 0$$

is necessary and sufficient for a solution $\underline{f}_\beta(z, \varepsilon)$ of equation (3.2) to be an eigenfunction of the operator $\tilde{\mathcal{L}}_\beta$ with eigenvalue λ_β.

Inserting $\tilde{\mathcal{L}}_\beta$ from (2.8) then into (3.2) one gets for an eigenfunction $\underline{f}_\beta$ of $\tilde{\mathcal{L}}_\beta$ the following equation

$$(3.7) \quad \lambda_\beta \left[\underline{f}_\beta(z, \varepsilon) - \chi^\Gamma(QT^\varepsilon Q) \underline{f}_\beta(z+1, \varepsilon) \right] - (\frac{1}{z+1})^{2\beta} \chi^\Gamma(QT^\varepsilon) \underline{f}_\beta(\frac{1}{z+1}, -\varepsilon) = 0.$$

From this follows that the transfer operator $\tilde{\mathcal{L}}_\beta$ does not have eigenvalue zero.

Equation (3.7) is a functional equation for $\underline{f}_\beta$ with a free parameter λ_β. The spectrum of the transfer operator $\tilde{\mathcal{L}}_\beta$ consists of the set of λ_β's for which a holomorphic solution $\underline{f}_\beta \in \oplus_{i=1\cdots2\mu} B(D)$ of equation (3.7) obeying the asymptotics (3.6) can be found. Since the operator $\tilde{\mathcal{L}}_\beta$ for $\beta \neq \beta_\kappa = \frac{1-\kappa}{2}$ is nuclear [**CM00**] only for a discrete set of λ_β there exist holomorphic solutions $\underline{f}_\beta$ in (3.7) fulfilling the asymptotic condition in (3.6).

Due to relation (2.13) the eigenvalue $\lambda_\beta = +1$ of $\tilde{\mathcal{L}}_\beta$ is responsible for the zeros of Selberg's zeta function. However, for our later discussion we consider here the more general case $\lambda_\beta = \pm 1$. First, let us define the quantity

$$Y(z, \varepsilon) \quad := \quad \lambda_\beta \, \underline{f}_\beta(z, \varepsilon) - \lambda_\beta \chi^\Gamma(QT^\varepsilon Q) \, \underline{f}_\beta(z + 1, \varepsilon)$$

$$(3.8) \qquad\qquad -(\frac{1}{z+1})^{2\beta} \, \chi^\Gamma(QT^\varepsilon) \, \underline{f}_\beta(\frac{1}{z+1}, -\varepsilon) \,,$$

which is nothing but the left-hand side of (3.7). Suppose $\underline{f}_\beta$ is a solution of (3.7), then $Y(z, \varepsilon)$ is identically zero. Inserting (3.8) into the identity

$$(\frac{1}{z+1})^{2\beta} \, \chi^\Gamma(QT^\varepsilon) \, Y(\frac{1}{z+1}, -\varepsilon)$$

$$(3.9) \qquad -\lambda_\beta \, (\frac{1}{z+2})^{2\beta} \, \chi^\Gamma(QT^\varepsilon QT^{-\varepsilon} QT^{-\varepsilon} Q) \, Y(\frac{-1}{z+2}, \varepsilon) + \lambda_\beta \, Y(z, \varepsilon) \equiv 0 \,,$$

we get

$$(\frac{1}{z+1})^{2\beta} \, \chi^\Gamma(QT^\varepsilon) \Big[\lambda_\beta \, \underline{f}_\beta(\frac{1}{z+1}, -\varepsilon)$$

$$-\lambda_\beta \chi^\Gamma(QT^{-\varepsilon} Q) \, \underline{f}_\beta(\frac{z+2}{z+1}, -\varepsilon) - (\frac{z+1}{z+2})^{2\beta} \, \chi^\Gamma(QT^{-\varepsilon}) \, \underline{f}_\beta(\frac{z+1}{z+2}, \varepsilon) \Big]$$

$$-\lambda_\beta \, (\frac{1}{z+2})^{2\beta} \, \chi^\Gamma(QT^\varepsilon QT^{-\varepsilon} QT^{-\varepsilon} Q) \Big[\lambda_\beta \, \underline{f}_\beta(\frac{-1}{z+2}, \varepsilon)$$

$$-\lambda_\beta \chi^\Gamma(QT^\varepsilon Q) \, \underline{f}_\beta(\frac{z+1}{z+2}, \varepsilon) - (\frac{z+2}{z+1})^{2\beta} \, \chi^\Gamma(QT^\varepsilon) \, \underline{f}_\beta(\frac{z+2}{z+1}, -\varepsilon) \Big]$$

$$+\lambda_\beta \Big[\lambda_\beta \, \underline{f}_\beta(z, \varepsilon) - \lambda_\beta \chi^\Gamma(QT^\varepsilon Q) \, \underline{f}_\beta(z + 1, \varepsilon)$$

$$-(\frac{1}{z+1})^{2\beta} \, \chi^\Gamma(QT^\varepsilon) \, \underline{f}_\beta(\frac{1}{z+1}, -\varepsilon) \Big] = 0 \,.$$

It is easy to see that the terms involving $\underline{f}_\beta(\frac{1}{z+1}, -\varepsilon)$, $\underline{f}_\beta(\frac{z+2}{z+1}, -\varepsilon)$ respectively $\underline{f}_\beta(\frac{z+1}{z+2}, \varepsilon)$ vanish. Using $QT^\varepsilon QT^{-\varepsilon} QT^{-\varepsilon} Q = QT^{2\varepsilon}$ which follows from the identity $(QT)^3 = id$ and setting $\lambda_\beta = \pm 1$ one finally gets the relation

$$(3.10) \quad \underline{f}_\beta(z, \varepsilon) - \chi^\Gamma(QT^\varepsilon Q) \, \underline{f}_\beta(z + 1, \varepsilon) - (\frac{1}{z+2})^{2\beta} \chi^\Gamma(QT^{2\varepsilon}) \, \underline{f}_\beta(\frac{-1}{z+2}, \varepsilon) = 0 \,.$$

Compared to equation (3.7) the parameter λ_β does not anymore appear and there is no sign change in the variable ε in (3.10). Every solution $\underline{f}_\beta$ of equation (3.7) solves equation (3.10) automatically. Conversely, it is easy to verify that under the condition

$$(3.11) \qquad \underline{f}_\beta(z - 1, \varepsilon) - \lambda_\beta \, z^{-2\beta} \, \chi^\Gamma(T^{-\varepsilon}) \, \underline{f}_\beta(\frac{1}{z} - 1, -\varepsilon) = 0, \quad \lambda_\beta = \pm 1 \,,$$

every solution $\underline{f}_\beta$ of equation (3.10) is also a solution of equation (3.7). To see this one only has to replace z in (3.11) by $1 - \frac{1}{z+2}$, leading to

$$(3.12) \qquad \underline{f}_\beta(\frac{-1}{z+2}, \varepsilon) = \lambda_\beta \, (\frac{z+1}{z+2})^{-2\beta} \, \chi^\Gamma(T^{-\varepsilon}) \, \underline{f}_\beta(\frac{1}{z+1}, -\varepsilon) \,,$$

and then to replace $\underline{f}_\beta$ in the last term of (3.10) by (3.12). On the other hand, every solution $\underline{f}_\beta$ of equation (3.7) is also a solution of (3.11), since (3.11) simply follows from the identity

$$(3.13) \qquad \chi^\Gamma(QT^\varepsilon QT^{-\varepsilon} Q) \, Y(z - 1, \varepsilon) - \lambda_\beta \, \chi^\Gamma(QT^{2\varepsilon} Q) z^{-2\beta} Y(\frac{1}{z} - 1, -\varepsilon) \equiv 0 \,.$$

For $\lambda_\beta = \pm 1$ this implies

$$\chi^\Gamma(QT^\varepsilon QT^{-\varepsilon}Q) \times$$

$$\left[\lambda_\beta \underline{f}_\beta(z-1,\varepsilon) - \lambda_\beta \chi^\Gamma(QT^\varepsilon Q)\underline{f}_\beta(z,\varepsilon) - (\frac{1}{z})^{2\beta}\chi^\Gamma(QT^\varepsilon)\underline{f}_\beta(\frac{1}{z},-\varepsilon)\right]$$

$$-\lambda_\beta\,\chi^\Gamma(QT^{2\varepsilon}Q)z^{-2\beta} \times$$

$$\left[\lambda_\beta \underline{f}_\beta(\frac{1}{z}-1,-\varepsilon) - \lambda_\beta\chi^\Gamma(QT^{-\varepsilon}Q)\underline{f}_\beta(\frac{1}{z},-\varepsilon) - z^{2\beta}\chi^\Gamma(QT^{-\varepsilon})\underline{f}_\beta(z,\varepsilon)\right] = 0.$$

The identity $(QT)^3 = id$ then leads to equation (3.11).

Define $\underline{\phi}(z,\varepsilon) := \underline{f}_\beta(z-1,\varepsilon)$, where we suppressed for the moment the dependence on β in the function $\underline{\phi}$. In terms of the function $\underline{\phi}$ the three functional equations (3.7), (3.10) and (3.11) can be written as

$$(I)' \quad \lambda_\beta\left[\underline{\phi}(z,\varepsilon) - \chi^\Gamma(QT^\varepsilon Q)\,\underline{\phi}(z+1,\varepsilon)\right]$$

(3.14)
$$-z^{-2\beta}\chi^\Gamma(QT^\varepsilon)\,\underline{\phi}(1+\frac{1}{z},-\varepsilon) = 0,$$

$$(II)' \quad \underline{\phi}(z,\varepsilon) - \chi^\Gamma(QT^\varepsilon Q)\,\underline{\phi}(z+1,\varepsilon)$$

(3.15)
$$-(z+1)^{-2\beta}\chi^\Gamma(QT^{2\varepsilon})\,\underline{\phi}(\frac{z}{z+1},\varepsilon) = 0, \quad \lambda_\beta = \pm 1,$$

(3.16)
$$(III)' \quad \underline{\phi}(z,\varepsilon) - \lambda_\beta\, z^{-2\beta}\,\chi^\Gamma(T^{-\varepsilon})\,\underline{\phi}(\frac{1}{z},-\varepsilon) = 0, \quad \lambda_\beta = \pm 1.$$

We call equation $(I)'$ the (generalized vector-valued) Lewis equation, because for the trivial one-dimensional representation $\chi^\Gamma = 1$ and $\lambda_\beta = +1$ equation (3.14) when supressing the parameter ε is the same as equation (2.41) of Lewis for $\Gamma(1)$ [**Lew97**]. We call equation $(II)'$ the (generalized vector-valued) master equation, because this equation for the trivial one-dimensional representation $\chi^\Gamma = 1$ when supressing the parameter ε was called master equation by Lewis and Zagier [**LZ97**]. Equation (3.7) respectively (3.10) corresponding to $(I)'$ respectively $(II)'$ are the Lewis respectively the master equation for the function $\underline{f}_\beta$.

REMARK 3.1. The three equations (3.14), (3.15) and (3.16) are equations for vector-valued functions. Every equation consists of μ equations, where μ is the dimension of the representation χ^Γ. The concrete form of these μ equations depends on the choice of the basis. For instance applying the matrix $\chi^\Gamma(Q)$ on the left to (3.14) and renaming $\chi^\Gamma(Q)\underline{\phi}$ by $\underline{\phi}$ one obtains a functional equation of the form:

$$\lambda_\beta\left[\underline{\phi}(z,\varepsilon) - \chi^\Gamma(T^\varepsilon)\,\underline{\phi}(z+1,\varepsilon)\right] - z^{-2\beta}\chi^\Gamma(T^\varepsilon Q)\,\underline{\phi}(1+\frac{1}{z},-\varepsilon) = 0.$$

Obviously, this equation is equivalent to (3.14).

If the induced representation χ^Γ in $\tilde{\mathcal{L}}_\beta$ in (3.1) has the property $\chi^\Gamma(T^2) = 1$, then the operator $\tilde{\mathcal{L}}_\beta$ can be decomposed as $\tilde{\mathcal{L}}_\beta = \begin{pmatrix} 0 & \mathcal{L}_\beta \\ \mathcal{L}_\beta & 0 \end{pmatrix}$ for $\Gamma(1)$ in (2.11).

In complete analogy to the above discussion for the operator $\tilde{\mathcal{L}}_\beta$ one can derive functional equations for the eigenfunctions of $\mathcal{L}_\beta$ similar to $(I)'$, $(II)'$ and $(III)'$:

$$(3.17)\,(I)\quad \lambda_\beta\left[\underline{\phi}(z)-\chi^\Gamma(QTQ)\underline{\phi}(z+1)\right]-z^{-2\beta}\chi^\Gamma(QT)\underline{\phi}(1+\frac{1}{z})=0,$$

$$(3.18)\,(II)\quad \underline{\phi}(z)-\chi^\Gamma(QTQ)\underline{\phi}(z+1)-(z+1)^{-2\beta}\chi^\Gamma(Q)\underline{\phi}(\frac{z}{z+1})=0,\quad \lambda_\beta=\pm 1,$$

$$(3.19)\,(III)\quad \underline{\phi}(z)-\lambda_\beta\, z^{-2\beta}\,\chi^\Gamma(T)\,\underline{\phi}(\frac{1}{z})=0,\quad \lambda_\beta=\pm 1.$$

Compared to equations $(I)'$, $(II)'$ and $(III)'$ the function $\underline{\phi}(z,\varepsilon)$ has been changed simply to $\underline{\phi}(z)$. Criterion (3.6) now reads

$$\lim_{z\to\infty}\left[\lambda_\beta\underline{f}_\beta(z)-\right.$$

$$(3.20)\qquad \left.\sum_{l=0}^{\kappa}\sum_{m=1}^{r}(\frac{1}{r})^{2\beta+l}\chi^\Gamma(QT^m)\frac{\underline{f}_\beta^{(l)}(0)}{l!}\zeta(2\beta+l,\frac{z+m}{r})\right]=0.$$

Summarizing, we showed that an eigenfunction of the operator $\tilde{\mathcal{L}}_\beta$ with the argument shifted from z to $z+1$ is always a solution of Lewis equation (3.14) with $\lambda_\beta=+1$. Conversely, a solution of Lewis equation (3.14) with parameter value $\lambda_\beta=+1$ is an eigenfunction of $\tilde{\mathcal{L}}_\beta$ with eigenvalue $\lambda_\beta=+1$ only if this solution satisfies the criterion (3.6). Next we will discuss this criterion for different regions of the variable $\beta\in\mathbb{C}$.

3.2. Asymptotic properties of the eigenfunctions of the transfer operator. Let us consider first the region $0<\Re\beta\le\frac{1}{2}$ which includes the two vertical lines $\Re\beta=\frac{1}{2}$ respectively $\Re\beta=\frac{1}{4}$ on which the spectral respectively the Riemann (if the Riemann hypothesis is true) zeros of $Z_S(\beta;\Gamma(1);\chi^\Gamma)$ lie. To obtain the analytic continuation of the transfer operator to this region one has to set $\kappa=0$ in (3.6). Then criterion (3.6) reads

$$(3.21)\qquad \lim_{z\to\infty}\left[\lambda_\beta\underline{f}_\beta(z,\varepsilon)-\sum_{m=1}^{r}(\frac{1}{r})^{2\beta}\chi^\Gamma(QT^{m\varepsilon})\underline{f}_\beta(0,-\varepsilon)\zeta(2\beta,\frac{z+m}{r})\right]=0.$$

For $\beta\ne\frac{1}{2}$ the function $\zeta(2\beta,\frac{z+m}{r})$ does not have any singularity. For large z with $|arg\,z|<\pi$ the Hurwitz zeta function behaves like [**MOS66**] p.25:

$$\zeta(s,z)\;=\;\frac{1}{\Gamma(s)}\left[z^{1-s}\Gamma(s-1)+\frac{1}{2}\Gamma(s)z^{-s}+\sum_{n=1}^{m-1}\frac{B_{2n}}{(2n)!}\Gamma(s+2n-1)z^{-2n-s+1}\right]$$

$$(3.22)\qquad +O(z^{-2m-s-1}),$$

where $\Gamma(s)$ is the gamma function and $B_n(z)$ is the n-th Bernoulli polynomial of degree n. That is, in the region $0<\Re s\le 1$ all the terms in (3.22) vanish in the limit $z\to\infty$, only the leading term $z^{1-s}\frac{\Gamma(s-1)}{\Gamma(s)}$ survives. Hence, in the region $0<\Re\beta\le\frac{1}{2}$ criterion (3.21) can be simplified to

$$\lim_{z\to\infty}\left[\lambda_\beta\underline{f}_\beta(z,\varepsilon)-\right.$$

$$(3.23)\qquad \left.\frac{\underline{f}_\beta(0,-\varepsilon)}{r}\frac{\Gamma(2\beta-1)}{\Gamma(2\beta)}\sum_{m=1}^{r}\chi^\Gamma(QT^{m\varepsilon})(z+m)^{1-2\beta}\right]=0.$$

To apply criterion (3.6) at the points $\beta=\beta_\kappa=\frac{1-\kappa}{2}$, $\kappa\in\mathbb{N}_0$, which in general are poles of the meromorphic operator $\tilde{\mathcal{L}}_\beta$ in (2.8), we recall first the properties of the eigenvalues and the eigenfunctions of the operators $\tilde{\mathcal{L}}_\beta$ and $\tilde{\mathcal{A}}_\beta^{(\kappa)}$ in the limit $\beta \to \beta_\kappa$: Since in general also $\tilde{\mathcal{A}}_\beta^{(\kappa)}$ is singular at $\beta = \beta_\kappa$, both $\tilde{\mathcal{L}}_\beta$ and $\tilde{\mathcal{A}}_\beta^{(\kappa)}$ can be singular and not defined at the critical points $\beta = \beta_\kappa$. As in the case $\Gamma(1)$ [**CM98**] the operators $\tilde{\mathcal{L}}_\beta$ and $\tilde{\mathcal{A}}_\beta^{(\kappa)}$ have besides the eigenvalues λ_β which are regular in the limit $\beta \to \beta_\kappa$ also eigenvalues $\hat{\lambda}_\beta$ becoming singular for $\beta \to \beta_\kappa$. The eigenfunctions of both λ_β and $\hat{\lambda}_\beta$ can however be rescaled by a multiplicative factor $(\beta - \beta_\kappa)^\delta$ with some $\delta \in \mathbb{Q}$ in such a way that the rescaled eigenfunctions remain regular in the limit $\beta \to \beta_\kappa$. Now if λ_β is a regular eigenvalue of $\tilde{\mathcal{L}}_\beta$ in (2.8) with regular eigenfunction $\underline{f}_\beta(z)$ for $\beta \to \beta_\kappa$, then the only singular term in the sum (2.9) respectively in (3.6), namely the term $l = \kappa$ must vanish. That is, either $\sum_{m=1}^r \chi^\Gamma(QT^{m\varepsilon})$ must be zero or $\underline{f}_\beta^{(\kappa)}(0, -\varepsilon)$ in (2.9) respectively (3.6) must vanish fast enough for $\beta \to \beta_\kappa$ to cancel the singularity in the Hurwitz function which for $\beta \to \beta_\kappa$ behaves like

$$\zeta(2\beta + \kappa, \frac{z+m}{r}) \sim \frac{1}{2}\frac{1}{\beta - \beta_\kappa}.$$

The remaining terms $l = 0, \cdots, \kappa - 1$ in (3.6) are polynomials of degree $\kappa - l$, because the zeta functions $\zeta(2\beta_\kappa + l, z)$, $l = 0, \cdots, \kappa - 1$ are for these β-values polynomials of degree $-(2\beta_\kappa + l) + 1 = \kappa - l$, since [**MOS66**]

$$(3.24) \qquad\qquad \zeta(-n, z) = \frac{B_{n+1}(z)}{n+1}, \; n \in \mathbb{N}_0$$

with B_n the Bernoulli polynomial of degree $n + 1$ Therefore, if λ_β and $\underline{f}_\beta$ fulfill the Lewis equation (3.7), are regular for $\beta \to \beta_\kappa$ and obey the asymptotics

$$\lim_{z \to \infty} \Big[\lambda_{\beta_\kappa} \underline{f}_{\beta_\kappa}(z, \varepsilon) -$$

$$(3.25) \qquad \sum_{l=0}^\kappa \sum_{m=1}^r (\frac{1}{r})^{2\beta_\kappa + l} \chi^\Gamma(QT^{m\varepsilon}) \frac{f_{\beta_\kappa}^{(l)}(0, -\varepsilon)}{l!} \frac{B_{k-l}(\frac{z+m}{r})}{k-l} \Big] = 0,$$

then they define a regular eigenvalue with eigenfunction $\underline{f}_\beta$ for $\tilde{\mathcal{L}}_\beta$ for $\beta = \beta_\kappa$. We call them simply regular eigenvalue and eigenfunction for $\mathcal{L}_{\beta_\kappa}$, although $\mathcal{L}_{\beta_\kappa}$ is not defined in the strict sense. Similar to the arguments in [**CM99**] one shows that, a vector-valued solution of Lewis equation (3.14) respectively (3.17) for $\beta = \beta_\kappa$ with components in $\oplus_{n=0}^\kappa \mathbb{C}z^n \oplus \frac{\mathbb{C}}{z}$ is always an eigenfunction of the transfer operator $\tilde{\mathcal{L}}_{\beta_\kappa}$ respectively $\mathcal{L}_{\beta_\kappa}$.

For $\Gamma(1)$ there are two kinds of eigenfunctions for $\mathcal{L}_{\beta_\kappa}$ [**CM99**]. One class lies in the polynomial space $\oplus_{n=0}^{\kappa-1} \mathbb{C}z^n$. An example of such a polynomial eigenfunction is

$$f_{-5}(z) = 10z + 45z^2 - 240z^3 - 1025z^4 - 1458z^5 - 915z^6 - 240z^7 + 10z^9 + z^{10}$$

for $\kappa = 11$ (corresponding to $\beta = -5$). The function $\phi(z) = f_{-5}(z - 1)$ is just the even part of the period polynomial of the holomorphic cusp form of $\Gamma(1)$ of weight 12. The other class consists of functions in $\oplus_{n=0}^\kappa \mathbb{C}z^n \oplus \frac{\mathbb{C}}{z+1}$. An example for such a solution is

$$f_{-1}(z) = \frac{1}{z+1} - 4 - 2z + 3z^2 + z^3$$

for $\kappa = 3$ (corresponding to $\beta = -1$), with the property $f(z) \sim_{z \to \infty} z^3$ and $f^{(3)}(0) = 0$. The function $\phi(z) = f_{-1}(z-1)$ is the odd part of the period function of the holomorphic Eisenstein series of weight 4 for the group $\Gamma(1)$.

We will next study solutions of the Lewis equation (3.14) for different subgroups Γ of $\Gamma(1)$ which are also eigenfunctions for the corresponding transfer operators.

4. Eigenfunctions of the transfer operator for subgroups of $\Gamma(1)$

Let Γ be one of the subgroups $\{\Gamma(1), \Gamma_2, \Gamma_0(2), \Gamma^0(2), \Gamma_\vartheta, \Gamma(2)\}$ of $\Gamma(1)$. Consider then the unitary irreducible representations of $\Gamma(1)$ defined through the following representations of the generators Q and T:

$$
\begin{array}{llll}
(4.1) & \chi_1(Q) = 1, & \chi_1(T) = 1, \\
(4.2) & \chi_{-1}(Q) = -1, & \chi_{-1}(T) = -1, \\
(4.3) & \chi_2(Q) = \begin{pmatrix} 1 & 0 \\ -1 & -1 \end{pmatrix}, & \chi_2(T) = \begin{pmatrix} 0 & 1 \\ 1 & 0 \end{pmatrix}.
\end{array}
$$

The induced representations χ^Γ of $\Gamma(1)$ induced from the trivial representation of the subgroup Γ can then be decomposed as follows [**CM00**]:

$$
\begin{array}{rcl}
\chi^{\Gamma(1)} & = & \chi_1, \\
\chi^{\Gamma_2} & = & \chi_1 \oplus \chi_{-1}, \\
\chi^{\Gamma'} & = & \chi_1 \oplus \chi_2, \quad \text{for } \Gamma' \text{ one of the subgroups } \{\Gamma_0(2), \Gamma^0(2), \Gamma_\vartheta\}, \\
(4.4) \quad \chi^{\Gamma(2)} & = & \chi_1 \oplus \chi_{-1} \oplus \chi_2 \oplus \chi_2.
\end{array}
$$

Since any of these representations χ^Γ has the property $\chi^\Gamma(T^2) = 1$, we can restrict the discussion of the operator $\tilde{\mathcal{L}}_\beta^{\Gamma(1),\chi^\Gamma}$ to the operator $\mathcal{L}_\beta^{\Gamma(1),\chi^\Gamma}$ since $\tilde{\mathcal{L}}_\beta^{\Gamma(1),\chi^\Gamma} = \begin{pmatrix} 0 & \mathcal{L}_\beta^{\Gamma(1),\chi^\Gamma} \\ \mathcal{L}_\beta^{\Gamma(1),\chi^\Gamma} & 0 \end{pmatrix}$ (see section 2.2.2). With

$$
(4.5) \qquad \mathcal{L}_\beta^{\chi_1} f_\beta(z) = \sum_{n=1}^{\infty} (\frac{1}{z+n})^{2\beta} f_\beta(\frac{1}{z+n}),
$$

$$
(4.6) \qquad \mathcal{L}_\beta^{\chi_{-1}} f_\beta(z) = \sum_{n=1}^{\infty} (-1)^{n+1} (\frac{1}{z+n})^{2\beta} f_\beta(\frac{1}{z+n}),
$$

$$
(4.7) \qquad \mathcal{L}_\beta^{\chi_2} \underline{f}_\beta(z) = \sum_{n=1}^{\infty} (\frac{1}{z+n})^{2\beta} \chi_2(QT^n) \underline{f}_\beta(\frac{1}{z+n})
$$

the transfer operators for the different subgroups Γ can be written as follows:

$$(4.8) \qquad \mathcal{L}_\beta^{\Gamma(1), \chi^{\Gamma(1)}} f(z) = \mathcal{L}_\beta^{\chi_1} f(z) \,,$$

$$(4.9) \qquad \mathcal{L}_\beta^{\Gamma(1), \chi^{\Gamma_2}} \underline{f}(z) = \begin{pmatrix} \mathcal{L}_\beta^{\chi_1} & 0 \\ 0 & \mathcal{L}_\beta^{\chi_{-1}} \end{pmatrix} \underline{f}(z) \,,$$

$$(4.10) \qquad \mathcal{L}_\beta^{\Gamma(1), \chi^\Gamma} \underline{f}(z) = \begin{pmatrix} \mathcal{L}_\beta^{\chi_1} & 0 \\ 0 & \mathcal{L}_\beta^{\chi_2} \end{pmatrix} \underline{f}(z) \,,$$

$$(4.11) \qquad \mathcal{L}_\beta^{\Gamma(1), \chi^{\Gamma(2)}} \underline{f}(z) = \begin{pmatrix} \mathcal{L}_\beta^{\chi_1} & & & 0 \\ & \mathcal{L}_\beta^{\chi_{-1}} & & \\ & & \mathcal{L}_\beta^{\chi_2} & \\ 0 & & & \mathcal{L}_\beta^{\chi_2} \end{pmatrix} \underline{f}(z) \,.$$

We can look for solutions $\varphi(z)$ of Lewis equation (3.17) for the two parameter values $\lambda_\beta = \pm 1$ where χ^Γ has been replaced by χ_i for $i = 1, -1, 2$. Then the functions $\underline{f}_\beta(z) = \varphi(z + 1)$ fulfilling criterion (3.20) (with χ^Γ replaced by χ_i) are eigenfunctions of the operator $\tilde{\mathcal{L}}_\beta^{\chi_i}$. Let us briefly recall for the following discussion the results for the transfer operator $\mathcal{L}_\beta^{\chi_1}$.

4.1. The eigenfunctions of the operator $\mathcal{L}_\beta^{\chi_1}$. The transfer operator (4.5)

$$\mathcal{L}_\beta^{\chi_1} f_\beta(z) = \sum_{n=1}^{\infty} \left(\frac{1}{z+n} \right)^{2\beta} f_\beta \left(\frac{1}{z+n} \right)$$

is nothing else but the transfer operator of the Gauß transformation [**May91a**]. The corresponding functional equations in (3.17), (3.18) and (3.19) are then

$$(4.12) \qquad (I) \quad \lambda_\beta \left[\phi(z) - \phi(z+1) \right] - z^{-2\beta} \phi\left(1 + \frac{1}{z}\right) = 0 \,,$$

$$(4.13) \qquad (II) \quad \phi(z) - \phi(z+1) - (z+1)^{-2\beta} \phi\left(\frac{z}{z+1}\right) = 0 \,, \quad \lambda_\beta = \pm 1 \,,$$

$$(4.14) \qquad (III) \quad \phi(z) - \lambda_\beta z^{-2\beta} \phi\left(\frac{1}{z}\right) = 0 \,, \quad \lambda_\beta = \pm 1 \,.$$

The functions $f_\beta(z)$ and $\phi(z+1)$ are scalar functions. Suppose $\phi(z)$ is a solution of (4.12). Then the function $f_\beta(z) = \phi(z+1)$ is an eigenfunction of the operator $\mathcal{L}_\beta^{\chi_1}$ if it satisfies condition (3.20) with $r = 1$ and $\chi^\Gamma = 1$ which hence reads

$$(4.15) \qquad \lim_{z \to \infty} \left[\lambda_\beta f_\beta(z) - \sum_{l=0}^{\kappa} \frac{f_\beta^{(l)}(0)}{l!} \zeta(2\beta + l, z + 1) \right] = 0 \,.$$

Defining $F(z) := z^\beta \phi(z)$ equation (4.14) can be rewritten as

$$(4.16) \qquad F(z) = \lambda_\beta F\left(\frac{1}{z}\right), \quad \lambda_\beta = \pm 1 \,.$$

That is, under the transformation $z \leftrightarrow \frac{1}{z}$ the function $F(z)$ is symmetric for $\lambda_\beta = +1$ respectively antisymmetric for $\lambda_\beta = -1$. If $F(z)$ is a solution of (4.16), then $\phi(z) = z^{-\beta} F(z)$ is a solution of (4.14).

It is well known [**Hej83**] that Selberg's zeta function $Z_S(\beta; \Gamma(1); 1)$ for the trivial representation of the modular group $\Gamma(1)$ has trivial zeros at $\beta = -N$,

$N \in \mathbb{N}$, spectral zeros on $\Re\beta = \frac{1}{2}$ and Riemann zeros on $\Re\beta = \frac{1}{4}$ (if the Riemann conjecture is true). Since [**CM00**]:

$$Z_S(\beta; \Gamma(1); 1) = \det(1 - \tilde{\mathcal{L}}_\beta^{\chi_1}) = \det(1 + \mathcal{L}_\beta^{\chi_1}) \det(1 - \mathcal{L}_\beta^{\chi_1}),$$

all the zeros β of the function $Z_S(\beta; \Gamma(1); 1)$ can be explained by the presence of an eigenvalue $\lambda_\beta = \pm 1$ for $\mathcal{L}_\beta^{\chi_1}$ at the corresponding β value. For such β's the eigenfunctions of $\mathcal{L}_\beta^{\chi_1}$ with $\lambda_\beta = \pm 1$ are explicitly related to the holomorphic modular forms and the Maaß wave forms of $\Gamma(1)$ as follows [**CM99**], [**CM98**]:

PROPOSITION 4.1. *Let* $\beta_\kappa = \frac{1-\kappa}{2}$ *and* $\kappa \in \{-1\} \cup \mathbb{N}_0$. *Then one has*

(i) $\mathcal{L}_1$ *(*$\kappa = -1$*) has the leading eigenvalue* $\lambda_1 = +1$ *with eigenfunction* $f_1(z) = \frac{1}{z+1}$. *The function* $\phi(z) = f_1(z-1) = \frac{1}{z}$ *is the Lewis transform (2.43) of the constant eigenfunction of* $-\triangle$ *with eigenvalue* $\rho = 0$. *(*$\frac{1}{\log 2}\frac{1}{z+1}$ *is the normalized Gauß measure on the unit interval* $[0, 1]$*).*

(ii) $\mathcal{L}_{\frac{1}{2}}$ *(*$\kappa = 0$*) has eigenvalue* $\lambda_{\frac{1}{2}} = -1$ *with eigenfunction* $f_{\frac{1}{2}}(z) = \frac{1}{z+1} - 1$.
$\mathcal{L}_0$ *(*$\kappa = 1$*) has eigenvalue* $\lambda_0 = +1$ *with eigenfunction* $f_0(z) = \frac{1}{z+1} - 2 + z$.

(iii) *For* $\kappa \geq 2$ *the operator* $\mathcal{L}_{\beta_\kappa}$ *has eigenvalue* $\lambda_{\beta_\kappa} = -1$ *with eigenfunction*

$$(4.17) \qquad\qquad f_{\beta_\kappa}(z) = (z+1)^{\kappa-1} - 1,$$

where $\phi(z) = f_{\beta_\kappa}(z-1) = p_{\kappa+1}^+(z)$ *is the even part of the period polynomial* $p_k^+(z)$ *in (2.36) for the holomorphic Eisenstein series of weight* $k = \kappa + 1$.

For $\kappa = 3, 5, 7, \cdots$ *the operator* $\mathcal{L}_{\beta_\kappa}$ *has eigenvalue* $\lambda_{\beta_\kappa} = +1$ *with eigenfunction*

$$(4.18) \qquad\qquad f_{\beta_\kappa}(z) = \sum_{-1 \leq n \leq \kappa, \, n \text{ odd}} \frac{B_{n+1}}{(n+1)!} \frac{B_{\kappa-n}}{(\kappa-n)!}(z+1)^n,$$

where $\phi(z) = f_{\beta_\kappa}(z-1) = p_{\kappa+1}^-(z)$ *is the odd part of the period polynomial* $p_k^-(z)$ *in (2.37) for the holomorphic Eisenstein series of weight* $k = \kappa + 1$.

(iv) $\mathcal{L}_{\beta_\kappa}$ *has eigenvalues* $\lambda_{\beta_\kappa} = -1$ *respectively* $\lambda_{\beta_\kappa} = +1$ *with polynomial eigenfunctions* $f_+(z)$ *respectively* $f_-(z)$ *in the space* $\oplus_{n=0}^{\kappa-1}\mathbb{C}z^n$ *for those* κ, *for which* $\phi_+(z) = f_+(z-1)$ *respectively* $\phi_-(z) = f_-(z-1)$ *is the even respectively odd part of the period polynomial of a holomorphic cusp form of* $\Gamma(1)$ *of weight* $k = \kappa + 1$ *(see section 2.3.1). The dimension of these eigenfunctions for fixed* κ *is equal to the dimension (2.15) of the cusp forms of weight* $k = \kappa + 1$.

(v) *Let* $\psi_\beta(z)$ *be the analytic continuation of the function given for* $\Re\beta > 1$ *as*

$$\psi_\beta(z) = \sum_{m,n \geq 1} (\frac{1}{mz+n})^{2\beta} + \frac{1}{2}\zeta(2\beta)(1 + (\frac{1}{z})^{2\beta}).$$

Then $f_\beta(z) = \psi_\beta(z+1)$ *is an eigenfunction of* $\mathcal{L}_\beta$ *with eigenvalue* $\lambda_\beta = +1$ *for those values of* β *for which* $\zeta(2\beta) = 0$.

4.2. Eigenfunctions of the operator $\mathcal{L}_\beta^{\chi_{-1}}$. The transfer oprator for the representation χ_{-1} in (4.6)

$$\mathcal{L}_\beta^{\chi_{-1}} f_\beta(z) = \sum_{n=1}^{\infty} (-1)^{n+1} (\frac{1}{z+n})^{2\beta} f_\beta(\frac{1}{z+n})$$

differs from $\mathcal{L}_\beta^{\chi_1}$ in (4.5) only in the additional factor $(-1)^{n+1}$ under the sum. The functional equations (3.17), (3.18) and (3.19) read now

(4.19)
$$(I) \quad \lambda_\beta \left[\phi(z) + \phi(z+1) \right] - z^{-2\beta} \phi(1 + \frac{1}{z}) = 0,$$

$$(II) \quad \phi(z) + \phi(z+1) + (z+1)^{-2\beta} \phi(\frac{z}{z+1}) = 0,$$

$$(III) \quad \phi(z) + \lambda_\beta z^{-2\beta} \phi(\frac{1}{z}) = 0, \quad \lambda_\beta = \pm 1.$$

Criterion (3.20) can be simplified in the region $\Re\beta > -\frac{\kappa}{2}$ with $\kappa \in \mathbb{N}_0$ and for $r = 2$ to
(4.20)
$$\lim_{z \to \infty} \left[\lambda_\beta f_\beta(z) - \sum_{l=0}^{\kappa} 2^{-2\beta-l} \frac{f_\beta^{(l)}(0)}{l!} \left(\zeta(2\beta + l, \frac{z+1}{2}) - \zeta(2\beta + l, \frac{z+2}{2}) \right) \right] = 0.$$

Let $\phi(z)$ be a solution of (4.19). Then $f_\beta(z) = \phi(z+1)$ is an eigenfunction of $\mathcal{L}_\beta^{\chi_{-1}}$ if $f_\beta(z)$ fulfills this criterion (4.20).

Since the singularities in the two zeta functions for $l = \kappa$ in (4.20) cancel each other for $\beta \to \beta_\kappa$, the operator $\mathcal{L}_\beta^{\chi_{-1}}$ is holomorphic at $\beta = \beta_\kappa$ and hence for all β in $\mathbb{C}$ [**CM00**]. The operator $\mathcal{L}_\beta^{\chi_{-1}}$ is nuclear in the entire complex β-plane with pure point spectrum [**CM00**]. This just reflects the fact that the representation χ_{-1} is regular (see section 2.3.2): the Laplace-Beltrami operator $-\triangle$ in (2.17) with this representation has only a discrete spectrum [**BV97**].

For $\beta = \frac{1-\kappa}{2}$ with $\kappa \in \mathbb{N}_0$ Lewis equation (4.19) has again solutions $\phi(z)$ in the space $\oplus_{n=0}^{\kappa-1} \mathbb{C} z^n$ of polynomials. It is straightforward to see that the corresponding polynomials $f_\beta(z-1)$ automatically fulfill the asymptotics in (4.20) and hence are eigenfunctions of the transfer operator $\mathcal{L}_\beta^{\chi_{-1}}$. These polynomial solutions can be determined exactly on the computer.

The number of polynomial solutions of the Lewis equation (4.19) as determined numerically for different $\beta = \beta_\kappa$ is shown in table 2 of section 2.

4.3. Eigenfunctions of the operator $\mathcal{L}_\beta^{\chi_2}$. With the two matrices

(4.21)
$$\chi_2(QT^n) = \begin{cases} \begin{pmatrix} 1 & 0 \\ -1 & -1 \end{pmatrix} & n \text{ even} \\ \begin{pmatrix} 0 & 1 \\ -1 & -1 \end{pmatrix} & n \text{ odd} \end{cases}$$

the transfer operator (4.7) for the representation χ_2 can be written as

$$\mathcal{L}_\beta^{\chi_2} \underline{f}_\beta(z) = \sum_{n=1}^{\infty} (\frac{1}{z+n})^{2\beta} \chi_2(QT^n) \underline{f}_\beta(\frac{1}{z+n})$$

$$= \begin{pmatrix} 0 & 1 \\ -1 & -1 \end{pmatrix} \sum_{n=1}^{\infty} \left(\frac{1}{z+(2n-1)} \right)^{2\beta} \underline{f}_\beta \left(\frac{1}{z+(2n-1)} \right)$$

(4.22)
$$+ \begin{pmatrix} 1 & 0 \\ -1 & -1 \end{pmatrix} \sum_{n=1}^{\infty} (\frac{1}{z+2n})^{2\beta} \underline{f}_\beta(\frac{1}{z+2n}).$$

The corresponding functional equations (3.17), (3.18) and (3.19) for χ_2 then read

$$(4.23) \quad (I) \quad \lambda_\beta \left[\underline{\phi}(z) - \begin{pmatrix} -1 & -1 \\ 0 & 1 \end{pmatrix} \underline{\phi}(z+1) \right] - z^{-2\beta} \begin{pmatrix} 0 & 1 \\ -1 & -1 \end{pmatrix} \underline{\phi}(1 + \frac{1}{z}) = 0,$$

$$(II) \ \underline{\phi}(z) - \begin{pmatrix} -1 & -1 \\ 0 & 1 \end{pmatrix} \underline{\phi}(z+1) - (z+1)^{-2\beta} \begin{pmatrix} 1 & 0 \\ -1 & -1 \end{pmatrix} \underline{\phi}(\frac{z}{z+1}), \ \lambda_\beta = \pm 1,$$

$$(4.24) \quad (III) \ \underline{\phi}(z) - z^{-2\beta} \lambda_\beta \begin{pmatrix} 0 & 1 \\ 1 & 0 \end{pmatrix} \underline{\phi}(\frac{1}{z}) = 0, \quad \lambda_\beta = \pm 1.$$

Criterion (3.20) for χ_2 in the region $\Re\beta > -\frac{\kappa}{2}$, $\kappa \in \mathbb{N}_0$ has the form

$$(4.25) \quad \lim_{z \to \infty} \left[\lambda_\beta \underline{f}_\beta(z) - \sum_{l=0}^{\kappa} \sum_{m=1}^{2} (\frac{1}{2})^{2\beta+l} \chi_2(QT^m) \frac{f_\beta^{(l)}(0)}{l!} \zeta(2\beta+l, \frac{z+m}{2}) \right] = 0.$$

Let $\underline{\phi} = \begin{pmatrix} \phi_1(z) \\ \phi_2(z) \end{pmatrix}$ then be a solution of the Lewis equation (4.23). The ϕ_i fulfill the equations

$$(4.26) \qquad \lambda_\beta \left[\phi_1(z) + \phi_1(z+1) + \phi_2(z+1) \right] - z^{-2\beta} \phi_2(1 + \frac{1}{z}) = 0,$$

$$(4.27) \qquad \lambda_\beta \left[\phi_2(z) - \phi_2(z+1) \right] + z^{-2\beta} \phi_1(1 + \frac{1}{z}) + z^{-2\beta} \phi_2(1 + \frac{1}{z}) = 0,$$

We denote the left-hand sides of (4.26) respectively (4.27) by $Y_1(z)$ respectively $Y_2(z)$. If $\underline{\phi}$ is a solution of (4.23), then $Y_1(z)$ and $Y_2(z)$ are identical zero and the identity

$$-\frac{z^{-2\beta}}{\lambda_\beta} Y_1(1 + \frac{1}{z}) + Y_2(z) + (z+1)^{-2\beta} Y_2(\frac{z}{z+1}) \equiv 0$$

implies

$$-z^{-2\beta} \left[\phi_1(\frac{z+1}{z}) + \phi_1(\frac{2z+1}{z}) + \phi_2(\frac{2z+1}{z}) \right] + \frac{1}{\lambda_\beta}(z+1)^{-2\beta} \phi_2(\frac{2z+1}{z+1})$$

$$+\lambda_\beta \left[\phi_2(z) - \phi_2(z+1) \right] + z^{-2\beta} \phi_1(\frac{z+1}{z}) + z^{-2\beta} \phi_2(\frac{z+1}{z})$$

$$+(z+1)^{-2\beta} \lambda_\beta \left[\phi_2(\frac{z}{z+1}) - \phi_2(\frac{2z+1}{z+1}) \right] + z^{-2\beta} \phi_1(\frac{2z+1}{z})$$

$$+z^{-2\beta} \phi_2(\frac{2z+1}{z}) = 0.$$

A simple calculation then gives

$$\lambda_\beta \left[\phi_2(z) - \phi_2(z+1) \right] + \lambda_\beta (z+1)^{-2\beta} \phi_2(\frac{z}{z+1}) + z^{-2\beta} \phi_2(\frac{z+1}{z})$$

$$(4.28) \qquad +(z+1)^{-2\beta} \phi_2(\frac{2z+1}{z+1}) (\frac{1}{\lambda_\beta} - \lambda_\beta) = 0,$$

where the function $\phi_1(z)$ has been eliminated. Since we are interested only in solutions for $\lambda_\beta = \pm 1$, the last term in (4.28) vanishes and one gets an equation for ϕ_2:

$$(4.29) \quad \lambda_\beta \left[\phi_2(z) - \phi_2(z+1) \right] + \lambda_\beta (z+1)^{-2\beta} \phi_2(\frac{z}{z+1}) + z^{-2\beta} \phi_2(\frac{z+1}{z}) = 0.$$

Similarly, adding $Y_1(z)$ and $Y_2(z)$ leads to the equation

$$(4.30) \qquad \lambda_\beta \left[\phi_1(z) + \phi_1(z+1) + \phi_2(z)\right] + z^{-2\beta} \phi_1(1 + \frac{1}{z}) = 0.$$

From this one can express the function $\phi_2(z)$ in terms of ϕ_1. Inserting this $\phi_2(z)$ into one of the relations (4.29), (4.26) or (4.27) one gets an equation for ϕ_1. That is, a solution $\underline{\phi}(z)$ of the Lewis equation (4.23) determines uniquely a solution $\phi_2(z)$ of equation (4.29). Conversely a solution $\phi_2(z)$ of (4.29) together with ϕ_1, determined from relation (4.27), gives a solution $\underline{\phi}(z) = \begin{pmatrix} \phi_1(z) \\ \phi_2(z) \end{pmatrix}$ of the Lewis equation (4.23). Hence equation (4.29) and criterion (4.25) determine the eigenfunctions of the transfer operator $\mathcal{L}_\beta^{\chi_2}$ with eigenvalue $\lambda_\beta = \pm 1$.

Let us next discuss some explicit solutions of these equations.

4.3.1. *Old solutions of Lewis' equation for the representation χ_2.* The first kind of solutions of the Lewis equation (4.23) for χ_2 are explicitly related to the solutions of the Lewis equation (4.12) for χ_1. Hence we call such solutions old solutions in analogy to the theory of old forms of Atkin-Lehner (see section 2.3.3).

PROPOSITION 4.2. *Let $\psi_\beta(z)$ be the analytic extension of Zagier's function in (2.45), which for $\Re\beta > 1$ reads*

$$(4.31) \qquad \psi_\beta(z) = \sum_{m,n \geq 1} (mz + n)^{-2\beta} + \frac{1}{2}\zeta(2\beta)(1 + z^{-2\beta}),$$

which for all β-values different from 1 is a solution of the Lewis equation (4.12) for χ_1 with $\lambda_\beta = 1$. Then

$$(4.32) \qquad \underline{\psi}_\beta(z) = \begin{pmatrix} \phi_1(z) \\ \phi_2(z) \end{pmatrix} = \begin{pmatrix} \frac{1}{3}(1 + 2^{1-2\beta}) & 0 & -2^{-2\beta} \\ \frac{1}{3}(1 + 2^{1-2\beta}) & -1 & 0 \end{pmatrix} \begin{pmatrix} \psi_\beta(z) \\ \psi_\beta(2z) \\ \psi_\beta(\frac{z}{2}) \end{pmatrix}$$

for β-values with $\beta \neq 1$ is a solution of the Lewis equation (4.23) for the representation χ_2 with $\lambda_\beta = 1$.

The proof of this proposition follows from Lemma 4.7 in section 4.5.1. We will argue next that $\underline{\psi}_\beta(z+1)$ is indeed an eigenfunction of $\mathcal{L}_\beta^{\chi_2}$ for β's with $\zeta(2\beta) = 0$.

COROLLARY 4.3. *Let $\underline{\psi}_\beta(z)$ be defined as in proposition 4.2. Then $\underline{\psi}_\beta(z+1)$ defines an eigenfunction of the transfer operator $\mathcal{L}_\beta^{\chi_2}$ for all β's with $\zeta(2\beta) = 0$.*

PROOF. The vector-valued function $\underline{\psi}_\beta(z) \in B(D) \oplus B(D)$ in (4.32) is obviously meromorphic in $\beta \in \mathbb{C}$, because $\psi_\beta(z)$ is meromorphic in the complex β-plane [**CM99**]. Since the function $\underline{\psi}_\beta(z)$ by Proposition 4.2 is for $\beta \neq 1$ a solution of the Lewis equation (4.23) with $\lambda_\beta = +1$, according to (3.7) the function $\underline{f}_\beta(z) = \underline{\psi}_\beta(z+1)$ is a solution of the following equation:

$$(4.33) \qquad \underline{f}_\beta(z) = \chi_2(QTQ)\underline{f}_\beta(z+1) + (\frac{1}{z+1})^{2\beta} \chi_2(QT)\underline{f}_\beta(\frac{1}{z+1}),$$

where we have set $\lambda_\beta = +1$. Iterating this equation N times one finds

$$\underline{f}_\beta(z) = \chi_2(QT^N Q)\underline{f}_\beta(z+N) + \sum_{n=1}^{N} (\frac{1}{z+n})^{2\beta} \chi_2(QT^n)\underline{f}_\beta(\frac{1}{z+n}),$$

where we used $\chi_2(Q^2) = 1$. Inserting the matrices (4.21) gives

$$(4.34) \qquad \underline{f}_\beta(z) = \begin{pmatrix} 1 & 0 \\ 0 & 1 \end{pmatrix} \underline{f}_\beta(z + N) + \sum_{n=1}^{N} (\frac{1}{z+n})^{2\beta} \chi_2(QT^n) \underline{f}_\beta(\frac{1}{z+n})$$

for even N and

$$(4.35) \qquad \underline{f}_\beta(z) = \begin{pmatrix} -1 & -1 \\ 0 & 1 \end{pmatrix} \underline{f}_\beta(z + N) + \sum_{n=1}^{N} (\frac{1}{z+n})^{2\beta} \chi_2(QT^n) \underline{f}_\beta(\frac{1}{z+n})$$

for odd N. However due to $\lim_{N\to\infty} \psi_\beta(z + 1 + N) = \frac{\zeta(2\beta)}{2}$ for $\Re\beta > 1$ we find

$$\lim_{N\to\infty} \underline{f}_\beta(z + N) = \lim_{N\to\infty} \underline{\psi}_\beta(z + 1 + N) = \begin{pmatrix} d_\beta \\ -2d_\beta \end{pmatrix} \quad \text{with } d_\beta = \frac{\cdot 1 - 2^{-2\beta}}{6} \zeta(2\beta) \,.$$

That is, in the limit $N \to \infty$ relations (4.34) and (4.35) reduce to

$$(4.36) \qquad \underline{f}_\beta(z) = \begin{pmatrix} 1 & 0 \\ 0 & 1 \end{pmatrix} \begin{pmatrix} d_\beta \\ -2d_\beta \end{pmatrix} + \mathcal{L}_\beta^{\chi_2} \underline{f}_\beta(z) \quad \text{for even } N$$

$$(4.37) \qquad \text{and} \quad \underline{f}_\beta(z) = \begin{pmatrix} -1 & -1 \\ 0 & 1 \end{pmatrix} \begin{pmatrix} d_\beta \\ -2d_\beta \end{pmatrix} + \mathcal{L}_\beta^{\chi_2} \underline{f}_\beta(z) \quad \text{for odd } N,$$

which obviously are identical to

$$(4.38) \qquad \underline{f}_\beta(z) = \begin{pmatrix} d_\beta \\ -2d_\beta \end{pmatrix} + \mathcal{L}_\beta^{\chi_2} \underline{f}_\beta(z) \,.$$

Since the function $\underline{f}_\beta(z) = \underline{\psi}_\beta(z+1)$ on the left-hand side of (4.38) is meromorphic for all $\beta \in \mathbb{C}$, the right-hand side must also be meromorphic in $\mathbb{C}$. For those $\beta \in \mathbb{C}$ with $\zeta(2\beta) = 0$, i.e., $d_\beta = 0$, the function $\underline{f}_\beta(z)$ is hence an eigenfunction of $\mathcal{L}_\beta^{\chi_2}$ with eigenvalue $\lambda_\beta = +1$. □

It is well known that Riemann's zeta function $\zeta(2\beta)$ has trivial zeros at $2\beta = -2, -4, -6, \cdots$ which correspond to $\beta = \beta_\kappa = \frac{1-\kappa}{2}$ with $\kappa = 3, 5, 7, \cdots$. For such β values the function $\psi_{\beta_\kappa}(z)$ coincides up to a factor with the odd part of the period polynomial $p_{\kappa+1}^-(z)$ in (2.37) for the holomorphic Eisenstein series of weight $k = \kappa + 1$ [**CM99**]. The Lewis equation (4.23) for $\lambda_\beta = +1$ has therefore the solution

$$(4.39) \qquad \underline{p}_{\kappa+1}^-(z) = \begin{pmatrix} \frac{1}{3}(1 + 2^\kappa) & 0 & -2^{\kappa-1} \\ \frac{1}{3}(1 + 2^\kappa) & -1 & 0 \end{pmatrix} \begin{pmatrix} p_{\kappa+1}^-(z) \\ p_{\kappa+1}^-(2z) \\ p_{\kappa+1}^-(\frac{z}{2}) \end{pmatrix}$$

which, up to a factor, is the same as the function $\underline{\psi}_\beta(z)$ at $\beta = \beta_\kappa$, which as we have shown is an eigenfunction if $\mathcal{L}_\beta^{\chi_2}$ for all β with $\zeta(2\beta) = 0$.

Another polynomial solution of the Lewis equation (4.23) for $\beta = \beta_\kappa$, $\kappa = 3, 5, 7, \cdots$ for the parameter value $\lambda_\beta = -1$ is

$$(4.40) \qquad \underline{p}_{\kappa+1}^+(z) = \begin{pmatrix} \frac{1}{3}(1 + 2^\kappa) & 0 & -2^{\kappa-1} \\ \frac{1}{3}(1 + 2^\kappa) & -1 & 0 \end{pmatrix} \begin{pmatrix} p_{\kappa+1}^+(z) \\ p_{\kappa+1}^+(2z) \\ p_{\kappa+1}^+(\frac{z}{2}) \end{pmatrix},$$

where $p_{\kappa+1}^+(z) = z^{\kappa-1} - 1$ is the even part of the period polynomial (2.36) of the holomorphic Eisenstein series of weight $k = \kappa + 1$. To verify this, one just inserts

$p^+_{\kappa+1}(z) = z^{\kappa-1} - 1$ into (4.40) and obtains

$$(4.41) \qquad \underline{p}^+_{\kappa+1}(z) = c_\kappa \begin{pmatrix} 2z^{\kappa-1} + 1 \\ -z^{\kappa-1} - 2 \end{pmatrix},$$

with $c_\kappa = \frac{2^{\kappa-1}-1}{3}$. As one can check immediately, the two components $\phi_1(z) = c_\kappa(2z^{\kappa-1} + 1)$ and $\phi_2(z) = c_\kappa(-z^{\kappa-1} - 2)$ in $\underline{p}^+_{\kappa+1}(z) = \begin{pmatrix} \phi_1(z) \\ \phi_2(z) \end{pmatrix}$ fulfill the two equations (4.26) and (4.27) which are equivalent to the Lewis equation (4.23). Therefore $\underline{p}^+_{\kappa+1}(z)$ is a solution of the Lewis equation (4.23) and the function $\underline{f}_{\beta_\kappa}(z) = \underline{p}^+_{\kappa+1}(z+1)$ is an eigenfunction of the operator $\mathcal{L}^{\chi_2}_\beta$ with eigenvalue $\lambda_\beta = -1$.

PROPOSITION 4.4. *Let $\underline{\psi}_\beta(z)$ be defined as in Proposition 4.2. Then the function $\underline{\psi}_\beta(z)$ is an eigenfunction of $\mathcal{L}^{\chi_2}_\beta$ for $\beta = 0$ with eigenvalue $\lambda_\beta = 1$.*

For the proof of this Proposition we refer to section 4.5.1.

Explicit examples for $\underline{p}^-_{\kappa+1}(z)$ and $\underline{p}^+_{\kappa+1}(z)$ for $\kappa = 3$ (corresponding to weight $k = 4$ and $\beta = -1$) are

$$\underline{p}^-_4(z) = \begin{pmatrix} -\frac{5}{z} - 5z + \frac{5}{2}z^3 \\ \frac{5}{2}\frac{1}{z} - 5z - 5z^3 \end{pmatrix} = \begin{pmatrix} 3 & 0 & -4 \\ 3 & -1 & 0 \end{pmatrix} \begin{pmatrix} p^-_4(z) \\ p^-_4(2z) \\ p^-_4(\frac{z}{2}) \end{pmatrix},$$

$$\underline{p}^+_4(z) = \begin{pmatrix} 1 + 2z^2 \\ -2 - z^2 \end{pmatrix} = \begin{pmatrix} 3 & 0 & -4 \\ 3 & -1 & 0 \end{pmatrix} \begin{pmatrix} p^+_4(z) \\ p^+_4(2z) \\ p^+_4(\frac{z}{2}) \end{pmatrix},$$

with the odd part of the period polynomial $p^-_4(z)$ respectively the even part of the period polynomial $p^+_4(z)$ of the Eisenstein series of weight 4 for $\Gamma(1)$ given as:

$$p^-_4(z) = \frac{1}{z} - 5z + z^3 \quad \text{respectively} \quad p^+_4(z) = z^2 - 1.$$

4.3.2. *Two classes of polynomial solutions for Lewis equation for χ_2.* Besides the solutions mentioned above, which are closely related to the nonholomorphic Eisenstein series respectively holomorphic Eisenstein series for β with $\zeta(2\beta) = 0$ for $\Gamma(1)$ the Lewis equation (4.23) has further polynomial solutions $\underline{\phi}(z) = \begin{pmatrix} \phi_1(z) \\ \phi_2(z) \end{pmatrix}$ for certain $\beta = \beta_\kappa$ with $\kappa \in \mathbb{N}_0$ and $\phi_1(z)$, $\phi_2(z) \in \oplus^{\kappa-1}_{n=0}\mathbb{C}z^n$ which obviously correspond to period polynomials of the holomorphic cusp forms of weight $k = \kappa+1$ for the group $\Gamma(1)$ (see section 2.4.3) respectively the corresponding polynomial eigenfunctions of the operator $\mathcal{L}^{\chi_1}_\beta$.

The two components of these solutions $\underline{\varphi}$ are linear combinations of the even respectively odd parts of the period polynomials $\varphi(z)$ of the group $\Gamma(1)$ with argument z, $2z$ and $\frac{z}{2}$ as shown in table 1. We call these solutions and the corresponding eigenfunctions old solutions respectively old eigenfunctions, because they can be constructed from the solutions and the eigenfunctions for the representation χ_1 of $\Gamma(1)$.

A second class of solutions $\underline{\phi}(z)$ has the property that the component $\phi_2(z)$ is proportional to the component $\phi_1(2z)$ plus the even part of the period polynomial of the holomorphic Eisenstein series in the case $\lambda_\beta = -1$ as shown in table 1. We call these solutions and eigenfunctions new solutions and new eigenfunctions, because they are not related to the solutions and the eigenfunctions for the representation χ_1 of $\Gamma(1)$. These solutions have been determined numerically

TABLE 1. Two classes of solutions of Lewis equation (4.23) for χ_2 with components in $\oplus_{n=0}^{\kappa-1}\mathbb{C}z^n$. $\varphi_o(z)$ respectively $\varphi_e(z)$ denote the odd respectively even part of a period polynomial of $\Gamma(1)$ and $p_{\kappa+1}^+$ is the even part of the period function of the holomorphic Eisenstein series of weight $\kappa+1$.

Class	λ_β	solutions in linear combinations
old	$+1$ (odd)	$\underline{\varphi}_o = \begin{pmatrix} c_1 & 0 & c_2 \\ c_1 & c_3 & 0 \end{pmatrix} \begin{pmatrix} \varphi_o(z) \\ \varphi_o(2z) \\ \varphi_o(\frac{z}{2}) \end{pmatrix}$
	-1 $(even)$	$\underline{\varphi}_e = \begin{pmatrix} c_1 & 0 & c_2 \\ c_1 & c_3 & 0 \end{pmatrix} \begin{pmatrix} \varphi_e(z) \\ \varphi_e(2z) \\ \varphi_e(\frac{z}{2}) \end{pmatrix} + \begin{pmatrix} d_1 & 0 & d_2 \\ d_1 & d_3 & 0 \end{pmatrix} \begin{pmatrix} p_{\kappa+1}^+(z) \\ p_{\kappa+1}^+(2z) \\ p_{\kappa+1}^+(\frac{z}{2}) \end{pmatrix}$
new	$+1$ (odd)	$\underline{\phi}_o = \begin{pmatrix} c_4\,\phi_o(z) \\ \phi_o(2z) \end{pmatrix}$
	-1 $(even)$	$\underline{\phi}_e = \begin{pmatrix} c_4\,\phi_e(z) \\ \phi_e(2z) \end{pmatrix} + \begin{pmatrix} d_5\,p_{\kappa+1}^+(z) \\ p_{\kappa+1}^+(2z) \end{pmatrix}$

Explicit examples for old solutions $\underline{\phi}(z)$ for $\kappa=11$ (corresponding to weight $k=12$ and $\beta=-5$) are

$$
\underline{\varphi}_o(z) = \begin{pmatrix} 52z - 85z^3 + 42z^5 - 10z^7 + z^9 \\ z - 10z^3 + 42z^5 - 85z^7 + 52z^9 \end{pmatrix} = \begin{pmatrix} 1 & 0 & 128 \\ 1 & \frac{1}{8} & 0 \end{pmatrix} \begin{pmatrix} \varphi_o(z) \\ \varphi_o(2z) \\ \varphi_o(\frac{z}{2}) \end{pmatrix},
$$

$$
\underline{\varphi}_e(z) = \begin{pmatrix} (t-3) + 22z^2 - 18z^4 + 6z^6 - z^8 + (2t+3)z^{10} \\ (-2t-3) + z^2 - 6z^4 + 18z^6 - 22z^8 + (3-t)z^{10} \end{pmatrix}
$$

$$
= \begin{pmatrix} 1 & 0 & 128 \\ 1 & \frac{1}{8} & 0 \end{pmatrix} \begin{pmatrix} \varphi_e(z) \\ \varphi_e(2z) \\ \varphi_e(\frac{z}{2}) \end{pmatrix} + \begin{pmatrix} 3 + \frac{683}{341}t & 0 & -\frac{1024}{341}t \\ 3 + \frac{683}{341}t & -\frac{1}{341}t & 0 \end{pmatrix} \begin{pmatrix} p_{12}^+(z) \\ p_{12}^+(2z) \\ p_{12}^+(\frac{z}{2}) \end{pmatrix},
$$

with $t \in \mathbb{C}$ a free parameter and

$$
\varphi_o(z) = \frac{4}{5}z - 5z^3 + \frac{42}{5}z^5 - 5z^7 + \frac{4}{5}z^9 \quad \text{respectively} \quad \varphi_e(z) = \frac{2}{3}z^2 - 2z^4 + 2z^6 - \frac{2}{3}z^8
$$

the odd respectively even part of the period polynomial of the holomorphic cusp form for $\Gamma(1)$ of weight 12 and $p_{12}^+(z) = z^{10} - 1$ the even part of the period function of the holomorphic Eisenstein series for $\Gamma(1)$ of weight 12.

Explicit examples of new solutions for $\kappa = 7$ (corresponding to weight $k = 8$ and $\beta = -3$) are

$$(4.42) \quad \underline{\phi}_o(z) = \begin{pmatrix} 4z - 5z^3 + z^5 \\ z - 5z^3 + 4z^5 \end{pmatrix} = \begin{pmatrix} 8\phi_o(\frac{z}{2}) \\ \phi_o(z) \end{pmatrix},$$

$$(4.43) \quad \underline{\phi}_e(z) = \begin{pmatrix} (t-1) + 6z^2 - 3z^4 + (2t+1)z^6 \\ (-2t-1) + 3z^2 - 6z^4 + (-t+1)z^6 \end{pmatrix}$$

$$= \begin{pmatrix} 8\phi_e(\frac{z}{2}) \\ \phi_e(z) \end{pmatrix} + \begin{pmatrix} 1 + \frac{43}{21}t & 0 & -\frac{64}{21}t \\ 1 + \frac{43}{21}t & -\frac{1}{21}t & 0 \end{pmatrix} \begin{pmatrix} p_8^+(z) \\ p_8^+(2z) \\ p_8^+(\frac{z}{2}) \end{pmatrix},$$

with $t \in \mathbb{C}$ is a free parameter and

$$\phi_o(z) = z - 5z^3 + 4z^5 \text{ respectively } \phi_e(z) = 3z^2 - 6z^4.$$

The polynomial $p_8^+(z) = z^6 - 1$ is the even part of the period function of the holomorphic Eisenstein series for $\Gamma(1)$ of weight 8. The components of the solutions in (4.42) respectively (4.43) lie in the polynomial space $\oplus_{n=0}^6 \mathbb{C}z^n$. The functions $\phi_o(z)$ and $\phi_e(z)$ for $\kappa = 7$ have obviously nothing to do with solutions of the Lewis equation (4.12) for the representation χ_1 of $\Gamma(1)$, because for $\kappa = 7$ equation (4.12) with $\lambda_\beta = \pm 1$ does not have any solution in the space $\oplus_{n=0}^6 \mathbb{C}z^n$ since there does not exist any cusp form for $\Gamma(1)$ of weight 8.

4.4. New and old solutions and new and old forms. The solutions of the Lewis equations for χ_1 and χ_{-1} in (4.12) and (4.19) in the space $\oplus_{n=0}^{\kappa-1} \mathbb{C}z^n$ of polynomials and the one for χ_2 in (4.23) with components in $\oplus_{n=0}^{\kappa-1} \mathbb{C}z^n$ can be determined easily numerically. The numbers of solutions found this way are summarized in table 2. Since 'odd' ($\lambda_\beta = +1$) and 'even' ($\lambda_\beta = -1$) solutions always arise in pairs, each pair will be counted as one solution. Numbers with the symbol $*$ in the row for χ_2 characterize the number of old solutions, the other number counts the new solutions.

According to the decompositions in (4.4)

$$\begin{aligned} \chi^{\Gamma(1)} &= \chi_1, \\ \chi^{\Gamma_2} &= \chi_1 \oplus \chi_{-1}, \\ \chi^{\Gamma'} &= \chi_1 \oplus \chi_2, \qquad \Gamma' \in \{\Gamma_0(2), \Gamma^0(2), \Gamma_\vartheta\}, \\ \chi^{\Gamma(2)} &= \chi_1 \oplus \chi_{-1} \oplus \chi_2 \oplus \chi_2 \end{aligned}$$

the number of solutions of the Lewis equation (3.17) with $\lambda_\beta = \pm 1$ for the groups $\Gamma(1), \Gamma_2, \Gamma_0(2), \Gamma^0(2), \Gamma_\vartheta$ and $\Gamma(2)$ are determined by the sum of solutions for χ_1, χ_{-1} and χ_2. They are also given in table 2, where the numbers with the symbol $*$ count the old solutions in the following sense:

The solutions of the Lewis equation (3.17) for χ^{Γ_2} are determined by the solutions of the Lewis equations for χ_1 and χ_{-1}. The first ones are called 'old' solutions since they are solutions of Lewis equation for the representation $\chi^{\Gamma(1)}$ of the larger group $\Gamma(1)$. The ones for χ_{-1} are 'new' solutions. The same notation is used for the other representations. For $\Gamma(2)$ the induced representation $\chi^{\Gamma(2)}$ can be decomposed into the representations χ_1, χ_{-1} and twice χ_2. The 'old' solutions for $\chi^{\Gamma(2)}$ then consist of the solutions for the representations χ_1 and for χ_2, since these representations appear also in the induced representations $\chi^{\Gamma(1)}$ respectively $\chi^{\Gamma_0(2)}$ for the larger groups $\Gamma(1)$ respectively $\Gamma_0(2)$ with $\Gamma(2) \subset \Gamma_0(2) \subset \Gamma(1)$.

TABLE 2. The number of solutions of the Lewis equation for different representations χ with components in the space $\oplus_{n=0}^{\kappa-1}\mathbb{C}z^n$, where $\Gamma' \in \{\Gamma_0(2), \Gamma^0(2), \Gamma_\vartheta\}$

β	-1	-2	-3	-4	-5	-6	-7	-8	-9	-10	-11
κ	3	5	7	9	11	13	15	17	19	21	23
χ_1	0	0	0	0	1	0	1	1	1	1	2
χ_{-1}	0	1	0	1	1	1	1	2	1	2	2
χ_2	0	0	1	1	1^*	2	$1+1^*$	$1+1^*$	$1+1^*$	$2+1^*$	$1+2^*$
$\chi^{\Gamma(1)}$	0	0	0	0	1	0	1	1	1	1	2
χ^{Γ_2}	0	1	0	1	$1+1^*$	1	$1+1^*$	$2+1^*$	$1+1^*$	$2+1^*$	$2+2^*$
$\chi^{\Gamma'}$	0	0	1	1	2^*	2	$1+2^*$	$1+2^*$	$2+2^*$	$2+2^*$	$1+4^*$
$\chi^{\Gamma(2)}$	0	1	2^*	$1+2^*$	$1+3^*$	$1+4^*$	$1+5^*$	$2+5^*$	$1+7^*$	$2+7^*$	$2+8^*$

Comparing this table with the table on p.296 of [**Miy89**] the numbers for the 'old' respectively the 'new' solutions for the representations $\chi^{\Gamma_0(2)}$ and $\chi^{\Gamma(2)}$ coincide exactly with the numbers given there for the old respectively the new cusp forms for the groups $\Gamma_0(2)$ and $\Gamma(2)$ (indeed $\Gamma(2)$ is conjugate to $\Gamma_0(4)$ [**BV97**] p.27). For the modular forms of the groups $\Gamma(1)$, $\Gamma_0(2)$, $\Gamma^0(2)$ and $\Gamma(2)$ the following result namely holds [**BV97**]:

LEMMA 4.5.

(i) If $g(z)$ is a modular form for $\Gamma(1)$, then

	the group $\Gamma_0(2)$	has the old forms $g(z)$ and $g(2z)$,
(4.44)	the group $\Gamma^0(2)$	has the old forms $g(z)$ and $g(\frac{z}{2})$,
	the group $\Gamma(2)$	has the old forms $g(z), g(2z)$ and $g(\frac{z}{2})$.

(ii) If $h(z)$ is a modular form of $\Gamma_0(2)$ respectively $\Gamma^0(2)$, then

	the group $\Gamma^0(2)$	has the modular form $h(\frac{z}{2})$,
(4.45)	the group $\Gamma(2)$	has the modular forms $h(z)$ and $h(\frac{z}{2})$

respectively

	the group $\Gamma_0(2)$	has the modular form $h(2z)$,
(4.46)	the group $\Gamma(2)$	has the modular forms $h(z)$ and $h(2z)$.

According to this lemma the space $M(\Gamma)^{old}$ of old modular forms for Γ is spanned by the following forms:

(4.47) $M(\Gamma_0(2))^{old} = \{c_1\, g_1(z) + c_2\, g_2(2z) \,|\, c_i \in \mathbb{C},\ g_i(z) \in M(\Gamma(1))\}$,

(4.48) $M(\Gamma^0(2))^{old} = \{c_1\, g_1(z) + c_2\, g_2(\frac{z}{2}) \,|\, c_i \in \mathbb{C},\ g_i(z) \in M(\Gamma(1))\}$,

(4.49) $M(\Gamma(2))^{old} = \{c_1\, g_1(z) + c_2\, g_2(2z) + c_3\, g_3(\frac{z}{2}) \,|\, c_i \in \mathbb{C},\ g_i(z) \in M(\Gamma(1))\}$

$$\oplus \{d_1\, h_1(z) + d_2\, h_2(\frac{z}{2}) \,|\, d_i \in \mathbb{C},\ h_i(z) \in M(\Gamma_0(2))\}.$$

The last subspace of $M(\Gamma(2))^{old}$ can be replaced by

$$\{d_1\, h_1(z) + d_2\, h_2(2z) \mid d_i \in \mathbb{C},\ h_i(z) \in M(\Gamma^0(2))\}\,.$$

The dimensions of these different spaces are hence related as follows:

$$
\begin{aligned}
(4.50)\quad dim(M(\Gamma_0(2))^{old}) &= dim(M(\Gamma^0(2))^{old}) = 2\,dim(M(\Gamma(1)))\,,\\
dim(M(\Gamma_0(2))) &= dim(M(\Gamma^0(2))^{old}) + dim(M(\Gamma(1))^{new})\,,\\
dim(M(\Gamma(2))) &= 3\,dim(M(\Gamma(1))) + 2\,dim(M(\Gamma_0(2))^{old})\\
(4.51)\quad & \quad + dim(M(\Gamma(2))^{new})\,.
\end{aligned}
$$

Obviously the numbers of solutions given in table 2 for the different groups fulfill relations (4.50) and (4.51). The form of old solutions for the Lewis equation for χ_2 in table 1 and in (4.32) is quite similar to the one of the old modular forms in (4.47), (4.48) and (4.49). The representation χ_2 appears in the induced representations $\chi^{\Gamma_0(2)}$, $\chi^{\Gamma^0(2)}$ and $\chi^{\Gamma(2)}$. This suggests that the transfer operator approach respects in a surprizing way the theory of old and new forms for such congruence subgroups which for instance in Eichlers period polynomials is not so obviously seen.

To study this in more detail, we restrict our discussion now to the group $\Gamma_0(2)$ and consider its induced representation $\chi^{\Gamma_0(2)}$ and the corresponding Lewis equation respectively transfer operator.

4.5. Solutions of the Lewis equation for $\Gamma_0(2)$. For the group $\Gamma_0(2)$ Lewis' equation in (3.17) reads

$$(4.52)\qquad \lambda_\beta\left[\underline{\phi}(z) - \chi^{\Gamma_0(2)}(QTQ)\,\underline{\phi}(z+1)\right] - z^{-2\beta}\chi^{\Gamma_0(2)}(QT)\,\underline{\phi}(1+\tfrac{1}{z}) = 0\,,$$

with $\chi^{\Gamma_0(2)}$ the induced representation defined by [**CM00**]:

$$(4.53)\qquad \chi^{\Gamma_0(2)}(Q) = \begin{pmatrix} 0 & 1 & 0 \\ 1 & 0 & 0 \\ 0 & 0 & 1 \end{pmatrix} \quad\text{and}\quad \chi^{\Gamma_0(2)}(T) = \begin{pmatrix} 1 & 0 & 0 \\ 0 & 0 & 1 \\ 0 & 1 & 0 \end{pmatrix}\,.$$

Let $\underline{\phi}(z) = \begin{pmatrix} \phi_1(z) \\ \phi_2(z) \\ \phi_3(z) \end{pmatrix}$ be a solution of equation (4.52). With

$$\chi^{\Gamma_0(2)}(QTQ) = \begin{pmatrix} 0 & 0 & 1 \\ 0 & 1 & 0 \\ 1 & 0 & 0 \end{pmatrix} \quad\text{and}\quad \chi^{\Gamma_0(2)}(QT) = \begin{pmatrix} 0 & 0 & 1 \\ 1 & 0 & 0 \\ 0 & 1 & 0 \end{pmatrix}$$

equation (4.52) is then equivalent to the following three equations:

$$(4.54)\qquad \lambda_\beta\left[\phi_1(z) - \phi_3(z+1)\right] - z^{-2\beta}\,\phi_3(1+\tfrac{1}{z}) = 0\,,$$

$$(4.55)\qquad \lambda_\beta\left[\phi_2(z) - \phi_2(z+1)\right] - z^{-2\beta}\,\phi_1(1+\tfrac{1}{z}) = 0\,,$$

$$(4.56)\qquad \lambda_\beta\left[\phi_3(z) - \phi_1(z+1)\right] - z^{-2\beta}\,\phi_2(1+\tfrac{1}{z}) = 0\,.$$

Summing up these three equations leads to

$$(4.57)\qquad \lambda_\beta\left[\hat{\phi}(z) - \hat{\phi}(z+1)\right] - z^{-2\beta}\,\hat{\phi}(1+\tfrac{1}{z}) = 0$$

with $\hat{\phi}(z) = \phi_1(z) + \phi_2(z) + \phi_3(z)$. This however is just Lewis' equation (4.12) for the representation χ_1. That is, if $\underline{\phi}$ is a solution of Lewis' equation for $\chi^{\Gamma_0(2)}$, then $\hat{\phi}(z)$ must be a solution of the Lewis equation (4.12) for χ_1, which can also be trivial. The three equations above can be rewritten as

$$(4.58) \qquad \phi_1(z) \;=\; \phi_3(z+1) + \frac{1}{\lambda_\beta}\, z^{-2\beta}\phi_3(1+\frac{1}{z})\,,$$

$$(4.59) \qquad \phi_1(z) \;=\; \lambda_\beta\left[\phi_2(\frac{1}{z-1}) - \phi_2(1+\frac{1}{z-1})\right](z-1)^{-2\beta}\,,$$

$$(4.60) \qquad \phi_3(z) \;=\; \phi_1(z+1) + \frac{1}{\lambda_\beta}\, z^{-2\beta}\phi_2(1+\frac{1}{z})\,.$$

Inserting (4.59) into (4.60) with $\lambda_\beta = \pm 1$ implies

$$\phi_3(z) \;=\; \lambda_\beta\left[\phi_2(\frac{1}{z}) - \phi_2(1+\frac{1}{z})\right] z^{-2\beta} + \frac{1}{\lambda_\beta}\, z^{-2\beta}\phi_2(1+\frac{1}{z})$$

$$(4.61) \qquad\qquad\;=\; \lambda_\beta\, z^{-2\beta}\, \phi_2(\frac{1}{z})\,.$$

Inserting $\phi_1(z)$ from (4.58) into relation (4.59) furthermore gives

$$\phi_3(z+1) + \frac{1}{\lambda_\beta}\, z^{-2\beta}\phi_3(1+\frac{1}{z}) = \lambda_\beta\left[\phi_2(\frac{1}{z-1}) - \phi_2(1+\frac{1}{z-1})\right](z-1)^{-2\beta}\,.$$

Due to expression (4.61) the function ϕ_3 can be replaced by $\phi_2(z)$, which leads finally to the following equation for $\phi_2(z)$:

$$(4.62) \qquad \lambda_\beta\left[\phi_2(z) - \phi_2(z+1)\right] = (2z+1)^{-2\beta}\left[\lambda_\beta\,\phi_2(\frac{z}{2z+1}) + \phi_2(\frac{z+1}{2z+1})\right]\,.$$

Hence, if $\phi(z)$ is a solution of the Lewis equation for $\chi^{\Gamma_0(2)}$, then $\phi_2(z)$ must fulfill equation (4.62). Conversely, any solution $\phi_2(z)$ of this equation defines a solution $\underline{\phi}$ of Lewis' equation (4.52) for $\lambda_\beta = \pm 1$ by means of the relations (4.59) and (4.61) as

$$(4.63) \qquad \underline{\phi}(z) = \begin{pmatrix}\phi_1(z)\\ \phi_2(z)\\ \phi_3(z)\end{pmatrix} = \begin{pmatrix}\lambda_\beta\left[\phi_2(\frac{1}{z-1}) - \phi_2(1+\frac{1}{z-1})\right](z-1)^{-2\beta}\\ \phi_2(z)\\ \lambda_\beta\, z^{-2\beta}\phi_2(\frac{1}{z})\end{pmatrix}\,,$$

This shows, that the solutions $\phi_2(z)$ of equation (4.62) and the solutions $\underline{\phi}(z)$ of Lewis' equation (4.52) are in $1-1$ correspondence.

Two special solutions of the form (4.63) can be obtained by choosing $\phi_2(z) = \phi(z)$ respectively $\phi_2(z) = \phi(2z)$, where $\phi(z)$ is an arbitrary solution of Lewis' equation (4.12) for the representation χ_1 and parameter value β. Indeed one shows

PROPOSITION 4.6. *Let $\phi(z)$ be a solution of the Lewis equation (4.12) for the representation χ_1 and parameter β with $\lambda_\beta = \pm 1$. Then the Lewis equation (4.52) for the representation $\chi^{\Gamma_0(2)}$ and the same β with $\lambda_\beta = \pm 1$ has the solutions*

$$(4.64) \qquad \underline{\phi}^{(1)}(z) = \begin{pmatrix}\phi(z)\\ \phi(z)\\ \phi(z)\end{pmatrix} \quad and \quad \underline{\phi}^{(2)}(z) = \begin{pmatrix}2^{-2\beta}\left[\phi(\frac{z-1}{2}) - z^{-2\beta}\phi(\frac{z-1}{2z})\right]\\ \phi(2z)\\ 2^{-2\beta}\phi(\frac{z}{2})\end{pmatrix}\,.$$

PROOF. One has to show $\phi(z)$ and $\phi(2z)$ to be solutions of equation (4.62). Then $\underline{\phi}^{(1)}(z)$ and $\underline{\phi}^{(2)}(z)$ in (4.64) correspond to the general solution (4.63) for $\phi_2(z) = \phi(z)$ respectively $\phi_2(z) = \phi(2z)$.

In the case $\phi_2(z) = \phi(z)$ we define

(4.65) $\qquad Y_1(z) \;:=\; \lambda_\beta \left[\phi(z) - \phi(z+1) \right] - z^{-2\beta} \phi(1 + \frac{1}{z}) ,$

(4.66) $\qquad Y_2(z) \;:=\; \phi(z) - \lambda_\beta \, z^{-2\beta} \phi(\frac{1}{z}) ,$

where $Y_1(z)$ respectively $Y_2(z)$ are just the left-hand sides of the functional equations (4.12) respectively (4.14). For $\phi(z)$ a solution of Lewis' equation (4.12) with $\lambda_\beta = \pm 1$, $Y_1(z) = Y_2(z) \equiv 0$. The trivial identity

$$Y_1(z) + \lambda_\beta \, z^{-2\beta} \, Y_1(1 + \frac{1}{z}) + z^{-2\beta} \, Y_2(2 + \frac{1}{z}) + \lambda_\beta \, Y_2(1 + \frac{z}{z+1}) \, (z+1)^{-2\beta} \equiv 0$$

then implies

$$\lambda_\beta \left[\phi(z) - \phi(z+1) \right] - z^{-2\beta} \phi(\frac{z+1}{z})$$
$$+ z^{-2\beta} \left[\phi(\frac{z+1}{z}) - \phi(\frac{2z+1}{z}) \right] - \lambda_\beta \, (z+1)^{-2\beta} \phi(\frac{2z+1}{z+1})$$
$$+ z^{-2\beta} \phi(\frac{2z+1}{z}) - \lambda_\beta \, (2z+1)^{-2\beta} \phi(\frac{z}{2z+1})$$
$$+ \lambda_\beta \, (z+1)^{-2\beta} \phi(\frac{2z+1}{z+1}) - (2z+1)^{-2\beta} \phi(\frac{z+1}{2z+1}) = 0 .$$

The terms proportional $\phi(\frac{z+1}{z})$, $\phi(\frac{2z+1}{z})$ and $\phi(\frac{2z+1}{z+1})$ cancel leading to equation (4.62). Hence $\phi(z)$ is a solution of this equation.

In the case $\phi_2(z) = \phi(2z)$ with $\phi(z)$ a solution of the Lewis equation for the representation χ_1 with $\lambda_\beta = \pm 1$, we define the quantities $Y_3(z)$ respectively $Y_4(z)$ as

(4.67) $\qquad Y_3(z) \;:=\; \lambda_\beta \left[\phi_2(\frac{z}{2}) - \phi_2(\frac{z+1}{2}) \right] - z^{-2\beta} \phi_2(\frac{1}{2} + \frac{1}{2z}) ,$

(4.68) $\qquad Y_4(z) \;:=\; \phi_2(\frac{z}{2}) - \lambda_\beta \, z^{-2\beta} \phi_2(\frac{1}{2z}) .$

Since $\phi(z)$ is a solution of the Lewis equation for the representation χ_1, $Y_3(z)$ and $Y_4(z)$ vanish identically in z. The trivial identity

(4.69) $\qquad Y_3(2z) + Y_3(2z+1) + (2z)^{-2\beta} \, Y_4(\frac{2z+1}{2z}) \equiv 0$

then implies

$$\lambda_\beta \left[\phi_2(z) - \phi_2(\frac{2z+1}{2}) \right] - (2z)^{-2\beta} \phi_2(\frac{1}{2} + \frac{1}{4z})$$
$$\lambda_\beta \left[\phi_2(\frac{2z+1}{2}) - \phi_2(z+1) \right] - (2z+1)^{-2\beta} \phi_2(\frac{1}{2} + \frac{1}{4z+2})$$
$$+ (2z)^{-2\beta} \phi_2(\frac{2z+1}{4z}) - \lambda_\beta \, (2z+1)^{-2\beta} \phi_2(\frac{z}{2z+1}) = 0 ,$$

which leads to equation (4.62). That is, $\phi_2(z) = \phi(2z)$ is also a solution of equation (4.62).

Replacing now in the general formula (4.63) $\phi_2(z)$ by $\phi(z)$ respectively $\phi(2z)$, one gets finally the two solutions:

$$(4.70) \qquad \underline{\phi}^{(1)}(z) = \begin{pmatrix} \lambda_\beta \left[\phi(\frac{1}{z-1}) - \phi(1 + \frac{1}{z-1}) \right] (z-1)^{-2\beta} \\ \phi(z) \\ \lambda_\beta \, z^{-2\beta} \phi(\frac{1}{z}) \end{pmatrix} = \begin{pmatrix} \phi(z) \\ \phi(z) \\ \phi(z) \end{pmatrix},$$

$$\underline{\phi}^{(2)}(z) = \begin{pmatrix} \lambda_\beta \left[\phi(\frac{2}{z-1}) - \phi(2 + \frac{2}{z-1}) \right] (z-1)^{-2\beta} \\ \phi(2z) \\ \lambda_\beta \, z^{-2\beta} \phi(\frac{2}{z}) \end{pmatrix}$$

$$(4.71) \qquad\qquad = \begin{pmatrix} 2^{-2\beta} \left[\phi(\frac{z-1}{2}) - z^{-2\beta} \phi(\frac{z-1}{2z}) \right] \\ \phi(2z) \\ 2^{-2\beta} \phi(\frac{z}{2}) \end{pmatrix},$$

where we used relations (4.12) and (4.14) to get (4.70) and (4.14) to get (4.71). □

The two solutions in (4.64) are called 'old' solutions of the Lewis equation for $\chi^{\Gamma_0(2)}$ since they are determined by solutions for $\chi^{\Gamma(1)}$. According to Proposition 4.6 the number of such old solutions of the Lewis equation for $\chi^{\Gamma_0(2)}$ is at least twice the number of the solutions for $\chi^{\Gamma(1)}$. This agrees with the numerical results in table 2, where we found that the number of old solutions of the Lewis equation for the representation $\chi^{\Gamma_0(2)} = \chi_1 \oplus \chi_2$, is at least for $\kappa \leq 23$, always twice the number of solutions for $\chi^{\Gamma(1)} = \chi_1$. This reflects also the fact that the dimension of the space of old forms for $\Gamma_0(2)$ is twice the dimension of the space of modular forms for $\Gamma(1)$. We obviously expect that the number of old solutions of the Lewis equation for $\chi^{\Gamma_0(2)}$ for all κ is exactly twice the number of solutions of the Lewis equation for $\chi^{\Gamma(1)}$. Since all polynomial solutions of Lewis' equation are also eigenfunctions of the corresponding transfer operator, an analogous result holds for the 'old' eigenfunctions of the transfer operator for $\chi^{\Gamma_0(2)}$.

4.5.1. *Old solutions of the Lewis equation for the representation χ_2.* Knowing the explicit form of two kinds of old solutions (4.64) of the Lewis equation for $\chi^{\Gamma_0(2)}$, one can ask how these solutions are related to the old solutions of the Lewis equation for the representation χ_2 as determined numerically in table 2. To answer this we rewrite the Lewis equation (4.52) for the representation $\chi^{\Gamma_0(2)}$ as

$$\lambda_\beta \left[M \underline{\phi}(z) - M \chi^{\Gamma_0(2)}(QTQ) M^{-1} M \underline{\phi}(z+1) \right]$$
$$(4.72) \qquad - z^{-2\beta} M \chi^{\Gamma_0(2)}(QT) M^{-1} M \underline{\phi}(1 + \frac{1}{z}) = 0,$$

where M denotes the matrix

$$M = \begin{pmatrix} \frac{1}{27} & \frac{1}{27} & \frac{1}{27} \\ \frac{1}{3} & \frac{1}{3} & -\frac{2}{3} \\ \frac{1}{3} & -\frac{2}{3} & \frac{1}{3} \end{pmatrix} \text{ and } M^{-1} = \begin{pmatrix} 9 & 1 & 1 \\ 9 & 0 & -1 \\ 9 & -1 & 0 \end{pmatrix} \text{ its inverse}.$$

The two matrices $M\chi^{\Gamma_0(2)}(QTQ)M^{-1}$ and $M\chi^{\Gamma_0(2)}(QT)M^{-1}$ in (4.72) then have the form

$$(4.73) \qquad \begin{pmatrix} \chi_1(QTQ) & 0 \\ 0 & \chi_2(QTQ) \end{pmatrix} \text{ and } \begin{pmatrix} \chi_1(QT) & 0 \\ 0 & \chi_2(QT) \end{pmatrix},$$

with χ_1 and χ_2 the representations given in (4.1) and (4.3). Equation (4.72) splits then into the two equations

$$(4.74) \qquad \lambda_\beta \left[\hat{\phi}_1(z) - \chi_1(QTQ)\, \hat{\phi}_1(z+1) \right] - z^{-2\beta} \chi_1(QT) \hat{\phi}_1(1 + \frac{1}{z}) = 0,$$

$$(4.75) \qquad \lambda_\beta \left[\hat{\underline{\phi}}_3(z) - \chi_2(QTQ)\, \hat{\underline{\phi}}_3(z+1) \right] - z^{-2\beta} \chi_2(QT) \hat{\underline{\phi}}_3(1 + \frac{1}{z}) = 0,$$

where $\hat{\phi}_1(z)$ respectively $\hat{\underline{\phi}}_3(z)$ denote the first respectively second component of $M\underline{\phi}(z) = \begin{pmatrix} \hat{\phi}_1(z) \\ \hat{\underline{\phi}}_3(z) \end{pmatrix}$. The equations (4.74) respectively (4.75) coincide however with the Lewis equations for the representations χ_1 respectively χ_2 as defined in (4.1) respectively (4.3).

For the solution $\underline{\phi}^{(1)}(z) = \begin{pmatrix} \phi(z) \\ \phi(z) \\ \phi(z) \end{pmatrix}$ in (4.64) one then finds

$$\begin{pmatrix} \hat{\phi}_1(z) \\ \hat{\underline{\phi}}_3(z) \end{pmatrix} = M\underline{\phi}^{(1)}(z) = \begin{pmatrix} \frac{1}{27} & \frac{1}{27} & \frac{1}{27} \\ \frac{1}{3} & \frac{1}{3} & -\frac{2}{3} \\ \frac{1}{3} & -\frac{2}{3} & \frac{1}{3} \end{pmatrix} \begin{pmatrix} \phi(z) \\ \phi(z) \\ \phi(z) \end{pmatrix} = \begin{pmatrix} \frac{1}{9}\phi(z) \\ 0 \\ 0 \end{pmatrix},$$

and hence $\hat{\phi}_1(z) = \frac{1}{9}\phi(z)$ and $\hat{\underline{\phi}}_3(z) = \begin{pmatrix} 0 \\ 0 \end{pmatrix}$. Obviously $\hat{\phi}_1(z)$ respectively $\hat{\underline{\phi}}_3(z)$ satisfy the Lewis equations for the representations χ_1 respectively χ_2.

For the solution $\underline{\phi}^{(2)}(z)$ in (4.64) on the other hand one gets

$$\begin{pmatrix} \hat{\phi}_1(z) \\ \hat{\underline{\phi}}_3(z) \end{pmatrix} = M\underline{\phi}^{(2)}(z) = \begin{pmatrix} \frac{1}{27} & \frac{1}{27} & \frac{1}{27} \\ \frac{1}{3} & \frac{1}{3} & -\frac{2}{3} \\ \frac{1}{3} & -\frac{2}{3} & \frac{1}{3} \end{pmatrix} \begin{pmatrix} 2^{-2\beta}\left[\phi(\frac{z-1}{2}) - z^{-2\beta}\phi(\frac{z-1}{2z})\right] \\ \phi(2z) \\ 2^{-2\beta}\phi(\frac{z}{2}) \end{pmatrix}$$

and hence

$$(4.76) \qquad \hat{\phi}_1(z) = \frac{1}{27}\left[2^{-2\beta}\left[\phi(\frac{z-1}{2}) - z^{-2\beta}\phi(\frac{z-1}{2z})\right] + \phi(2z) + 2^{-2\beta}\phi(\frac{z}{2}) \right],$$

according to (4.57) is a solution of the Lewis equation for the representation χ_1. The second component $\hat{\underline{\phi}}_3(z)$, which is a solution of Lewis equation for the representation χ_2, can be rewritten as

$$(4.77) \quad \hat{\underline{\phi}}_3(z) = \begin{pmatrix} \frac{1}{3} & \frac{1}{3} & -\frac{2}{3} \\ \frac{1}{3} & -\frac{2}{3} & \frac{1}{3} \end{pmatrix} \underline{\phi}^{(2)}(z) = \left[\begin{pmatrix} \frac{1}{3} & \frac{1}{3} & -\frac{2}{3} \\ \frac{1}{3} & -\frac{2}{3} & \frac{1}{3} \end{pmatrix} N^{-1} \right] \left[N\,\underline{\phi}^{(2)}(z) \right],$$

with

$$N = \begin{pmatrix} \frac{1}{3\delta} & \frac{1}{3\delta} & \frac{1}{3\delta} \\ 0 & 1 & 0 \\ 0 & 0 & 2^{2\beta} \end{pmatrix} \text{ respectively } N^{-1} = \begin{pmatrix} 3\delta & -1 & -2^{-2\beta} \\ 0 & 1 & 0 \\ 0 & 0 & 2^{-2\beta} \end{pmatrix}$$

for arbitrary $\delta \neq 0$. After a simple calculation one finds

$$(4.78) \qquad \hat{\underline{\phi}}_3(z) = \begin{pmatrix} \delta & 0 & -2^{-2\beta} \\ \delta & -1 & 0 \end{pmatrix} \begin{pmatrix} \frac{9}{\delta}\hat{\phi}_1(z) \\ \phi(2z) \\ \phi(\frac{z}{2}) \end{pmatrix},$$

with $\hat{\phi}_1(z)$ given in (4.76). The form of these old solutions (4.78) of the Lewis equation for the representation χ_2 explains completely the form of the old solutions given in table 1.

As shown in (4.57) the functions $\hat{\phi}_1(z)$ in (4.76) as well as $\phi(z)$ from which it is constructed are solutions of the Lewis equation for the representation χ_1. In general $\hat{\phi}_1(z)$ is not necessarily proportional to $\phi(z)$ and lies only in the vector space spanned by the solutions of the Lewis equation for χ_1 for a fixed β. Hence, if for some β-value with $\Re\beta = \frac{1}{2}$ the functions $\hat{\phi}_1(z)$ and $\phi(z)$ are not proportional to each other and satisfy criterion (4.15), then the transfer operator $\mathcal{L}_\beta$ has a degenerate eigenvalue $\lambda_\beta = \pm 1$ and also the Laplace-Beltrami operator $-\triangle$ has a degenerate eigenvalue $\beta(1-\beta)$ and vice versa. This could be a possibility to test if the operator $-\triangle$ on the modular surface for $\Gamma(1)$ has degenerate eigenvalues. For β-values with $\zeta(2\beta) = 0$ and $\Re\beta > 0$ the solutions of the Lewis equation for χ_1 fulfilling criterion (4.15) are related to the nonholomorphic Eisenstein series. Their multiplicity is one and hence $\hat{\phi}_1(z)$ must be proportional to $\phi(z)$.

Indeed for $\phi(z) = \psi_\beta(z)$ defined in (4.31) one finds with $\delta = \frac{1}{3}(1 + 2^{1-2\beta})$ for (4.78):

$$(4.79) \qquad \underline{\hat{\phi}}_3(z) = \begin{pmatrix} \frac{1}{3}(1 + 2^{1-2\beta}) & 0 & -2^{-2\beta} \\ \frac{1}{3}(1 + 2^{1-2\beta}) & -1 & 0 \end{pmatrix} \begin{pmatrix} \hat{\psi}_\beta(z) \\ \psi_\beta(2z) \\ \psi_\beta(\frac{z}{2}) \end{pmatrix},$$

with

$$(4.80) \qquad \hat{\psi}_\beta(z) = \frac{2^{-2\beta}\left[\psi_\beta(\frac{z-1}{2}) - z^{-2\beta}\psi_\beta(\frac{z-1}{2z})\right] + \psi_\beta(2z) + 2^{-2\beta}\psi_\beta(\frac{z}{2})}{1 + 2^{1-2\beta}}.$$

Then one shows:

LEMMA 4.7. $\hat{\psi}_\beta(z) = \psi_\beta(z)$.

PROOF. Using definition (4.31)

$$\psi_\beta(z) = \sum_{m,n\geq 1}(mz+n)^{-2\beta} + \frac{1}{2}\zeta(2\beta)(1 + z^{-2\beta}), \qquad \Re\beta > 1$$

one gets for $\Re\beta > 1$

$$2^{-2\beta}\left[\psi_\beta(\frac{z-1}{2}) - z^{-2\beta}\psi_\beta(\frac{z-1}{2z})\right]$$

$$= \sum_{m,n\geq 1}(m(z-1)+2n)^{-2\beta} + \frac{1}{2}\zeta(2\beta)(2^{-2\beta} + (z-1)^{-2\beta})$$

$$- \sum_{m,n\geq 1}(m(z-1)+2nz)^{-2\beta} - \frac{1}{2}\zeta(2\beta)((2z)^{-2\beta} + (z-1)^{-2\beta})$$

$$= \sum_{m,n\geq 1}(mz + (2n-m))^{-2\beta} - \sum_{m,n\geq 1}((m+2n)z-m)^{-2\beta}$$

$$(4.81) \qquad + \frac{1}{2}\zeta(2\beta)(2^{-2\beta} - (2z)^{-2\beta}).$$

Writing

$$\sum_{m,n\geq 1} (mz + (2n-m))^{-2\beta}$$

$$= \Bigg[\sum_{m,n\geq 1,\, 2n>m} + \sum_{m,n\geq 1,\, 2n=m} + \sum_{m,n\geq 1,\, 2n<m} \Bigg] (mz + (2n-m))^{-2\beta}$$

$$= \Bigg[\sum_{m_o,n_0\geq 1} (m_o z + n_0)^{-2\beta} + \sum_{m_e,n_e\geq 1} (m_e z + n_e)^{-2\beta} \Bigg] + \sum_{n\geq 1}(2nz)^{-2\beta}$$

$$+ \sum_{n,\, m'=m-2n\geq 1} ((m' + 2n)z - m')^{-2\beta},$$

where m_o, n_0 respectively m_e, n_e run over the odd respectively the even natural numbers, (4.81) can be simplified to

$$2^{-2\beta}\Big[\psi_\beta(\frac{z-1}{2}) - z^{-2\beta}\psi_\beta(\frac{z-1}{2z})\Big]$$

$$= \sum_{m_o,n_0\geq 1} (m_o z + n_0)^{-2\beta} + \sum_{m_e,n_e\geq 1} (m_e z + n_e)^{-2\beta}$$

(4.82)
$$+ \frac{1}{2}\zeta(2\beta)\Big[2^{-2\beta} + (2z)^{-2\beta}\Big].$$

Furthermore one finds

$$\psi_\beta(2z) = \sum_{m,n\geq 1} (2mz + n)^{-2\beta} + \frac{1}{2}\zeta(2\beta)(1 + (2z)^{-2\beta})$$

(4.83)
$$= \sum_{m_e,n\geq 1} (m_e z + n)^{-2\beta} + \frac{1}{2}\zeta(2\beta)[1 + (2z)^{-2\beta}]$$

respectively

$$2^{-2\beta}\psi_\beta(\frac{z}{2}) = \sum_{m,n\geq 1} (mz + 2n)^{-2\beta} + \frac{1}{2}\zeta(2\beta)(2^{-2\beta} + z^{-2\beta})$$

(4.84)
$$= \sum_{m,n_e\geq 1} (mz + n_e)^{-2\beta} + \frac{1}{2}\zeta(2\beta)[2^{-2\beta} + z^{-2\beta}].$$

Adding (4.82), (4.83) and (4.84) the numerator of the function $\hat\psi_\beta(z)$ in (4.80) is equal to

$$\sum_{m_o,n_0\geq 1} (m_o z + n_0)^{-2\beta} \sum_{m_e,n_e\geq 1} (m_e z + n_e)^{-2\beta}$$

$$\sum_{m_e,n_0\geq 1} (m_e z + n_0)^{-2\beta} \sum_{m_e,n_e\geq 1} (m_e z + n_e)^{-2\beta}$$

$$\sum_{m_o,n_e\geq 1} (m_o z + n_e)^{-2\beta} \sum_{m_e,n_e\geq 1} (m_e z + n_e)^{-2\beta}$$

$$+ \frac{1}{2}\zeta(2\beta)[(1 + z^{-2\beta})(1 + 2^{1-2\beta})]$$

$$= (1 + 2^{1-2\beta})\,\psi_\beta(z),$$

where we used $\sum_{m_e, n_e \geq 1}(m_e z + n_e)^{-2\beta} = 2^{-2\beta}\sum_{m,n \geq 1}(mz + n)^{-2\beta} = 2^{-2\beta}\psi_\beta(z)$ for $\Re\beta > 1$. This implies immediately $\hat{\psi}_\beta(z) = \psi_\beta(z)$ for $\Re\beta > 1$ and by analytic continuation for arbitrary β. $\square$

This Lemma together with the definition of $\hat{\underline{\phi}}_3(z)$ in (4.79) proves also Proposition 4.2 and Proposition 4.4 in section 4.3.1.

In [**CM99**] we showed that the function $\varphi_0(z) = \frac{1}{z} + z - 3$ is a solution of Lewis equation (4.12) for $\Gamma(1)$. The function $\psi_\beta(z)$ on the other hand takes for $\beta = 0$ the form $\psi_0(z) = \frac{1}{12}\left(\frac{1}{z} + z\right)$. Inserting φ_0 for ϕ in (4.78) hence leads to a solution $\hat{\underline{\phi}}_3$ of Lewis equation for χ_2. By accident the same solution $\hat{\underline{\phi}}_3$ is obtained when inserting instead of $\varphi_0(z)$ the function $\psi_0(z)$ since the constant term in $\varphi_0(z)$ just cancels when inserted into (4.78).

4.5.2. *New solutions of the Lewis equation for $\Gamma_0(2)$.* Besides the old solutions given in (4.64) the Lewis equation (4.52) for the representation $\chi^{\Gamma_0(2)}$ has other solutions, which obviously are not related to solutions of the Lewis equation for the representation χ_1. These solutions obviously must be related to the new forms of the group $\Gamma_0(2)$. For $\beta = \frac{1-\kappa}{2}$, $\kappa = 3, 5, \cdots 21$ we found numerically, that these new solutions have the structure as shown in table 3.

TABLE 3. New solutions of the Lewis equation (4.52) with components in $\oplus_{n=0}^{\kappa-1}\mathbb{C}z^n$, where $c \in \mathbb{Q}$, $e = \pm 1$ and $p_{\kappa+1}^+$ denotes the even part of the period polynomial of the holomorphic Eisenstein series of weight $\kappa + 1$.

λ_β	Structure of the solutions
$+1 (odd)$	$\underline{\phi}_o = \begin{pmatrix} -\phi_o(z) - e\phi_o(\frac{z}{2})2^{\frac{\kappa-1}{2}} \\ \phi_o(z) \\ e\phi_o(\frac{z}{2})2^{\frac{\kappa-1}{2}} \end{pmatrix}$
$-1 (even)$	$\underline{\phi}_e = \begin{pmatrix} -\phi_e(z) - e\phi_e(\frac{z}{2})2^{\frac{\kappa-1}{2}} + cp_{\kappa+1}^+(z) \\ \phi_e(z) \\ e\phi_e(\frac{z}{2})2^{\frac{\kappa-1}{2}} \end{pmatrix}$

Explicit examples of such solutions $\underline{\phi}(z)$ for $\kappa = 7$ (corresponding to weight $k = 8$ and $\beta = -3$) are

$$\underline{\phi}_o(z) = \begin{pmatrix} -\phi_o(z) - \phi_o(\frac{z}{2})2^{\frac{\kappa-1}{2}} \\ \phi_o(z) \\ \phi_o(\frac{z}{2})2^{\frac{\kappa-1}{2}} \end{pmatrix} = \begin{pmatrix} 5z - 10z^3 + 5z^5 \\ -z + 5z^3 - 4z^5 \\ -4z + 5z^3 - z^5 \end{pmatrix}$$

$$\underline{\phi}_e(z) = \begin{pmatrix} -\phi_e(z) - \phi_e(\frac{z}{2})2^{\frac{\kappa-1}{2}} + p_8^+(z) \\ \phi_e(z) \\ \phi_e(\frac{z}{2})2^{\frac{\kappa-1}{2}} \end{pmatrix} = \begin{pmatrix} 1 - 3z^2 + 3z^4 - z^6 \\ z^2 - 2z^4 \\ 2z^2 - z^4 \end{pmatrix}.$$

We expect the polynomials

$$\phi_o = -z + 5z^3 - 4z^5 \text{ respectively } \phi_e = z^2 - 2z^4$$

to be related to a new cusp form of weight 8 for the group $\Gamma_0(2)$ since there are no such forms for $\Gamma(1)$ for this weight. The polynomial $p_8^+(z) = z^6 - 1$ is the even part of the period polynomial of the holomorphic Eisenstein series of weight 8 for $\Gamma(1)$.

For $\underline{\phi}(z) = \begin{pmatrix} \phi_1(z) \\ \phi_2(z) \\ \phi_3(z) \end{pmatrix}$ a general solution of the Lewis equation (4.52) for the

representation $\chi^{\Gamma_0(2)}$, we have shown that $\hat{\phi}(z) = \phi_1(z) + \phi_2(z) + \phi_3(z)$ must be a solution of the Lewis equation for the representation χ_1 of $\Gamma(1)$. Our numerical results show that for the odd new solutions $\underline{\phi}_o(z)$ one always has $\hat{\phi}(z) \equiv 0$ for $\beta = -N$, $N \in \mathbb{N}$. For the even new solutions $\underline{\phi}_e(z)$ however one finds $\hat{\phi}(z) \not\equiv 0$ proportional to the even part of the period polynomial of the Eisenstein form.

4.5.3. *Solutions of the Lewis equations and period functions of $\Gamma_0(2)$.* The old solutions of Lewis equation for $\Gamma_0(2)$ have the form

$$(4.85) \qquad \underline{\phi}^{(1)}(z) = \begin{pmatrix} \phi(z) \\ \phi(z) \\ \phi(z) \end{pmatrix} \quad \text{and} \quad \underline{\phi}^{(2)}(z) = \begin{pmatrix} 2^{-2\beta}\left[\phi(\frac{z-1}{2}) - z^{-2\beta}\phi(\frac{z-1}{2z})\right] \\ \phi(2z) \\ 2^{-2\beta}\phi(\frac{z}{2}) \end{pmatrix}$$

According to Lemma 4.5 in section 4.5 we know on the other hand that for $g(z)$ a modular form for $\Gamma(1)$ $g(z)$ and $g(2z)$ are old forms for $\Gamma_0(2)$ respectively $g(z)$ and $g(\frac{z}{2})$ are old forms for $\Gamma^0(2)$. Consider then the Eichler period polynomial for the modular form $u(z)$ of weight k given in (2.30)

$$r(z) = \int_0^{i\infty} (t-z)^{k-2}u(t)\, dt$$

respectively the period function (2.43) for an even Maaß wave form $u_s(z)$ with eigenvalue $\lambda = s(1-s)$

$$\psi_s(z) = z \int_0^\infty y^s\, u_s(iy)\, (z^2 + y^2)^{-s-1}\, dy\,, \quad \Re(z) > 0\,.$$

A trivial calculation then shows, that

$$r(2z) = 2^{k-1} \int_0^{i\infty} (t-z)^{k-2}u(2t)\, dt \quad \text{and} \quad r(\frac{z}{2}) = 2^{1-k} \int_0^{i\infty} (t-z)^{k-2}u(\frac{t}{2})\, dt$$

and similarly $\psi_s(2z)$ respectively $\psi_s(\frac{z}{2})$ is related through the above integral to $u(2z)$ and $u(\frac{z}{2})$.

These relations obviously are also reflected in the form of the old solutions of the Lewis equation for $\Gamma_0(2)$ respectively in the old eigenfunctions of the transfer operator $\tilde{\mathcal{L}}_\beta^{\chi^{\Gamma_0(2)}}$. The third component of both the old solutions in (4.85) and the new solutions in Table 3 seem to be directly related to the old and new forms for the group $\Gamma_0(2)$. One could then also expect that the first component of these solutions is related to forms of the group Γ_ϑ which is also conjugate to the two groups $\Gamma_0(2)$ and $\Gamma^0(2)$.

There arises immediately the question how the above polynomial solutions are related to Eichlers period polynomials for the group $\Gamma_0(2)$ respectively $\Gamma^0(2)$. Besides for the polynomial

$$\Omega_{e_3}(z) = z^{k-2}\Omega_Q(-\frac{2z+1}{z}) + \Omega_Q(z)$$

in (2.54) with $k = \kappa + 1$ and Ω_Q the period polynomial of $\Gamma(1)$ we could not find an explicit relation between the polynomial solutions of Lewis equation (4.58) for $\Gamma_0(2)$ and the solutions of Eichler's cocycle relation (2.53) for the period polynomials for the modular cusp forms of the group $\Gamma_0(2)$. Numerical calculations however show that the number of polynomial solutions of the cocycle relation (2.53) and the number of polynomial solutions of equation (4.62) for the parameter values $\beta = \beta_\kappa = \frac{1-\kappa}{2}$ coincide at least for the κ values we considered. From this we expect that there exists a close relation between the two equations.

An obvious remarkable fact concerning our approach via Lewis equation (4.61) or the transfer operator is, that contrary to Eichler's cocycle relations (2.53) the theory of old and new forms for $\Gamma_0(2)$ can be recognized immediately also in the solutions of this equation respectively the eigenfunctions of the transfer operator $\tilde{\mathcal{L}}_\beta^{\chi^{\Gamma_0(2)}}$. Interestingly enough the old solutions of Lewis's equation respectively the old eigenfunctions of the transfer operator are obtained just like the automorphic forms of $\Gamma_0(2)$ from those of the larger group $\Gamma(1)$. The same seems to hold true for the group $\Gamma(2)$. Unfortunately, we did not succeed up to now to relate also the new solutions of Lewis equation for $\Gamma_0(2)$ or $\Gamma(2)$ to the corresponding new automorphic forms for these groups. The old solutions respectively eigenfunctions however seem to be related to the old automorphic forms through the same transformations Lewis found for $\Gamma(1)$. This is true at least for the groups $\Gamma_0(2)$ and $\Gamma(2)$.

5. Conclusion

From the transfer operator $\tilde{\mathcal{L}}_\beta^\Gamma$ for subgroups of finite index of the modular group $\Gamma(1) = PSL(2,\mathbb{Z})$ one can derive a functional equation for the eigenfunctions of $\tilde{\mathcal{L}}_\beta^\Gamma$ which generalize the functional equation of Lewis for $\Gamma(1)$. For the group $\Gamma_0(2)$ we discussed in more detail this equation and found explicit solutions. They can be characterized as old ones and new ones in complete analogy to the well known theory of old and new automorphic forms for subgroups of $\Gamma(1)$. The old polynomial solutions seem to be related to the old cusp forms by Eichler's integral transformation for $\Gamma(1)$, the same holds true for the old nonpolynomial solutions which seem to be related to the old Maaß wave forms through Lewis transformations for $\Gamma(1)$. From this one should expect that this holds true also for the new solutions and the new automorphic forms.

References

[AL70] A. O. L. Atkin and J. Lehner. Hecke operators on $\Gamma_0(m)$. *Math. Ann.*, 185:134–160, 1970.

[BV97] E. Balslev and A.B. Venkov. The Weyl law for subgroups of the modular group. Preprint, 1997.

[CM98] C.-H. Chang and D. Mayer. The period function of the nonholomorphic Eisenstein series for $PSL(2,\mathbb{Z})$. *Math. Phys. Elec. J.*, 4(6), 1998.

[CM99] C.-H. Chang and D. Mayer. The transfer operator approach to Selberg's zeta function and modular and Maass wave forms for $PSL(2,\mathbb{Z})$. In D. Hejhal and M. Gutzwiller et al, editors, *IMA Volumes 109 'Emerging applications of number theory'*, pages 72–142. Springer-Verlag, 1999.

[CM00] C.-H. Chang and D. Mayer. Thermodynamic Formalism and Selberg's zeta function for modular groups. *Regular & Chaotic Dynamics*, 5(3):281–312, 2000.

[Eic57] M. Eichler. Eine Verallgemeinerung der Abelschen Integrale. *Math. Zeitschr.*, Bd. 67:267–298, 1957.

[Hej76]   D. Hejhal. The Selberg trace formula and the Riemann zeta function. *Duke Math. J.*,
           43(3):441–482, Sep. 1976.

[Hej83]   D. Hejhal. *The Selberg Trace Formula for PSL(2, ℝ)*. Springer-Verlag, 1983. L. N. in
           Math. 1001.

[Iwa95]   H. Iwaniec. *Introduction to the spectral theory of automorphic forms*. Revista Matem-
           atica Iberoamericana. 1995. Biblioteca de la Revista Matematica Iberoamericana ISSN
           0213-2230.

[Kno78]   M. Knopp. Rational period functions of the modular group. *Duke Math. J.*, 45(1):47–62,
           March 1978.

[Kob93]   N. Koblitz. *Introduction to Elliptic Curves and Modular Forms*. Springer-Verlag, 1993.

[Kub73]   T. Kubota. *Elementary Theory of Eisenstein Series*. Wiley N.Y., 1973.

[Lan76]   S. Lang. *Introduction to Modular Forms*. Springer-Verlag, 1976.

[Lew97]   J. Lewis. Spaces of holomorphic functions equivalent to the even Maass cusp forms.
           Invent. Math., 127(2):271–306, 1997.

[LZ]   J. Lewis and D. Zagier. Period functions for Maaß wave forms. Preprint 1999.

[LZ97]   J. Lewis and D. Zagier. Period functions and the Selberg zeta function for the modular
           group. In *The Mathematical Beauty of Physics*, Adv. Series in Math. Physics 24, pages
           83–97. World Scientific, Singapore, 1997.

[May91a]   D. Mayer. Continued fractions and related transformations. In *Ergodic Theory, Symbolic
           Dynamics and Hyperbolic Spaces*, chapter 7, pages 175–222. Oxford Univ. Press, Oxford,
           1991.

[May91b]   D. Mayer. The thermodynamic formalism approach to Selberg's zeta function for
           PSL(2, ℤ). *Bull. Am. Math. Soc.*, 25:55–60, 1991.

[Miy89]   T. Miyake. *Modular Forms*. Springer-Verlag, 1989.

[MOS66]   W. Magnus, F. Oberhettinger, and R. Soni. *Formulas and Theorems for the Special
           Functions of Mathematical Physics*. Springer-Verlag, 1966.

[Ran77]   R. Rankin. *Modular Forms and Functions*. Cambridge University Press, 1977.

[Sar90]   P. Sarnak. *Some applications of modular forms*. Cambridge University Press, Cam-
           bridge, 1990.

[Shi71]   G. Shimura. *Introduction to the Arithmetic Theory of Automorphic Functions*. Prince-
           ton University Press, 1971.

[Ter85]   A. Terras. *Harmonic Analysis on Symmetric Spaces and Applications I*. Springer-
           Verlag, 1985.

[Zag91]   D. Zagier. Periods of modular forms and Jacobi theta functions. *Invent. Math.*, pages
           449–465, 1991.

[Zag92a]   D. Zagier. Introduction to Modular Forms. In M. Waldschmidt et al, editor, *'From
           Number Theory to Physics'*, chapter 4, pages 238–291. Springer-Verlag, Heidelberg,
           1992.

[Zag92b]   D. Zagier. Periods of modular forms, traces of Hecke operators and multiple zeta values.
           Studies of automorphic forms and L-functions, RIMS Kyoto, pages 162–170, 1992.

FACHBEREICH MATHEMATIK, ABTEILUNG MATHEMATISCHE PHYSIK, SEKRETARIAT MA 7-2,
TU BERLIN, D-10623 BERLIN, GERMANY
    E-mail address: cchang@math.tu-berlin.de

INSTITUT FÜR THEORETISCHE PHYSIK, TU CLAUSTHAL, ARNOLD-SOMMERFELD-STRASSE 6,
D-38678 CLAUSTHAL-ZELLERFELD, GERMANY
    E-mail address: dieter.mayer@tu-clausthal.de

Contemporary Mathematics
Volume **290**, 2001

A Note on Dynamical Trace Formulas

Christopher Deninger and Wilhelm Singhof

ABSTRACT. We prove a dynamical Lefschetz trace formula for certain flows respecting a one-codimensional foliation.

0. Introduction

In their book [**GS**] Ch. VI Guillemin and Sternberg established a trace formula for certain flows on closed manifolds with coefficients in a vector bundle V. It expresses a suitable distributional trace of the induced flow on the global sections of V as a sum of local contributions coming from the periodic orbits and the fixed points.

In the presence of a foliation respected by the flow, this formula has been used to heuristically derive formulas for the alternating sum of distributional traces on leafwise cohomologies. The step which is difficult to justify rigorously is the passage from an alternating sum of traces on global sections to the alternating sum of traces on cohomology. In fact in [**DS**] it is shown that in general for codimension ≥ 2 this passage is not possible. However in [**G**] an example is given where this heuristic method leads to a correct result, namely the Selberg trace formula. Moreover in [**D**] § 4 the first author pointed out that the resulting formula for codimension one foliations were similar to the "explicit formulas" of analytic number theory.

The aim of the present note is to show that in the case of a codimension one foliation which is everywhere transversal to the flow the heuristic

---

1991 *Mathematics Subject Classification.* 37Cxx, 37C27, 53C12.
Key words and phrases. Flow, foliation, trace formula, leafwise cohomology.

argument can be made to work in certain cases. This uses the leaf-wise Hodge decomposition theorem of [**AK1**] and a crude estimate for the wave front sets of the Schwartz kernels of the Hodge projectors. Our first proof of this estimate was unneccessarily complicated. Y. Kordyukov kindly pointed out to us that it already followed from a simple application of the principle of propagation of singularities. An earlier application due to Hörmander of this principle to the definition of the index of a transversally elliptic operator was worked out in [**S**] and [**NZ**].

Our C^∞-approach to the trace formula requires some artificial conditions at the moment which could probably be overcome with more effort. We refrain from this however since complete results have been obtained in the meantime by Alvárez López and Kordyukov [**AK2**]. They also use harmonic forms along the leaves but base their arguments on heat kernel asymptotics.

We would like to thank J. Alvárez López, U. Bunke and Y. Kordyukov for helpful explanations. The first author is also grateful for the hospitality and support of the University of Santiago de Compostela and to J. Alvárez López for inviting him there.

1. Estimates for wave front sets of certain Hodge projectors

We begin by reviewing the Hodge decomposition theorem of Alvárez López and Kordyukov [**AK1**] Cor. 1.3. Consider a closed manifold X with a foliation $\mathcal{F}$. We assume that $\mathcal{F}$ is Riemannian. This means that there is a Riemannian metric g on X such that the foliation is locally defined by Riemannian submersions. A metric with this property is called bundle-like. Alternatively a metric is bundle-like for a foliation if any geodesic which is perpendicular to a leaf at one point remains perpendicular to the leaves at all other points. For example any one-codimensional foliation given by a closed one-form without singularities is Riemannian.

Let V be a Riemannian vector bundle on X with a flat, Riemannian connection $d_\mathcal{F}$ along the leaves of the foliation. Let

$$\mathcal{A}^\bullet(\mathcal{F}, V) = \Gamma(X, \Lambda^\bullet T^*\mathcal{F} \otimes V)$$

be the de Rham complex of V-valued forms along the leaves of $\mathcal{F}$ on X with the differential $d_\mathcal{F}$. The graded Fréchet space $\mathcal{A}^\bullet(\mathcal{F}, V)$ carries a canonical inner product $(,)$. Since the foliation is Riemannian and the metric bundle-like the operator $\delta_{\mathcal{F}}$ defined by the de Rham coderivative on the leaves is adjoint to $d_{\mathcal{F}}$ with respect to $(,)$. The Laplace operator along the leaves

$$\Delta_{\mathcal{F}} = d_{\mathcal{F}}\delta_{\mathcal{F}} + \delta_{\mathcal{F}}d_{\mathcal{F}}$$

is leafwise elliptic and formally self adjoint.

In [**AK1**] Cor. 1.3 the following remarkable result is proved:

THEOREM 1.1. *There is an orthogonal Hodge-decomposition:*

$$\mathcal{A}^{\bullet}(\mathcal{F}, V) = \ker \Delta_{\mathcal{F}} \oplus \overline{\operatorname{im} d_{\mathcal{F}}} \oplus \overline{\operatorname{im} \delta_{\mathcal{F}}} \ .$$

In particular the reduced leafwise cohomology

$$\mathcal{H}^{\bullet}(\mathcal{F}, V) = \ker d_{\mathcal{F}}/\overline{\operatorname{im} d_{\mathcal{F}}}$$

is isomorphic to $\ker \Delta_{\mathcal{F}}$.

REMARK The example in [**DS**] shows that this Hodge decomposition does not hold in general for non-Riemannian foliations.

Let us write

$$P_{\Delta}, P_d, P_{\delta} : \mathcal{A}^{\bullet}(\mathcal{F}, V) \longrightarrow \mathcal{A}^{\bullet}(\mathcal{F}, V)$$

for the projectors to $\ker \Delta_{\mathcal{F}}, \overline{\operatorname{im} d_{\mathcal{F}}}$ and $\overline{\operatorname{Im} \delta_{\mathcal{F}}}$ defined by the above decomposition. If P is anyone of them we may view its Schwartz kernel K_P as a distributional section of $\operatorname{End}(\Lambda^{\bullet}T^*\mathcal{F} \otimes V)$ on $X \times X$ with the defining property that for all test functions α, β in $\mathcal{A}^{\bullet}(\mathcal{F}, V)$ we have

(1.1) $$\langle K_P, \alpha \otimes \beta \rangle = (P(\beta), \alpha) \ .$$

Let $N^*\mathcal{F}$ be the conormal bundle to the foliation, and let $N^*\Delta$ be the conormal bundle to the diagonal $\Delta \subset X \times X$. For any vector bundle F set $\tilde{F} := F \smallsetminus 0$.

PROPOSITION 1.2. *The following estimates for the wave front set of K_P in $\tilde{T}^*(X \times X)$ hold:*

$$WF(K_{P_{\Delta}}) \subset (N^*\mathcal{F} \times N^*\mathcal{F}) \smallsetminus 0 \ .$$

Moreover $WF(K_{P_d})$ and $WF(K_{P_{\delta}})$ are contained in the union of

$$(N^*\mathcal{F} \times N^*\mathcal{F}) \smallsetminus 0 \text{ with } \tilde{N}^*\Delta \ .$$

PROOF. First note that because each of the projectors $P = P_\Delta, P_d,$ P_δ is self-adjoint we have for all $\alpha, \beta \in \mathcal{A}^\bullet(\mathcal{F}, V)$ that:

$$\begin{aligned} \langle K_P, \alpha \otimes \beta \rangle &= (P(\beta), \alpha) \\ &= (\beta, P(\alpha)) = \langle K_P, \beta \otimes \alpha \rangle . \end{aligned}$$

Hence

$$K_P = s^* K_P$$

where $s : X \times X \to X \times X$ is defined by $s(x, y) = (y, x)$.

Thus

$$WF(K_P) = s^* WF(K_P)$$

and therefore it suffices to show the estimates:

$$(1.2) \quad \begin{aligned} WF(K_{P_\Delta}) &\subset (N^*\mathcal{F} \times T^*X) \smallsetminus 0 \\ WF(K_{P_d}) \cup WF(K_{P_\delta}) &\subset \tilde{N}^*\Delta \cup (N^*\mathcal{F} \times T^*X) \smallsetminus 0 . \end{aligned}$$

Since $\Delta_\mathcal{F}$ is formally self adjoint, we have:

$$\begin{aligned} \langle (\Delta_\mathcal{F} \hat\otimes \mathrm{id})(K_P), \alpha \otimes \beta \rangle &= \langle K_P, \Delta_\mathcal{F} \alpha \otimes \beta \rangle \\ &= (P \Delta_\mathcal{F} \alpha, \beta) . \end{aligned}$$

Hence

$$(\Delta_\mathcal{F} \hat\otimes \mathrm{id})(K_P) = K_{P\Delta_\mathcal{F}} .$$

By the principle of propagation of singularities [**H**] Theorem 8.3.1 we have:

$$WF(K_P) \subset WF(K_{P\Delta_\mathcal{F}}) \cup \mathrm{char}(\Delta_\mathcal{F} \hat\otimes \mathrm{id}) .$$

Here char denotes the charakteristic set of a differential operator. Since $\Delta_\mathcal{F}$ is leafwise elliptic we get

$$\mathrm{char}(\Delta_\mathcal{F} \hat\otimes \mathrm{id}) \subset (N^*\mathcal{F} \times T^*X) \smallsetminus 0 .$$

For $P = P_\Delta, P_d, P_\delta$ the composition $P\Delta_\mathcal{F}$ equals respectively $0, d_\mathcal{F}\delta_\mathcal{F},$ $\delta_\mathcal{F} d_\mathcal{F}$. Since the wave front set of the Schwartz kernel of a differential operator is contained in $\tilde{N}^*\Delta$ we get (1.2) and hence the proposition.
$\square$

In the interesting case where all leaves are dense we don't know how to improve the estimate in the proposition which doesn't impose any restriction on the singular support of K_P if $\dim \mathcal{F} < \dim X$. In general we can do a little better by the following observation. It is known that the closures $\overline{L}$ of the leaves L of the foliation form a partition of X. Let $\mathcal{R}_{\overline{\mathcal{F}}} \subset X \times X$ denote the corresponding equivalence relation. It is a closed subset of $X \times X$. We then have the following basic information about the kernel K_{P_Δ}:

PROPOSITION 1.3. *The support of K_{P_Δ} is contained in $\mathcal{R}_{\overline{\mathcal{F}}}$.*

PROOF. By [**AK1**] Cor. 1.3 we have for any $\alpha \in \mathcal{A}(\mathcal{F}, V)$ that

$$P_\Delta(\alpha) = \lim_{t \to \infty} e^{-t\Delta_\mathcal{F}}(\alpha)$$

in $\mathcal{A}(\mathcal{F}, V)$. Now:

$$u_\alpha(x, t) = e^{-t\Delta_\mathcal{F}}(\alpha)$$

is the unique smooth solution to the leafwise heat equation

$$\frac{\partial u_\alpha}{\partial t} = \Delta_\mathcal{F} u_\alpha \quad \text{with } u_\alpha(0, x) = \alpha(x).$$

Since heat evolves along the leaves it follows that for all $t \geq 0$ the support of $u_\alpha(t, \_)$ is contained in the closure of the union of all leaves which have a non-empty intersection with supp α. Hence supp $P_\Delta(\alpha)$ is contained in this set as well. Using (1.1) we find that supp $K_{P_\Delta} \subset \mathcal{R}_{\overline{\mathcal{F}}}$. $\qquad\square$

1.4. We now give a corollary of proposition 1.2 which is relevant for the C^∞-approach to the trace formula in section 2. Let X be a closed manifold with a flow $\phi : X \times \mathbb{R} \to X$ which is everywhere transversal to a one-codimensional foliation $\mathcal{F}$ and such that ϕ^t maps leaves to leaves for all t. Then $\mathcal{F}$ is Riemannian and we fix a bundle-like metric. The flow doesn't have fixed points. It should be non-degenerate in the following sense:

If x lies on a periodic orbit γ of length $l(\gamma)$ then for all integers $k \geq 1$ (equivalently: for all nonzero integers k) the 1-eigenspace of $T_x\phi^{kl(\gamma)}$ is one-dimensional. Note here that $T_x\phi^{kl(\gamma)}$ always has $Y_{\phi,x}$ as an eigenvector with eigenvalue 1, Y_ϕ being the vector field generated by the flow.

We also assume that we are given a Riemannian bundle V with a connection as before. Moreover there should be a smooth action

$$\psi^t : \phi^{t*}V \longrightarrow V$$

which is compatible with the connection in the evident sense. The induced action ψ^{t*} on $\mathcal{A}^\bullet(\mathcal{F}, V)$ respects the differential $d_\mathcal{F}$. For the bundle $E = \Lambda^\bullet T^*\mathcal{F} \otimes V$ consider the composition

$$\psi^* : \Gamma(X, E) \xrightarrow{(\phi|_{X \times \mathbb{R}^*})^{-1}} \Gamma(X \times \mathbb{R}^*, \phi^*E) \xrightarrow{\psi} \Gamma(X \times \mathbb{R}^*, p^*E)$$

where $p : X \times \mathbb{R}^* \to X$ is the projection.

COROLLARY 1.5. *For $P = P_\Delta, P_d, P_\delta$ the wave front set of the Schwartz kernels $K_{\psi^* \circ P}$ and $K_{(P \hat{\otimes} \mathrm{id}) \circ \psi^*}$ and $K_{(P \hat{\otimes} \mathrm{id}) \circ \psi^* \circ P}$ on $X \times \mathbb{R}^* \times X$ is disjoint from the conormal bundle to the "diagonal"*

$$\tilde{\Delta} : X \times \mathbb{R}^* \longrightarrow X \times \mathbb{R}^* \times X$$

defined by $\tilde{\Delta}(x, t) = (x, t, x)$. In particular the pullbacks $\tilde{\Delta}^* \mathrm{tr}_V K_{\psi^* \circ P}$ and $\tilde{\Delta}^* \mathrm{tr}_V K_{(P \hat{\otimes} \mathrm{id}) \circ \psi^*}$ and $\tilde{\Delta}^* \mathrm{tr}_V K_{(P \hat{\otimes} \mathrm{id}) \circ \psi^* \circ P}$ are defined, [**H**] *Cor. 8.2.7.*

PROOF. We consider only $K_{\psi^* \circ P}$. The other cases are similar. The Schwartz kernel of ψ^* is given by a smooth density on the graph $\Gamma_\phi \subset X \times \mathbb{R}^* \times X$ of $\phi|_{X \times \mathbb{R}^*}$. By [**H**] Ex. 8.2.5 this implies that

$$WF(K_{\psi^*}) \subset \tilde{N}^* \Gamma_\phi .$$

By definition the conormal bundle $N^* \Gamma_\phi$ consists of all cotangent vectors

$$(\xi_x, \mu \, dt|_t, \eta_{\phi^t(x)}) \quad \text{in } T^*(X \times \mathbb{R}^* \times X)$$

which satisfy:

$$\xi_x = -(T_x \phi^t)^* (\eta_{\phi^t(x)}) \quad \text{and} \quad \mu = -\langle Y_{\phi, \phi^t(x)}, \eta_{\phi^t(x)} \rangle .$$

Recall the following notations for distributions K on $X \times Y$:

$$WF'(K) = \{(\xi_x, \eta_y) \in T^*X \times T^*Y \quad \text{such that } (\xi_x, -\eta_y) \in WF(K)\}$$

and

$$WF(K)_X = \{\xi_x \in T^*X \quad \text{such that } (\xi_x, 0_y) \in WF(K) \text{ for some } y \in Y\}$$

and

$$WF'(K)_Y = \{\eta_y \in T^*X \quad \text{s. t. } (0_x, \eta_y) \in WF'(K) \text{ for some } x \in X\} .$$

Thus $WF'(K_{\psi^*})$ is contained in the set

$$\{(\xi_x, \mu \, dt|_t, \eta_{\phi^t(x)}) \, | \, \xi_x = (T_x \phi^t)^* (\eta_{\phi^t(x)}) \text{ and } \mu = \langle Y_{\phi, \phi^t(x)}, \eta_{\phi^t(x)} \rangle\} .$$

In particular the sets

$$WF(K_{\psi^*})_{X \times \mathbb{R}^*} \quad \text{and} \quad WF'(K_{\psi^*})_X \quad \text{are empty} .$$

Hence the generalization of [**H**] Th. 8.2.14 to manifolds gives the estimate:

$$WF'(K_{\psi^* \circ P}) \subset WF'(K_{\psi^*}) \circ WF'(K_P) \cup 0_{X \times \mathbb{R}^*} \times \tilde{T}^*X$$

where $\circ$ denotes the product of correspondences.

By proposition 1.2 we know that

$$WF'(K_P) \subset (\text{diagonal in } \tilde{T}^*X \times \tilde{T}^*X) \cup (N^*\mathcal{F} \times N^*\mathcal{F}) \smallsetminus 0 .$$

Hence $WF'(K_{\psi^* \circ P})$ is contained in the union of $WF'(K_{\psi^*})$ with $WF'(K_{\psi^*}) \circ ((N^*\mathcal{F} \times N^*\mathcal{F}) \smallsetminus 0)$ and $0_{X \times \mathbb{R}^*} \times \tilde{T}^*X$. Thus $WF(K_{\psi^* \circ P})$ is contained in the union of $\tilde{N}^*\Gamma_\phi$ with $0_{X \times \mathbb{R}^*} \times \tilde{T}^*X$ and with the set of elements

$$0 \neq (\xi_x, \mu\, dt\,|_t, \zeta_z) \in T^*X \times T^*\mathbb{R} \times N^*\mathcal{F}$$

such that there exists some $\eta_{\phi^t(x)} \in N^*_{\phi^t(x)}\mathcal{F}$ with

$$\xi_x = (T_x\phi^t)^*(\eta_{\phi^t(x)}) \quad \text{and} \quad \mu = \langle Y_{\phi,\phi^t(x)}, \eta_{\phi^t(x)} \rangle \ .$$

The fact that the flow is non-degenerate is equivalent to the intersection of $\tilde{N}^*\Gamma_\phi$ with

$$N^*\tilde{\Delta} = \{(\xi_x, 0_t, -\xi_x) \,|\, \xi_x \in T^*_x X\}$$

being empty. Obviously

$$(0_{X \times \mathbb{R}^*} \times \tilde{T}^*X) \cap N^*\tilde{\Delta} = \emptyset \ .$$

Finally, if an element $(\xi_x, \mu\, dt\,|_t, \zeta_z)$ as above were in $N^*\tilde{\Delta}$ then $\mu = 0$ and hence $\eta_{\phi^t(x)}$ would have to vanish since the flow is transversal to the foliation. Thus $\xi_x = 0$, and because of $\zeta_z = -\xi_x$ also $\zeta_z = 0$. Contradiction. Hence we have seen that:

$$WF(K_{\psi^* \circ P}) \cap N^*\tilde{\Delta} = \emptyset \ .$$

$\square$

2. Application to a dynamical trace formula

The starting point is a trace formula in differential topology due to Guillemin and Sternberg [**GS**] Ch. VI, p. 311.

Consider a smooth compact manifold X with a flow ϕ^t without fixed points. We assume that ϕ^t is non-degenerate in the sense of 1.4.
Let $T^0 = \mathbb{R} \cdot Y_\phi \subset TX$ be the one-dimensional bundle of tangents to the flow. Let V be a smooth vector bundle with a contravariant action

$$\psi^t = \psi^t_V : \phi^{t*}V \longrightarrow V \ .$$

For any point x on a periodic orbit γ of length $l(\gamma)$ we get endomorphisms

$$\psi_x^{kl(\gamma)} : V_{\phi^{kl(\gamma)}(x)} = V_x \longrightarrow V_x$$

for any integer k. Note that the trace $\text{Tr}(\psi_x^{kl(\gamma)} \,|\, V_x)$ is independent of the chosen point $x \in \gamma$.

Guillemin and Sternberg define the distributional trace of the composition

$$\psi^* : \Gamma(X, V) \xrightarrow{(\phi|_{X \times \mathbb{R}^*})^{-1}} \Gamma(X \times \mathbb{R}^*, \phi^* V) \xrightarrow{\psi^*} \Gamma(X \times \mathbb{R}^*, p^* V)$$

by the formula

$$\mathrm{Tr}(\psi^* \,|\, \Gamma(X, V)) = \pi_* \mathrm{tr}_V \tilde{\Delta}^* K_{\psi^*} \quad \text{in } \mathcal{D}'(\mathbb{R}^*) \,.$$

Here $p : X \times \mathbb{R}^* \to X$ and $\pi : X \times \mathbb{R}^* \to \mathbb{R}^*$ are the projections and K_{ψ^*} is the Schwartz kernel of ψ^*. The pullback of K_{ψ^*} by the "diagonal" $\tilde{\Delta} : X \times \mathbb{R}^* \to X \times \mathbb{R}^* \times X, \tilde{\Delta}(x, t) = (x, t, x)$ is defined since the wave front set of K_{ψ^*} is disjoint from $N^* \tilde{\Delta}$, the flow being non-degenerate. Then according to [**GS**] Ch. VI, p. 311 the following is true:

PROPOSITION 2.1. *With assumptions as above the following formula holds in* $\mathcal{D}'(\mathbb{R}^*)$:

$$\mathrm{Tr}(\psi^* \,|\, \Gamma(X, V)) = \sum_\gamma l(\gamma) \sum_{k \in \mathbb{Z} \smallsetminus 0} \frac{\mathrm{Tr}(\psi_x^{kl(\gamma)} \,|\, V_x)}{|\det(1 - T_x \phi^{kl(\gamma)} \,|\, T_x X / T_x^0)|} \delta_{kl(\gamma)} \,.$$

Here γ *runs over the periodic orbits and in the sum* x *denotes any point on* γ. *Note that non-degeneracy of the flow (1.4) implies that there are only finitely many orbits* $\gamma \subset X$ *of length bounded by a given constant* C.

Now assume that in addition X carries a smooth foliation $\mathcal{F}$ such that ϕ^t maps leaves to leaves for any t. Hence $T_x \phi^t$ maps $T_x \mathcal{F}$ to $T_{\phi^t(x)} \mathcal{F}$ and hence the dual map gives rise to a contravariant action:

$$T^* \phi^t : \phi^{t*} T^* \mathcal{F} \longrightarrow T^* \mathcal{F} \,.$$

Now recall the formulas:

$$\sum_i (-1)^i \mathrm{Tr}(\Lambda^i f \,|\, \Lambda^i E) = \det(1 - f \,|\, E)$$

and

$$\mathrm{Tr}(f \otimes g \,|\, E \otimes F) = \mathrm{Tr}(f \,|\, E) \mathrm{Tr}(g \,|\, E)$$

for endomorphisms f, g of finite dimensional $\mathbb{R}$-vector spaces E resp. F. Applying (2.1) to the bundles $\Lambda^i T^* \mathcal{F} \otimes V$ with $\psi^t = \psi_V^t \otimes \Lambda^i T^* \phi^t$ and taking the alternating sum over i we get in $\mathcal{D}'(\mathbb{R}^*)$:

$$(2.1) \quad \sum_i (-1)^i \mathrm{Tr}(\psi^* \,|\, \mathcal{A}^i(\mathcal{F}, V))$$

$$= \sum_\gamma{}' l(\gamma) \sum_{k \in \mathbb{Z} \smallsetminus 0} \varepsilon_\gamma(k) \frac{\mathrm{Tr}(\psi_x^{kl(\gamma)} \,|\, V_x)}{\det(1 - T_x \phi^{kl(\gamma)} \,|\, T_x X / (T_x \mathcal{F} \oplus T_x^0))} \delta_{kl(\gamma)} \,.$$

Here the prime at the sum over periodic orbits means that we only sum over those γ's which are not contained in a leaf i.e. those for which $Y_{\phi,x} \notin T_x\mathcal{F}$ for one (equivalently: for every) point $x \in \gamma$. Observe here that for the other closed orbits we have

$$\det(1 - T_x\phi^{kl(\gamma)} \,|\, T_x\mathcal{F}) = 0 \quad \text{for } x \in \gamma, k \in \mathbb{Z} \smallsetminus 0$$

since $T_x\phi^{l(\gamma)}(Y_{\phi,x}) = Y_{\phi,x}$. Finally $\varepsilon_\gamma(k)$ denotes the sign of the determinant

$$\det(1 - T_x\phi^{kl(\gamma)} \,|\, T_xX/T_x^0) \,.$$

It is independent of the chosen point $x \in \gamma$.

Let us now assume in addition that V carries a flat connection $d_\mathcal{F}$ along the leaves which is compatible with ψ. One would like to know conditions under which the alternating sum in (2.1) can be replaced by an alternating sum of traces on the reduced leafwise cohomologies $\mathcal{H}^i(\mathcal{F}, V)$ of $(V, d_\mathcal{F})$.

If $\mathcal{F}$ has codimension zero this is possible since the resulting formula is just Hopf's formula.

For $\mathcal{F}$ of codimension two this is not always possible as the example in [**DS**] §2 shows.

For Riemannian foliations and Riemannian connections a Hodge theoretic approach to the problem makes use of the leafwise Hodge decomposition 1.1. If the wave front sets of the Schwartz kernels $K_{\psi^* \circ P}$ are disjoint from $N^*\tilde{\Delta}$ for $P = P_\Delta, P_d, P_\delta$ the traces

$$(2.2) \quad \begin{aligned} \mathrm{Tr}(\psi^* \,|\, \ker \Delta_\mathcal{F}) &:= \pi_* \tilde{\Delta}^* \mathrm{tr}_V K_{\psi^* \circ P_\Delta} \\ \mathrm{Tr}(\psi^* \,|\, \overline{\mathrm{im}\, d_\mathcal{F}}) &:= \pi_* \tilde{\Delta}^* \mathrm{tr}_V K_{\psi^* \circ P_d} \\ \mathrm{Tr}(\psi^* \,|\, \overline{\mathrm{im}\, \delta_\mathcal{F}}) &:= \pi_* \tilde{\Delta}^* \mathrm{tr}_V K_{\psi^* \circ P_\delta} \end{aligned}$$

are well defined and we have for all i:

$$\mathrm{Tr}(\psi^* \,|\, \mathcal{A}^i(\mathcal{F}, V)) = \mathrm{Tr}(\psi^* \,|\, \ker \Delta_\mathcal{F}^i) + \mathrm{Tr}(\psi^* \,|\, \overline{\mathrm{im}\, d_\mathcal{F}^{i-1}}) + \mathrm{Tr}(\psi^* \,|\, \overline{\mathrm{im}\, \delta_\mathcal{F}^i}) \,.$$

If one can show that

$$(2.3) \qquad \sum_i (-1)^i \mathrm{Tr}(\psi^* \,|\, \overline{\mathrm{im}\, d_\mathcal{F}^i}) = \sum_i (-1)^i \mathrm{Tr}(\psi^* \,|\, \overline{\mathrm{im}\, \delta_\mathcal{F}^i})$$

e.g. if

$$(2.4) \qquad \mathrm{Tr}(\psi^* \,|\, \overline{\mathrm{im}\, d_\mathcal{F}^i}) = \mathrm{Tr}(\psi^* \,|\, \overline{\mathrm{im}\, \delta_\mathcal{F}^i}) \quad \text{for all } i$$

then it follows that:

$$(2.5) \qquad \sum_i (-1)^i \mathrm{Tr}(\psi^* \mid \mathcal{A}^*(\mathcal{F}, V)) = \sum_i (-1)^i \mathrm{Tr}(\psi^* \mid \ker \Delta_{\mathcal{F}}^i) \,.$$

We will now discuss this approach in the particular codimension one situation described in section 1.4. According to Corollary 1.5 the problem of defining the individual traces (2.2) is solved in this case. Moreover the traces $\pi_* \tilde{\Delta}^* \mathrm{tr}_V K_{(P \hat{\otimes} \mathrm{id}) \circ \psi^*}$ and $\pi_* \tilde{\Delta}^* \mathrm{tr}_V K_{(P \hat{\otimes} \mathrm{id}) \circ \psi^* \circ P}$ are defined as well. It can be shown by a somewhat lengthy calculation that these traces all agree i.e. that:

$$(2.6) \qquad \mathrm{Tr}(\psi^* \mid \ker \Delta_{\mathcal{F}}^i) \;=\; \pi_* \tilde{\Delta}^* \mathrm{tr}_V K_{(P_\Delta \hat{\otimes} \mathrm{id}) \circ \psi^*}$$
$$\;=\; \pi_* \tilde{\Delta}^* \mathrm{tr}_V K_{(P_\Delta \hat{\otimes} \mathrm{id}) \circ \psi^* \circ P_\Delta}$$

etc. In particular, because of the commutative diagram

$$\begin{array}{ccc}
\ker \Delta_{\mathcal{F}}^i & \xrightarrow{\;\sim\;} & \mathcal{H}^i(\mathcal{F}, V) \\
{\scriptstyle P \circ \psi^{t*}} \downarrow & & \downarrow {\scriptstyle \psi^{t*}} \\
\ker \Delta_{\mathcal{F}}^i & \xrightarrow{\;\sim\;} & \mathcal{H}^i(\mathcal{F}, V)
\end{array}$$

we may view $\mathrm{Tr}(\psi^* \mid \ker \Delta_{\mathcal{F}}^i)$ as a distributional trace of the flow on $\mathcal{H}^i(\mathcal{F}, V)$. Of course there is still the problem to express $\mathrm{Tr}(\psi^* \mid \ker \Delta_{\mathcal{F}}^i)$ purely in terms of the induced flow on $\mathcal{H}^i(\mathcal{F}, V)$ without any reference to a metric. In **2.5** below this is achieved for flows which are isometric for some bundle like metric.

We have been able to establish equation (2.3) in two special cases only:

2.2. If the leafwise cohomologies are Hausdorff then $d_{\mathcal{F}}$ induces an isomorphism

$$d_{\mathcal{F}} : \mathrm{im}\, \delta_{\mathcal{F}} \xrightarrow{\;\sim\;} \mathrm{im}\, d_{\mathcal{F}}$$

which commutes with the flow. This may be used to prove (2.4). In general $d_{\mathcal{F}}$ induces only an isomorphism

$$d_{\mathcal{F}} : \overline{\mathrm{im}\, \delta_{\mathcal{F}}} \xrightarrow{\;\sim\;} \mathrm{im}\, d_{\mathcal{F}} \,.$$

2.3. Assume ϕ^t and ψ^t are isometric and $n = \dim X$ is even. The Hodge star operator along the leaves $*_{\mathcal{F}}$ c.f. [**AK1**] §3 commutes with ψ^{t*}. It maps $\overline{\mathrm{im}\, d_{\mathcal{F}}^i}$ isomorphically onto $\overline{\mathrm{im}\, \delta_{\mathcal{F}}^{n-i-2}}$ where $n = \dim X$. From this one may deduce that

$$\mathrm{Tr}(\psi^* \mid \overline{\mathrm{im}\, d_{\mathcal{F}}^i}) = \mathrm{Tr}(\psi^* \mid \overline{\mathrm{im}\, \delta_{\mathcal{F}}^{n-i-2}}) \,.$$

Hence (2.3) follows since $\dim X$ was supposed to be even.

THEOREM 2.4. *In the situation of 1.4 assuming 2.2 or 2.3 the following trace formula holds as an equality in $\mathcal{D}'(\mathbb{R}^*)$:*

$$\sum_i (-1)^i \mathrm{Tr}(\psi^* \mid \ker \Delta^i_{\mathcal{F}}) = \sum_\gamma l(\gamma) \sum_{k \in \mathbb{Z} \smallsetminus 0} \varepsilon_\gamma(k) \mathrm{Tr}(\psi^{kl(\gamma)}_x \mid V_x) \delta_{kl(\gamma)} \; .$$

REMARK Heat kernel methods lead to a stronger version of this result. In [**AK2**] Alvárez López and Kordyukov prove the trace formula in the setting of 1.4 without any restrictions. Moreover they can prove a version in $\mathcal{D}'(\mathbb{R})$: Quite beautifully there appears the contribution $\chi_{\mathrm{Co}}(\mathcal{F}, V) \cdot \delta_0$ from the origin where $\chi_{\mathrm{Co}}(\mathcal{F}, \_)$ is Connes' Euler characteristic for foliated bundles.

In the case of isometric flows their formula also follows from the transverse index theorem of Lazarov [**L**] Theorem 2.10 combined with the leafwise Hodge decomposition of [**AK1**].

2.5. We now assume that ϕ^t and ψ^t are isometric. In this case the distributional trace $\mathrm{Tr}(\psi^* \mid \ker \Delta^i_{\mathcal{F}})$ is completely determined by the action of ψ^{t*} on the reduced leafwise cohomology $\mathcal{H}^i(\mathcal{F}, V)$ as follows. For $\alpha \in \mathbb{C}$ let $E_\alpha \subset \mathcal{H}^i(\mathcal{F}, V)$ denote the subspace of $\mathcal{H}^i(\mathcal{F}, V)$ where ψ^{t*} acts by multiplication with $e^{\alpha t}$ for all $t \in \mathbb{R}$. Since ψ^{t*} is an isometry on $L^2(\ker \Delta_{\mathcal{F}})$ the commutative diagram:

(2.7)

$$
\begin{array}{ccc}
\mathcal{H}^i(\mathcal{F}, V) & \xrightarrow{\;\sim\;} & \ker \Delta^i_{\mathcal{F}} \\
\psi^{t*} \downarrow & & \downarrow \psi^{t*} \\
\mathcal{H}^i(\mathcal{F}, V) & \xrightarrow{\;\sim\;} & \ker \Delta^i_{\mathcal{F}}
\end{array}
$$

implies that $E_\alpha = 0$ unless α is purely imaginary.

THEOREM 2.6. *The multiplicity* $\dim E_\alpha$ *is finite for all* α *and non-zero for countably many values only. We have:*

$$\mathrm{Tr}(\psi^* \mid \ker \Delta^i_{\mathcal{F}}) = \sum_\alpha \dim E_\alpha \cdot e^{\alpha t} \quad in \; \mathcal{D}'(\mathbb{R}^*) \; .$$

Here the functions $e^{\alpha t}$ *are viewed as distributions on* $\mathbb{R}^*$ *and the sum converges in* $\mathcal{D}'(\mathbb{R}^*)$.

As an example for such kinds of sums note that the sum over $\nu \geq 1$ of the functions $e^{i\nu t}$ converges in $\mathcal{D}'(\mathbb{R})$ since by partial integration for any $\varphi \in \mathcal{D}(\mathbb{R})$

$$\langle e^{i\nu t}, \varphi \rangle = O(\nu^{-N}) \quad \text{for all } N \geq 1 \; .$$

On the other hand the sum $\sum e^{i\nu t}$ does not converge for a single real t.

PROOF OF 2.6 Consider the orthogonal decomposition with respect to the bundle like metric

$$TX = T\mathcal{F}^\perp \oplus T\mathcal{F} \quad \text{with } T\mathcal{F}^\perp = \mathbb{R} \cdot Y_\phi \ .$$

It allows us to view $\Lambda^\bullet T^*\mathcal{F}$ as a direct summand of $\Lambda^\bullet T^*X$. Using a corresponding decomposition of V we may view $\mathcal{A}^\bullet(\mathcal{F}, V)$ as a direct summand of the space $\mathcal{A}^\bullet(V)$ of all smooth V-valued forms on X. Let Θ be the infinitesimal generator of the induced flow ψ^{t*} on $\ker \Delta^i_\mathcal{F}$. Then we have the equality

(2.8) $$-\Theta^2 = \Delta^i |_{\ker \Delta^i_\mathcal{F}} \ .$$

Here Δ^i is the ordinary Laplacian on $\mathcal{A}^i(V)$. In particular Δ^i maps $\ker \Delta^i_\mathcal{F}$ into itself. The proof of (2.8) goes as follows. The orthogonal decomposition

$$TX = T\mathcal{F} \oplus \mathbb{R} \cdot Y_\phi$$

corresponds to a decomposition $d = d_\mathcal{F} + d^0$ of the exterior differential. We set

$$\Delta^0 = d^0 d^{0*} + d^{0*} d^0 \ .$$

Using the fact that $\Delta_\mathcal{F} \omega = 0$ if and only if $d_\mathcal{F} \omega = 0 = d^*_\mathcal{F} \omega$ we find that for a form $\omega \in \mathcal{A}^i(V)$ in $\ker \Delta_\mathcal{F}$ we have:

$$\Delta\omega = \Delta^0 \omega + d_\mathcal{F} d^{0*}\omega + d^*_\mathcal{F} d^0 \omega \ .$$

By isometry of the flow the function $x \mapsto \|Y_{\phi,x}\|$ is constant and we may therefore assume that the metric on X is normalized such that $\|Y_{\phi,x}\| = 1$ for all x. Let ω_ϕ be the 1-form on X which is zero on $T\mathcal{F}$ and such that $\langle \omega_\phi, Y_\phi \rangle = 1$. Then $\|\omega_{\phi,x}\| = 1$ for all x and for any form α with values in $\Lambda^p T^*\mathcal{F} \otimes V$ we have

$$d^0 \alpha = \Theta\alpha \wedge \omega_\phi \quad \text{and} \quad d^{0*}(\alpha \wedge \omega_\phi) = -\Theta\alpha \ .$$

Here Θ is the infinitesimal generator of the induced flow on $\mathcal{A}^\bullet(V)$. Note that Θ is skew-symmetric since the flow is isometric. It follows that $\Delta^0 = -\Theta^2$. Moreover

$$d_\mathcal{F} d^{0*}(\alpha \wedge \omega_\phi) = -d_\mathcal{F} \Theta\alpha = -\Theta d_\mathcal{F} \alpha = 0$$

if $d_\mathcal{F}(\alpha \wedge \omega_\phi) = 0$. Hence for $\omega \in \ker \Delta_\mathcal{F}$ we have

$$d_\mathcal{F} d^{0*}\omega = 0 \ .$$

By isometry of the flow Θ commutes with the $*$-operator along the leaves, $*_\mathcal{F}$ up to sign. Hence $d^*_\mathcal{F}\Theta = \pm\Theta d^*_\mathcal{F}$ since $d^*_\mathcal{F} = \pm *_\mathcal{F} d_\mathcal{F} *_\mathcal{F}$ and therefore

$$\begin{aligned}
d_{\mathcal{F}}^* d^0 \omega &= d_{\mathcal{F}}^* (\Theta \omega \wedge \omega_\phi) \\
&= \pm \Theta d_{\mathcal{F}}^* \omega \wedge \omega_\phi \\
&= 0 \quad \text{if } \omega \in \ker \Delta_{\mathcal{F}} .
\end{aligned}$$

This completes the proof of equation (2.8).

Since E_α is the eigenspace of Θ the first assertions of the theorem follow from the spectral theory of the ordinary Laplacian e.g. [**BGV**] ch. 2. Moreover it follows that there is an orthonormal basis $\{\omega_\nu\}$ of $L^2(\ker \Delta_{\mathcal{F}}^i)$ of eigenvectors of Θ with $\omega_\nu \in \ker \Delta_{\mathcal{F}}^i$. If $\Theta \omega_\nu = \alpha_\nu \omega_\nu$ then $\psi^{t*} \omega_\nu = e^{t \alpha_\nu} \omega_\nu$. A short calculation now shows that:

$$K_{(P_\Delta \hat{\otimes} \mathrm{id}) \circ \psi^* \circ P_\Delta} = \sum_\nu e^{t \alpha_\nu} \omega_\nu \otimes \overline{\omega}_\nu ,$$

where the sum is a convergent sum of distributional sections on $X \times \mathbb{R}^* \times X$.

For any test function $\varphi \in \mathcal{D}'(\mathbb{R}^*)$ define a distributional section K_φ on $X \times X$ by "contraction":

$$(2.9) \quad K_\varphi = \langle K_{(P_\Delta \hat{\otimes} \mathrm{id}) \circ \psi^* \circ P_\Delta}, \varphi \rangle = \sum_\nu \left(\int_{\mathbb{R}} e^{t \alpha_\nu} \varphi(t) \, dt \right) \omega_\nu \otimes \overline{\omega}_\nu .$$

We claim that the sum does not only converge in the distributional sense but even in the smooth topology. In particular K_φ is a smooth section on $X \times X$. For this it suffices to show that all the series:

$$(2.10) \quad \sum_\nu \left| \int_{\mathbb{R}} e^{t \alpha_\nu} \varphi(t) \, dt \right| \| \omega_\nu \otimes \overline{\omega}_\nu \|_{2,k} \quad \text{for } k \geq 1$$

converge. Here $\|\eta\|_{2,k}$ denotes the Sobolev norm given by the L^2-norm of $(\mathrm{id} + \Delta_{X \times X}^i)^k(\eta)$. Since

$$\Delta_{X \times X}^i = \Delta_X^i \otimes \mathrm{id} + \mathrm{id} \otimes \Delta_X^i$$

and because of (2.8) we have

$$\| \omega_\nu \otimes \overline{\omega}_\nu \|_{2,k} = (1 - \alpha_\nu^2 - \overline{\alpha}_\nu^2)^{k/2} = (1 + 2|\alpha_\nu|^2)^{k/2} .$$

On the other hand partial integration shows that for any $N \geq 1$ we have

$$\int_{\mathbb{R}} e^{ty} \varphi(t) \, dt = O(y^{-N})$$

as $y \to \infty$. Hence up to finitely many terms the sum in (2.10) is majorised by a constant times

$$\sum_{\alpha_\nu \neq 0} |\alpha_\nu|^{-N} (1 + 2|\alpha_\nu|^2)^{k/2} .$$

Ths sum converges for large enough $N = N(k)$ since the numbers $|\alpha_\nu|^2$ are among the eigenvalues λ of the ordinary Laplacian Δ on V-valued forms on X and since the sum $\sum_{\lambda \neq 0} \lambda^{-m}$ converges for $m \gg 0$. We have therefore seen that the series (2.10) converge for all $k \geq 1$.

A straightforeward argument using the definition of the pullback of distributions in [H] Theorem 8.2.4 shows that if $\Delta : X \to X \times X$ denotes the diagonal:

$$\begin{aligned} \langle \pi_* \tilde{\Delta}^* \mathrm{tr}_V K_{(P_\Delta \hat{\otimes} \mathrm{id}) \circ \psi^* \circ P_\Delta}, \varphi \rangle &= \langle \tilde{\Delta}^* \mathrm{tr}_V K_{(P_\Delta \hat{\otimes} \mathrm{id}) \circ \psi^* \circ P_\Delta}, \varphi \circ \pi \rangle \\ &= \langle \Delta^* \mathrm{tr}_V K_\varphi, 1_X \rangle . \end{aligned}$$

Since K_φ is a smooth section we have

$$\langle \Delta^* \mathrm{tr}_V K_\varphi, 1_X \rangle = \int_X \mathrm{tr}_V K_\varphi(x, x) \, dx .$$

Using (2.6) and equation (2.9) with its convergence in the smooth topology we find:

$$\langle \mathrm{Tr}(\psi^* \mid \ker \Delta^i_{\mathcal{F}}), \varphi \rangle = \sum_\nu \int_{\mathbb{R}} e^{t\alpha_\nu} \varphi(t) \, dt ,$$

as desired. $\qquad \square$

References

[AK1] J.A. Alvárez López, Y. Kordyukov, Long time behaviour of leafwise heat flow for Riemannian foliations. Preprint dg-ga/9612010,1996. To appear in Compositio Math.

[AK2] J.A. Alvárez López, Y. Kordyukov, Distributional Betti numbers of transitive foliations of codimension one. Preprint 2000.

[BGV] N. Berline, E. Getzler, M. Vergne, Heat kernels and Dirac operators. Springer 1992

[D] C. Deninger, Some analogies between number theory and dynamical systems on foliated spaces. Doc. Math. J. DMV Extra volume ICM I (1998), 23–46

[DS] C. Deninger, W. Singhof, A counterexample to smooth leafwise Hodge decomposition for general foliations and to a type of dynamical trace formula. To appear in Ann. Inst. Fourier 2001

[G] V. Guillemin, Lectures on spectral theory of elliptic operators. Duke Math. J. **44** (1977), 485–517

[GS] V. Guillemin, S. Sternberg, Geometric asymptotics, Math. Surveys **14**, Amer. Math. Soc., Providence, R.I. 1977

[H]   L. Hörmander, The analysis of linear partial differential operators I. Springer
       1983
[L]   C. Lazarov, Transverse index and periodic orbits. GAFA **10** (2000), 124–159
[NZ]   A. Neske, F. Zickermann, The index of transversally elliptic complexes. Pro-
       ceedings of the 13th winter school on abstract analysis (Srni, 1985). Rend.
       Circ. Mat. Palermo (2) Suppl. No. **9** (1986), 165–175
[S]   I.M. Singer, Index theory for elliptic operators, Proc. Symp. Pure Math. **28**
       (1973), 11–31

MATHEMATISCHES INSTITUT, WWU MÜNSTER, EINSTEINSTR. 62, 48149 MÜN-
STER, GERMANY

E-mail address: deninge@math.uni-muenster.de

MATHEMATISCHES INSTITUT, UNIVERSITÄTSSTR. 1, 40225 DÜSSELDORF, GER-
MANY

E-mail address: singhof@cs.uni-duesseldorf.de

Contemporary Mathematics
Volume **290**, 2001

Small Eigenvalues and Hausdorff Dimension of Sequences of Hyperbolic Three-Manifolds

Carol E. Fan and Jay Jorgenson

ABSTRACT. This paper uses heat kernel convergence techniques to prove continuity of the small eigenvalues and small eigenfunctions of the Laplacian for sequences of strongly convergent hyperbolic three-manifolds with geometrically finite limit manifold. As a consequence, we conclude that if, in addition, the Hausdorff dimension of the limit set of the limit manifold is larger than 1, then the Hausdorff dimension of the associated limit sets is also continuous with respect to the strong topology.

1. Introduction

In this article, we apply the heat kernel convergence techniques developed in [**8**] and [**13**] to prove continuity of the small eigenvalues and small eigenfunctions of the Laplacian for sequences of strongly convergent hyperbolic three-manifolds with geometrically finite limit manifold. As a consequence, we conclude that if, in addition, the Hausdorff dimension of the limit set of the limit manifold is larger than 1, then the Hausdorff dimension of the associated limit sets is also continuous with respect to the strong topology.

The paper is organized as follows. In §2 we establish necessary notation and discuss two different notions of convergence of hyperbolic three-manifolds—geometric convergence and strong convergence. In §3 we study the heat kernel associated to an infinite volume hyperbolic three-manifold and prove convergence of heat kernels in the strong topology. As a corollary of this heat kernel convergence theorem, we prove continuity of the small eigenvalues and small eigenfunctions, as well as continuity of the Hausdorff dimension for strongly convergent sequences of hyperbolic three-manifolds under additional conditions.

In [**4**], [**7**], and [**18**], the authors consider continuity of the smallest eigenvalue and also the Hausdorff dimension of the limit set under strong convergence. The proof in Canary-Taylor [**4**] involves studying the smallest eigenvalue by minimizing a Rayleigh quotient. In [**7**], Comar and Taylor show that the eigenfunctions associated to the lowest eigenvalues converge to an eigenfunction on the limit manifold.

2000 *Mathematics Subject Classification.* Primary 58J35, 30F40; Secondary 58J50.

The second named author gratefully acknowledges support from NSF grant DMS-96-22535. The authors would like to thank Dick Canary, Timothy Comar and Hee Oh for helpful conversations.

In [18] McMullen studies the behavior of the Patterson-Sullivan measure under the strong topology, which itself can be related to a Rayleigh quotient. The advantage of our heat kernel approach is that we prove convergence of all small eigenvalues and associated eigenfunctions. Then, as in [4, 18], we use the relationship between the smallest eigenvalue and the Hausdorff dimension of the limit set to prove continuity of the Hausdorff dimension. For related results on heat kernel convergence, see Judge [15].

2. Convergent sequences

There are many references for the basic geometry of hyperbolic three-manifolds, so we refer the reader to any of the texts [1], [2] or [17] for a complete discussion of background material. We assume this material to be known and instead focus on defining different notions of convergence of hyperbolic three-manifolds.

A hyperbolic three-manifold $M = \Gamma \backslash \mathbf{h}_3$ is a complete Riemannian three-manifold of constant curvature -1, where Γ is a discrete, torsion-free subgroup of $\mathrm{PSL}_2(\mathbb{C})$, the orientation-preserving isometry group of $\mathbf{h}_3$. To compare two hyperbolic three-manifolds, we introduce the concept of an approximate isometry. Let (M_1, x_1, e_1) and (M_2, x_2, e_2) be two hyperbolic three-manifolds with base points x_1, x_2 and orthonormal frames e_1, e_2 in the tangent spaces at the base points. We say that $f : M_1 \to M_2$ is a *framed (K, r)-approximate isometry* if:

1. For $k = 1, 2$, there exist subsets B_k of M_k, each of which contains the ball of radius r based at x_k, such that f is a diffeomorphism from B_1 to B_2;
2. We have that $f(x_1) = x_2$, and the differential Df_{x_1} at x_1 maps e_1 to e_2;
3. For any points $x, y \in B_1$, we have

$$\frac{d_{M_1}(x, y)}{K} \leq d_{M_2}(f(x), f(y)) \leq K d_{M_1}(x, y).$$

DEFINITION 2.1. A sequence $\{(M_k, x_k, e_k)\}$ of hyperbolic three-manifolds with base frame *converges geometrically* to $(M_\infty, x_\infty, e_\infty)$ if there exists a sequence of (K_k, r_k)-approximate isometries $\{f_k : (M_k, x_k, e_k) \to (M_\infty, x_\infty, e_\infty)\}$ such that as $k \to \infty$, we have $K_k \to 1$ and $r_k \to \infty$.

Alternatively, one can define geometric convergence in terms of convergence of the underlying Kleinian groups (see [2] or [3]).

DEFINITION 2.2. A sequence of Kleinian groups $\{\Gamma_k\}$ *converges geometrically* to Γ_∞ if:

1. If $\gamma \in \mathrm{PSL}_2(\mathbf{C})$ is an accumulation point of a sequence $\{\gamma_k \in \Gamma_k\}$, then $\gamma \in \Gamma_\infty$;
2. If $\gamma \in \Gamma_\infty$, then there exists a sequence $\{\gamma_k \in \Gamma_k\}$ such that $\gamma_k \to \gamma_\infty$.

A hyperbolic three-manifold can be related to a Kleinian group in the following canonical way. Choose a base point p in $\mathbf{h}_3$ and an orthonormal frame $e_p = e_{\mathbf{h}_3, p}$ in the tangent space of $\mathbf{h}_3$ based at p. Given a manifold M with base point x and an orthonormal frame $e_x = e_{M,x}$ in the tangent space of M based at x, there exists a unique torsion-free Kleinian group Γ such that Γ is the group of isometries acting on $\mathbf{h}_3$ for which $M = \Gamma \backslash \mathbf{h}_3$, where p is a Γ-lift of x and the frame e_p at p on $\mathbf{h}_3$ is a Γ-lift of the frame e_x at x on M. Therefore, we can state that a sequence $\{(M_k, x_k, e_k)\}$ of hyperbolic three-manifolds with base frame converges geometrically to $(M_\infty, x_\infty, e_\infty)$ if and only if the sequence $\{\Gamma_k\}$ of their associated Kleinian groups converges geometrically to Γ_∞.

REMARK 2.3. Consider a sequence $\{(M_k, x_k, e_k)\}$ of hyperbolic three-manifolds with base frame which converges geometrically to $(M_\infty, x_\infty, e_\infty)$. For a point $x \in M_\infty$, by abuse of notation, we will use x to refer to the point $x \in M_\infty$, as well as the point $f_k^{-1}(x) \in M_k$ (for k sufficiently large). It will be clear from context on which manifold x lies. Also, we will abbreviate our notation by dropping the references to the base point x_k, as well as the frame e_k based at x_k.

For a fixed hyperbolic three-manifold $M = \Gamma \backslash \mathbf{h}_3$ and points $x, y \in M$, let

$$\mathbf{L}_M(x, y; T) = \{d_{\mathbf{h}_3}(\tilde{x}, \gamma \tilde{y}) : d_{\mathbf{h}_3}(\tilde{x}, \gamma \tilde{y}) < T \text{ for } \gamma \in \Gamma\}$$

be the set of geodesic distances between $\tilde{x}$ and $\tilde{y}$ bounded above by T, where $\tilde{x}, \tilde{y} \in \mathbf{h}_3$ are any pair of Γ-lifts of $x, y \in M$. A straightforward application of the definitions yields the following result about convergence of sets of geodesic distances.

PROPOSITION 2.4. *Let $\{M_k = \Gamma_k \backslash \mathbf{h}_3\}$ be a sequence of hyperbolic three-manifolds which converges geometrically to $M_\infty = \Gamma_\infty \backslash \mathbf{h}_3$. Then the sequence of sets of geodesic distances $\{\mathbf{L}_{M_k}(x, y; T)\}$ converges to $\mathbf{L}_{M_\infty}(x, y; T)$. The convergence is uniform on compact subsets of $M_\infty \times M_\infty \times \mathbf{R}^+$.*

In addition to geometric convergence, we have the notion of strong convergence:

DEFINITION 2.5. A sequence $\{M_k = \Gamma_k \backslash \mathbf{h}_3\}$ of hyperbolic three-manifolds *converges strongly* to $M_\infty = \Gamma_\infty \backslash \mathbf{h}_3$ if:

1. The sequence $\{\Gamma_k\}$ converges geometrically to Γ_∞;
2. There exists a sequence of surjective homomorphisms $\{\chi_k : \Gamma_\infty \to \Gamma_k\}$ such that $\lim_{k \to \infty} \chi_k(\gamma_\infty) = \gamma_\infty$ for all $\gamma_\infty \in \Gamma_\infty$.

REMARK 2.6. Not all geometrically convergent sequences are strongly convergent. Consider the example described in Remark 1, Section 3 of Comar-Taylor [7] in which each Γ_k is a Klein-Maskit combination of geometrically finite groups Γ_A and Γ_B, where Γ_B is gradually conjugated off to infinity. The geometric limit in this case is Γ_A. This sequence of hyperbolic three-manifolds is geometrically convergent, but not strongly convergent, because Part 2 of Definition 2.5 is not satisfied.

REMARK 2.7. Sequences of manifolds obtained using Thurston's Dehn Surgery Theorem are examples of strongly convergent finite volume manifolds. Comar's extension [6] of Thurston's Dehn Surgery Theorem allows one to construct a strongly convergent sequence of infinite volume hyperbolic three-manifolds as well.

REMARK 2.8. Note that Definition 2.5 coincides with the definition of strong convergence given in McMullen [18]. Our definition is slightly weaker than the definition for strong convergence given in Canary-Taylor [4], which requires that the surjective homomorphisms $\{\chi_k\}$ be faithful as well as surjective. The definition in [4] excludes the Dehn surgery examples referred to in Remark 2.7.

3. Heat kernel convergence

Let us recall a few basic properties of the heat kernel; for further details, see [5]. Let M be a hyperbolic three-manifold, and let Δ_M denote the Friedrichs' extension of the Laplacian which acts on the space of L^2 functions on M. The heat kernel $K_M(t, x, y)$ is the integral kernel which is a real-valued function of $x, y \in M$ and $t > 0$ which inverts the operator $\Delta_M + \partial/\partial t$. Specifically, the heat kernel is uniquely characterized by the following two local properties:

1. (Differential Equation): If $\Delta_{M,x}$ denotes the Laplacian which acts on the variable x on M, then

$$\left(\Delta_{M,x} + \frac{\partial}{\partial t}\right) K_M(t,x,y) = 0;$$

2. (Dirac property): For any smooth and compactly supported function $f :$ $M \to \mathbf{R}$, we have

$$f(x) = \lim_{t \to 0} \int_M K_M(t,x,y) f(y) d\mu_M(y).$$

In our setting, the heat kernel exists, is unique, and is symmetric in x and y (p.188, [5]). By realizing the manifold M as $M = \Gamma \backslash \mathbf{h}_3$, the heat kernel on M can be written as

$$K_M(t,x,y) = \sum_{\gamma \in \Gamma} K_{\mathbf{h}_3}(t, d_{\mathbf{h}_3}(\tilde{x}, \gamma \tilde{y})),$$

where

$$K_{\mathbf{h}_3}(t,\rho) = \frac{e^{-t}}{(4\pi t)^{3/2}} \frac{\rho}{\sinh \rho} e^{-\rho^2/4t}.$$

In other words, the heat kernel on M is the Γ-periodization of the heat kernel on its universal cover $\mathbf{h}_3$ (p.150, [5]). Note that the above formulae are such that the heat kernel extends to allow the time variable t to assume complex values z provided $\mathrm{Re}(z) > 0$.

If M is geometrically finite, the spectrum of the Laplacian is discrete below one (Theorem 1.1, [20]). Thus, one can express the heat kernel via the partial spectral expansion

$$K_M(t,x,y) = \sum_{\lambda_{M,n} < 1} e^{-\lambda_{M,n}t} \phi_{M,n}(x)\phi_{M,n}(y) + F_{M,1}(t,x,y)$$

where:

1. $\{\lambda_{M,n}\}$ is the set of eigenvalues of the Laplacian Δ_M which are less than one;
2. $\{\phi_{M,n}\}$ is the set of L^2-eigenfunctions of norm one corresponding to the eigenvalues $\{\lambda_{M,n}\}$;
3. $F_{M,1}(t,x,y) : \mathbf{R} \times M \times M \to \mathbf{R}$ is a smooth function which satisfies the following asymptotic bounds:
   (a) for any $\varepsilon > 0$ we have

   $$F_{M,1}(t,x,y) = O\left(e^{-(1-\varepsilon)t}\right) \quad \text{as } t \to \infty;$$

   (b)

   $$F_{M,1}(t,x,y) = O\left(t^{-3/2} e^{-d_M^2(x,y)/4t}\right) \quad \text{as } t \to 0.$$

Note that the asymptotic expansion as t approaches zero can be further developed (p.154, [5]).

THEOREM 3.1. *Let $\{M_k\}$ be a strongly convergent sequence of hyperbolic three-manifolds with limit manifold M_∞. For any points $x, y \in M_\infty$ and $z \in \mathbf{C}$ with $\mathrm{Re}(z) > 0$, we have*

$$\lim_{k \to \infty} K_{M_k}(z,x,y) = K_{M_\infty}(z,x,y).$$

The convergence is uniform on compact subsets of $\mathbf{C}_{\mathrm{Re}(z)>0} \times M_\infty \times M_\infty$.

PROOF. The ideas in this proof are taken from Lemma 4 of [13] which dealt with the sequences of finite area hyperbolic surfaces. This result is analogous to Theorem 2.1(i) of [8] which considered sequences of finite volume hyperbolic three-manifolds, as constructed by Thurston's Dehn Surgery Theorem. For the sake of completeness, let us outline the proof.

For any discrete subgroup Γ of $\mathrm{PSL}_2(\mathbf{C})$, let

$$N_\Gamma(x, y; \rho) = \mathrm{card}\{\gamma \in \Gamma : d_{\mathbf{h}_3}(\tilde{x}, \gamma\tilde{y}) < \rho\};$$

that is, $N_\Gamma(x, y; \rho)$ is the counting function associated to the set $\mathbf{L}_{\Gamma\backslash\mathbf{h}_3}(x, y; \rho)$.

Assume that the injectivity radius at y in M is $> r$. Our first claim is that for $r < \delta < \rho$, we have

$$N_\Gamma(x, y; \rho) \leq N_\Gamma(x, y; \delta) + \frac{V(\rho + r) - V(\delta - r)}{V(r)}$$

where $V(\rho) = \pi(\sinh 2\rho - 2\rho)$ denotes the volume of a ball of radius ρ in $\mathbf{h}_3$ (see also Sec 3, [14]). By definition, the injectivity radius at y in $M = \Gamma\backslash\mathbf{h}_3$ is $inj_M(y) = \frac{1}{2}\inf\{d_{\mathbf{h}_3}(y, \gamma y) : \gamma \in \Gamma - \{id\}\}$. The first term of our inequality is the number of preimages of y contained in the ball of radius δ based at x. The second term is an upper bound on the number of preimages contained in the region bounded by the spheres centered at x of radius ρ and δ. Because $inj_M(y) > r$, balls of radius r centered at preimages of y are disjoint and are contained in the larger region between spheres centered at x of radius $\rho + r$ and $\delta - r$. The number of these disjoint balls (and hence the number of preimages) is less than or equal to the volume of this larger region divided by the volume of a ball of radius r. Thus our first claim is proved.

Our second claim is that if $f(\rho)$ is a smooth, positive function on $\mathbf{R}^+$ which is eventually decreasing and continuous at δ, then we have the following inequality of distributional integrals

$$\int_0^\infty f(\rho) dN_\Gamma(x, y; \rho) \leq \int_0^\delta f(\rho) dN_\Gamma(x, y; \rho)$$

$$+ f(\delta)\frac{V(\delta + r)}{V(r)} + \frac{4\pi}{V(r)} \int_\delta^\infty f(\rho) \sinh^2(\rho + r) d\rho$$

We begin by defining a function

$$U(\rho) = \begin{cases} N_\Gamma(x, y, \rho) & \rho < \delta \\ N_\Gamma(x, y, \delta) + \frac{V(\rho + r) - V(\delta - r)}{V(r)} & \delta \leq \rho \end{cases}$$

so that $N_\Gamma(x, y, \rho) \leq U(\rho)$. We can calculate the (distributional) differential

$$dU(\rho) = \begin{cases} dN_\Gamma(x, y, \rho) & \rho < \delta \\ \frac{V'(\rho + r)}{V(r)} = \frac{\pi(2\cosh 2\rho - 2)}{V(r)} = \frac{4\pi \sinh^2(\rho + r)}{V(r)} & \delta < \rho \end{cases}$$

where $dU(\delta)$ is a Dirac mass of weight $\frac{V(\delta + r) - V(\delta - r)}{V(r)} \leq \frac{V(\delta + r)}{V(r)}$ since $V(r) > 0$. Then by Lemma 2.2 of [16], we know that our inequality of integrals holds.

Our final claim is that for fixed points $x, y \in M_\infty$ and $z \in \mathbf{C}$ with $\mathrm{Re}(z) > 0$, we have

$$\lim_{k \to \infty} K_{M_k}(z, x, y) = K_{M_\infty}(z, x, y).$$

or, letting $\rho = d_{\mathbf{h}_3}(\tilde{x}, \gamma\tilde{y})$, we have

$$\lim_{k\to\infty} \sum_{\gamma\in\Gamma_k} K_{\mathbf{h}_3}(z,\rho) = \sum_{\gamma\in\Gamma_\infty} K_{\mathbf{h}_3}(z,\rho)$$

To prove our final claim, first we note that for fixed x, y, there exist positive constants c_1 and c_2 such that

$$|K_{\mathbf{h}_3}(z,\rho)| = \left| \frac{e^{-t}}{(4\pi t)^{3/2}} \frac{\rho}{\sinh\rho} e^{-\rho^2/4t} \right| \le c_1 \exp(-c_2\rho^2)$$

Furthermore, $K_{\mathbf{h}_3}(z,\rho)$ is a smooth, positive function on $\mathbf{R}^+$ which is eventually decreasing and continuous at δ.

Next we define $A_{T,k} = \{\gamma \in \Gamma_k : d_{\mathbf{h}_3}(\tilde{x}, \gamma\tilde{y}) < T\}$ so that $\mathbf{L}_M(x,y;T)$ is the number of elements in $A_{T,k}$. Let $\tilde{x}, \tilde{y} \in \mathbf{h}_3$. By strong convergence, if the injectivity radius of $p_\infty(\tilde{y})$ in M_∞ is > 0, then there exists $r > 0$ such that the injectivity radius at $p_k(\tilde{y})$ in M_k is $> 2r$ for all k. Using our second claim, we know that for each $k > 0$ or $k = \infty$, we have

$$\left| \sum_{\gamma\in\Gamma_k - A_{T,k}} K_{\mathbf{h}_3}(z,\rho) \right| \le \sum_{\gamma\in\Gamma_k - A_{T,k}} |K_{\mathbf{h}_3}(z,\rho)| \le \sum_{\gamma\in\Gamma_k - A_{T,k}} c_1 \exp(-c_2\rho^2)$$

$$\le \int_T^\infty c_1 \exp(-c_2\rho^2) dN_{\Gamma_k}(x,y,\rho)$$

$$+ c_1 \exp(-c_2\rho^2) \frac{V(r+T)}{V(r)}$$

$$+ \frac{4c_1\pi}{V(r)} \int_T^\infty \exp(-c_2\rho^2) \sinh^2(\rho+r) d\rho$$

Because for large ρ we have that both $V(\rho) = O(e^{2\rho})$ and $V'(\rho) = O(e^{2\rho})$, then for large enough T, we can guarantee that the last expression in the inequality is smaller than $\epsilon/4$.

Strong convergence implies that $A_{T,k}$ converges to $A_{T,0}$ so that with T fixed as desired above, there exists $k_\infty > 0$ such that for all $k \ge k_\infty$, we have

$$\left| \sum_{\gamma\in A_{T,\infty}} K_{\mathbf{h}_3}(z,\rho) - \sum_{\gamma\in A_{T,k}} K_{\mathbf{h}_3}(z,\rho) \right| < \epsilon/2$$

Thus we have pointwise convergence. The convergence here is uniform for x, y on compact subsets of M_∞ and for z in compact subsets of $\mathbf{C}_{\mathrm{Re}(z)>0}$, because all of the above bounds can be made uniform.

Further details of the above argument are given in Section 2 of [8], thus concluding the proof of this lemma. $\qquad\square$

COROLLARY 3.2. *Let $\{M_k\}$ be a strongly convergent sequence of hyperbolic three-manifolds with geometrically finite limit manifold M_∞.*

1. *For any $x, y \in M_\infty$, $T < 1$, and $p > 0$, we have*

$$\lim_{k\to\infty} \sum_{\lambda_{M_k,n}<T} (T - \lambda_{M_k,n})^p \phi_{M_k,n}(x)\phi_{M_k,n}(y)$$

$$= \sum_{\lambda_{M_\infty,n}<T} (T - \lambda_{M_\infty,n})^p \phi_{M_\infty,n}(x)\phi_{M_\infty,n}(y).$$

2. *For any $T < 1$ which is not equal to an eigenvalue of M_∞, we have*

$$\lim_{k \to \infty} \sum_{\lambda_{M_k,n} < T} 1 = \sum_{\lambda_{M_\infty,n} < T} 1.$$

3. *For any $x \in M_\infty$ and $T < 1$ which is not equal to an eigenvalue of M_∞, we have*

$$\lim_{k \to \infty} \sum_{\lambda_{M_k,n} < T} \phi_{M_k,n}(x) = \sum_{\lambda_{M_\infty,n} < T} \phi_{M_\infty,n}(x).$$

The convergence is uniform on compact subsets of M_∞.

PROOF. For Part 1, we argue as in [**10**] and Section 15 of [**8**]. For any function f on $\mathbf{R}^+$, let $\mathcal{L}(f)$ denote the Laplace transform, defined by

$$\mathcal{L}(f)(z) = \int_0^\infty e^{-zu} f(u)du.$$

Assume also that f is continuous, $f \in L^1(0, R)$ for every $R > 0$, and $\mathcal{L}(f)(t + i\sigma) \in L^1(t)$ for some $\sigma > 0$. Then by Theorem 3.1 and the dominated convergence theorem, we have for $T > 0$ and $x, y \in M_k$ not on a pinching geodesic,

$$\lim_{k \to \infty} \frac{1}{2\pi i} \int_{t-i\infty}^{t+i\infty} K_{M_k}(z, x, y)\mathcal{L}(f)(z)e^{Tz}dz = \frac{1}{2\pi i} \int_{t-i\infty}^{t+i\infty} K_{M_\infty}(z, x, y)\mathcal{L}(f)(z)e^{Tz}dz$$

(see also Theorem 15.2, [**8**]). Since $T < 1$, standard formulae for the complex inversion formula for the Laplace transform (see Theorem 7.3, [**23**]) give the equality

$$\frac{1}{2\pi i} \int_{t-i\infty}^{t+i\infty} K_{M_k}(z, x, y)\mathcal{L}(f)(z)e^{Tz}dz = \sum_{\lambda_{M_k,n} \leq T} f(T - \lambda_{M_k,n})\phi_{M_k,n}(x)\phi_{M_k,n}(y).$$

To finish, simply note that if $f_p(t) = t^p$ with $p > 0$, we have

$$\mathcal{L}(f_p)(z) = \frac{\Gamma(p+1)}{z^{p+1}},$$

from which Part 1 of the corollary follows.

For Part 2, choose T so that $\lambda_{M_\infty,1} < T < \lambda_{M_\infty,2}$. Then by taking $p = 1$ and $x = y$, we have the following formula from Part 1,

$$\lim_{k \to \infty} \sum_{\lambda_{M_k,n} < T} (T - \lambda_{M_k,n})\phi_{M_k,n}(x)^2 = (T - \lambda_{M_\infty,1}) \sum_{\lambda_{M_\infty,n} < T} \phi_{M_\infty,n}(x)^2$$

where the formula takes into account the possibility that the eigenspace associated to $\lambda_{M_\infty,1}$ may have dimension greater than 1. Note that the right hand side is a degree one polynomial in T; by taking T sufficiently close to $\lambda_{M_\infty,1}$ and k sufficiently large, we may assume that the terms on the left hand side are also eventually, degree one polynomials. Parts 2 and 3 of the Corollary then follow by equating coefficients of these sequences of polynomials. This concludes the proof of Parts 2 and 3 in the case that $T < \lambda_{M_\infty,2}$. The general result follows by induction when T is take to lie in the range $\lambda_{M_\infty,j} < T < \lambda_{M_\infty,j+1}$. For more details of such arguments, see Sections 14 and 15 of [**8**] and in [**10**]. $\square$

REMARK 3.3. If the limit manifold is not geometrically finite, then the continuous spectrum can extend below one, in which case one can adjust Corollary 3.2 by taking T sufficiently small. Further, as discussed in [8], one can take arbitrary T in Corollary 3.2, in which case one sees the continuous spectrum, in particular the Eisenstein series, appear through degeneration.

THEOREM 3.4. *Let $\{M_k\}$ be a strongly convergent sequence of hyperbolic three-manifolds with geometrically finite limit manifold M_∞. If the Hausdorff dimension $\delta_{M_\infty} \geq 1$, then $\delta_{M_k} \to \delta_{M_\infty}$ as $k \to \infty$.*

PROOF. By Corollary 3.2(2), the smallest eigenvalue is continuous for strongly convergent sequences of manifolds with geometrically finite limit manifold. Since the limit manifold is geometrically finite, for large k, the manifolds in the strongly convergent sequence are also geometrically finite (see [22] or [18]). Because

$$1 \leq \delta_{M_\infty} \leq \lim\inf \delta_{M_k},$$

for large k, the Hausdorff dimension δ_{M_k} is related to the smallest eigenvalue $\lambda_{M_k,0}$ through the formula

$$\lambda_{M_k,0} = \delta_{M_k}(2 - \delta_{M_k}) \quad \text{or} \quad \delta_{M_k} = 1 + \sqrt{(1 - \lambda_{M_k,0})}$$

(see [19] or [21]), which concludes the proof.   □

REMARK 3.5. Theorem 3.4 is the same as Corollary 7.6 of [18], which is proved via the same methods, namely by first proving continuity of the smallest eigenvalue. The Main Theorem of [4] states that the Hausdorff dimension varies continuously in the strong topology over the space of discrete, faithful representations of non-handlebody, hyperbolic three-manifolds. If one follows the proof of their Main Theorem, one then can extend Theorem 3.4 to the Main Theorem of [4].

REMARK 3.6. For real s sufficiently large, we can define the function

$$G_M(s,x,y) = (2s - 2) \int_0^\infty K_M(t,x,y)e^{-s(s-2)t}dt.$$

Using the group expansion of the heat kernel, we have

$$G_M(s,x,y) = \sum_{\gamma \in \Gamma} \frac{(2s-2)}{(4\pi)^{3/2}} \frac{d(\tilde{x},\gamma\tilde{y})}{\sinh d(\tilde{x},\gamma\tilde{y})} \int_0^\infty e^{-d^2(\tilde{x},\gamma\tilde{y})/4t} e^{-(s-1)^2 t} t^{-1/2} \frac{dt}{t}.$$

Standard formulae for the K-Bessel function (see [11]) allow us to explicitly evaluate the above integral, giving the formula

$$G_M(s,x,y) = \frac{1}{4\pi} \sum_{\gamma \in \Gamma} \frac{d(\tilde{x},\gamma\tilde{y})}{\sinh d(\tilde{x},\gamma\tilde{y})} e^{-(s-1)d(\tilde{x},\gamma\tilde{y})}.$$

Using the spectral expansion of the heat kernel cited in the beginning of Section 3, it can be shown that $G_M(s,x,y)$ admits a meromorphic continuation to a region slightly larger than the half-plane of convergence of its series expansion. Furthermore, the series expansion for $G_M(s,x,y)$ can be bounded above and below by constant multiples of the series $\sum e^{-sd(\tilde{x},\gamma\tilde{y})}$, which is the usual Patterson-Sullivan series studied in the present context. Using Theorem 3.1, we have continuity of $G_{M_k}(s,x,y)$ for strongly convergent sequences with geometrically finite limit manifold. Further analysis of the function $G_M(s,x,y)$, including its full meromorphic continuation and functional equation, is certainly warranted.

References

[1] A. Beardon, *On the Geometry of Discrete Groups,* Graduate Texts in Mathematics **91** New York: Springer-Verlag, (1983).

[2] R. Benedetti, and C. Petronio, *Lectures on Hyperbolic Geometry,* Universitext, New York: Springer-Verlag, (1992).

[3] R. Canary, D. Epstein, and P. Green, "Notes on notes of Thurston," in *Analytical and Geometrical Aspects of Hyperbolic Spaces*, ed. by D. Epstein. Cambridge: Cambridge University Press, (1987), 3–92.

[4] R. Canary and E. Taylor, "Hausdorff dimension and limits of Kleinian groups," *Geom. Funct. Anal.* **9** (1999), no. 2, 283–297.

[5] I. Chavel, *Eigenvalues in Riemannian Geometry,* Graduate Texts in Mathematics **91** New York: Academic Press, (1984).

[6] T. Comar, "Hyperbolic Dehn surgery and convergence of Kleinian groups," Ph.D. Thesis, University of Michigan, (1996).

[7] T. Comar and E. Taylor, "Geometrically convergent Kleinian groups and the lowest eigenvalue of the Laplacian," *Indiana Univ. Math. J.* **47** (1998), no. 2, 601–623.

[8] J. Dodziuk and J. Jorgenson, "Spectral asymptotics on degenerating hyperbolic 3-manifolds," *Mem. Amer. Math. Soc.* **135** (1998), no. 643.

[9] D. Hejhal, "Regular b-groups, degenerating Riemann surfaces, and spectral theory," *Mem. Amer. Math. Soc.* **88** (1990), no. 437.

[10] J. Huntley, J. Jorgenson, and R. Lundelius, "Continuity of small eigenfunctions on degenerating hyperbolic Riemann surfaces with hyperbolic cusps," *Bol. Soc. Mat. Mexicana (3)* **1** (1995) 119–125.

[11] J. Jorgenson and S. Lang, "Extension of analytic number theory and the theory of regularized harmonic series from Dirichlet series to Bessel series," *Math. Ann.* **306** (1996) 75–124.

[12] T. Jorgensen and A. Marden, "Algebraic and Geometric Convergence of Kleinian Groups," *Math Scand.* **66** (1990) 47–72.

[13] J. Jorgenson and R. Lundelius, "Convergence of the heat kernel and the resolvent kernel on degenerating hyperbolic Riemann surfaces of finite volume," *Quaestiones Math.* **18** (1995), no. 4, 345–363.

[14] J. Jorgenson and R. Lundelius, "Continuity of relative hyperbolic spectral theory through metric degeneration," *Duke Math. J.* **84** (1996), no. 1, 47–81.

[15] C. Judge, "Heating and Stretching Riemannian manifolds," in *Spectral Problems in Geometry and Arithmetic,* Contemporary Mathematics **237** (1999).

[16] R. Lundelius, "Asymptotics of the determinant of the Laplacian on hyperbolic surfaces of finite volume," *Duke Math. J.* **71** (1993) 211–242.

[17] B. Maskit, *Kleinian Groups*. Grundlehren der mathematischen Wissenschaften **287** New York: Springer Verlag (1988).

[18] C. McMullen "Hausdorff dimension and conformal dynamics I: Strong convergence of Kleinian groups," Preprint, (1997) (http://math.berkeley.edu/~ctm/papers.html).

[19] S. Patterson, "Lectures on measures on limit sets of Kleinian groups," in *Analytical and Geometrical Aspects of Hyperbolic Spaces,* ed. by D. Epstein. Cambridge: Cambridge University Press (1987) 281–323.

[20] R. Phillips and P. Sarnak, "The Laplacian for domains in hyperbolic space and limit sets of Kleinian groups," *Acta Math.* **155** (1985) 173–241.

[21] D. Sullivan "Related aspects of positivity in Riemannian geometry," *J. Diff. Geom.* **25** (1987) 327–351.

[22] E. Taylor, "Geometric finiteness and the convergence of Kleinian groups," *Comm. Anal. Geom.* **5** (1997) 497–533.

[23] D. Widder, *The Laplace Transform,* Princeton, NJ: Princeton University Press, (1941).

DEPARTMENT OF MATHEMATICS, OKLAHOMA STATE UNIVERSITY, STILLWATER, OK 74075
  Current address: Department of Mathematics MC-8130, Loyola Marymount University, Los
Angeles, CA 90045-8130
  E-mail address: `cfan@lmumail.lmu.edu`

DEPARTMENT OF MATHEMATICS, OKLAHOMA STATE UNIVERSITY, STILLWATER, OK 74075
  Current address: Department of Mathematics, City College of New York, New York, NY
10031
  E-mail address: `jjorgen@math0.sci.ccny.cuny.edu`

Contemporary Mathematics
Volume **290**, 2001

Dynamical zeta functions and asymptotic expansions in Nielsen theory

Alexander Fel'shtyn

ABSTRACT. We prove that the Nielsen zeta function is a rational function or a radical of a rational function for orientation preserving homeomorphisms on closed orientable 3-dimensional manifolds which are special Haken or Seifert manifolds. In the case of pseudo-Anosov homeomorphisms of surfaces, we obtain an asymptotic expansion for the number of twisted conjugacy classes or for the number of Nielsen fixed point classes whose norm is at most x.

0. Introduction

Before discussing the main results of the paper, we briefly describe the few basic notions of Nielsen fixed point theory which will be used. We assume X to be a connected, compact polyhedron and $f : X \to X$ to be a continuous map. Let $p : \tilde{X} \to X$ be the universal cover of X and $\tilde{f} : \tilde{X} \to \tilde{X}$ a lifting of f, i.e. $p \circ \tilde{f} = f \circ p$. Two liftings $\tilde{f}$ and $\tilde{f}'$ are called *conjugate* if there is a $\gamma \in \Gamma \cong \pi_1(X)$ such that $\tilde{f}' = \gamma \circ \tilde{f} \circ \gamma^{-1}$. The subset $p(\mathrm{Fix}(\tilde{f})) \subset \mathrm{Fix}(f)$ is called *the fixed point class of f determined by the lifting class $[\tilde{f}]$.* Two fixed points x_0 and x_1 of f belong to the same fixed point class iff there is a path c from x_0 to x_1 such that $c \cong f \circ c$ (homotopy relative endpoints). This fact can be considered as an equivalent definition of a non-empty fixed point class. Every map f has only finitely many non-empty fixed point classes, each a compact subset of X.

A fixed point class is called *essential* if its index is nonzero. The number of lifting classes of f (and hence the number of fixed point classes, empty or not) is called the *Reidemeister number* of f, denoted $R(f)$. This is a positive integer or infinity. The number of essential fixed point classes is called the *Nielsen number* of f, denoted by $N(f)$.

The Nielsen number is always finite. $R(f)$ and $N(f)$ are homotopy invariants. In the category of compact, connected polyhedra, the Nielsen number of a map is, apart from certain exceptional cases, equal to the least number of fixed points of maps with the same homotopy type as f.

Taking a dynamical point of view, we consider the iterates of f, and we may define several zeta functions connected with Nielsen fixed point theory (see [**4, 5, 7**]). The Nielsen zeta function of f is defined as the following power series:

2000 *Mathematics Subject Classification.* Primary 58F20, Secondary 20F34.

$$N_f(z) \quad := \quad \exp\left(\sum_{n=1}^{\infty} \frac{N(f^n)}{n} z^n\right).$$

The Nielsen zeta function $N_f(z)$ is a homotopy invariant. The function $N_f(z)$ has a positive radius of convergence which has a sharp estimate in terms of the topological entropy of the map f [7].

We begin this article by proving in Section 1 that the Nielsen zeta function is a rational function or a radical of a rational function for orientation preserving homeomorphisms of special Haken or special Seifert 3-manifolds.

In Section 2, we obtain an asymptotic expansion for the number of twisted conjugacy classes or for the number of Nielsen fixed point classes whose norm is at most x in the case of pseudo-Anosov homeomorphisms of surfaces.

The author would like to thank M. Gromov, Ch. Epstein, R. Hill, L. Potyagailo, R. Sharp, V.G. Turaev for stimulating discussions. In particular, Turaev proposed to the author in 1987 the conjecture that the Nielsen zeta function is a rational function or a radical of a rational function for homeomorphisms of Haken or Seifert manifolds.

1. The Nielsen zeta function and homeomorphisms of 3-manifolds

1.1. Periodic maps and homeomorphisms of hyperbolic manifolds.

We prove in Corollary 1 of this subsection that the Nielsen zeta function is a radical of a rational function for any homeomorphism of a compact hyperbolic 3-manifold. Lemma 1 and Corollary 1 play also an important role in the proof of the main theorem of this section.

We denote $N(f^n)$ by N_n. We shall say that $f : X \to X$ is a periodic map of period m, if f^m is the identity map $id_X : X \to X$. Let $\mu(d), d \in N$, be the Möbius function from number theory. As is known, it is given by the following equations: $\mu(d) = 0$ if d is divisible by a square different from one ; $\mu(d) = (-1)^k$ if d is not divisible by a square different from one, where k denotes the number of prime divisors of d; $\mu(1) = 1$.

We give the proof of the following key lemma for completeness.

LEMMA 1. [7] Let f be a periodic map of least period m of the connected compact polyhedron X. Then the Nielsen zeta function is equal to

$$N_f(z) = \prod_{d|m} \sqrt[d]{(1 - z^d)^{-P(d)}},$$

where the product is taken over all divisors d of the period m, and $P(d)$ is the integer

$$P(d) = \sum_{d_1|d} \mu(d_1) N_{d|d_1}.$$

PROOF Since $f^m = id$, for each $j, N_j = N_{m+j}$. If $(k, m) = 1$, then there exist positive integers t and q such that $kt = mq + 1$. So $(f^k)^t = f^{kt} = f^{mq+1} = f^{mq} f = (f^m)^q f = f$. Consequently, $N((f^k)^t) = N(f)$. Let two fixed points x_0 and x_1 belong to the same fixed point class. Then there exists a path α from x_0 to x_1 such that $\alpha * (f \circ \alpha)^{-1} \simeq 0$. We have $f((\alpha * f \circ \alpha)^{-1}) = (f \circ \alpha) * (f^2 \circ \alpha)^{-1} \simeq 0$ and a product $\alpha * (f \circ \alpha)^{-1} * (f \circ \alpha) * (f^2 \circ \alpha)^{-1} = \alpha * (f^2 \circ \alpha)^{-1} \simeq 0$. It follows that $\alpha * (f^k \circ \alpha)^{-1} \simeq 0$ by iterating this process. So x_0 and x_1 belong to the same fixed point class of f^k. If two points belong to different fixed point classes of f, then they belong to different fixed point classes of f^k. So, each essential class (i.e., class with nonzero index) for f is an essential class for f^k; in addition, different essential classes for f are different essential classes for f^k. So $N(f^k) \geq N(f)$. Analogously, $N(f) = N((f^k)^t) \geq N(f^k)$. Consequently, $N(f) = N(f^k)$. One can prove completely analogously that $N_d = N_{di}$, if $(i, m/d) = 1$, where d is a divisor of m. Using these sequences of equal Nielsen numbers, one can regroup the terms of the series in the exponential of the Nielsen zeta function so as to get logarithmic functions by adding and subtracting missing terms with necessary coefficients. We show how to do this first for period $m = p^l$, where p is a prime number. We have the following sequence of equal Nielsen numbers:

$$N_1 = N_k, \ (k, p^l) = 1 \ \text{(i.e., no } N_{ip}, N_{ip^2}, \ldots, N_{ip^l}, \ i = 1, 2, 3, \ldots),$$

$$N_p = N_{2p} = N_{3p} = \cdots = N_{(p-1)p} = N_{(p+1)p} = \ldots \ \text{(no } N_{ip^2}, N_{ip^3}, \ldots, N_{ip^l})$$

etc.; finally,

$$N_{p^{l-1}} = N_{2p^{l-1}} = \ldots \ \text{(no } N_{ip^l})$$

and separately the number N_{p^l}.
Further,

$$
\begin{aligned}
\sum_{i=1}^{\infty} \frac{N_i}{i} z^i &= \sum_{i=1}^{\infty} \frac{N_1}{i} z^i + \sum_{i=1}^{\infty} \frac{(N_p - N_1)}{p} \frac{z^{pi}}{i} + \\
&+ \sum_{i=1}^{\infty} \frac{(N_{p^2} - (N_p - N_1) - N_1)}{p^2} \frac{z^{p^2 i}}{i} + \cdots \\
&+ \sum_{i=1}^{\infty} \frac{(N_{p^l} - \cdots - (N_p - N_1) - N_1)}{p^l} \frac{z^{p^l i}}{i} \\
&= -N_1 \cdot \log(1 - z) + \frac{N_1 - N_p}{p} \cdot \log(1 - z^p) + \\
&+ \frac{N_p - N_{p^2}}{p^2} \cdot \log(1 - z^{p^2}) + \cdots \\
&+ \frac{N_{p^{l-1}} - N_{p^l}}{p^l} \cdot \log(1 - z^{p^l}).
\end{aligned}
$$

For an arbitrary period m, we get completely analogously,

$$
\begin{aligned}
N_f(z) &= \exp\left(\sum_{i=1}^{\infty} \frac{N(f^i)}{i} z^i\right) \\
&= \exp\left(\sum_{d|m} \sum_{i=1}^{\infty} \frac{P(d)}{d} \cdot \frac{z^{di}}{i}\right) \\
&= \exp\left(\sum_{d|m} \frac{P(d)}{d} \cdot \log(1 - z^d)\right) \\
&= \prod_{d|m} \sqrt[d]{(1 - z^d)^{-P(d)}},
\end{aligned}
$$

where the integers $P(d)$ are calculated recursively by the formula

$$P(d) = N_d - \sum_{d_1 | d; d_1 \neq d} P(d_1).$$

Moreover, if the last formula is rewritten in the form

$$N_d = \sum_{d_1 | d} \mu(d_1) \cdot P(d_1)$$

and one uses the Möbius Inversion law for real functions in number theory, then

$$P(d) = \sum_{d_1 | d} \mu(d_1) \cdot N_{d/d_1},$$

where $\mu(d_1)$ is the Möbius function in number theory. The lemma is proved.

COROLLARY 1. Let $f : M^n \to M^n$, $n \geq 3$, be a homeomorphism of a compact hyperbolic manifold. Then by Mostow's rigidity theorem, f is homotopic to a periodic homeomorphism g. So Lemma 1 applies and the Nielsen zeta function $N_f(z)$ is equal to

$$N_f(z) = N_g(z) = \prod_{d | m} \sqrt[d]{(1 - z^d)^{-P(d)}},$$

where the product is taken over all divisors d of the least period m of g, and $P(d)$ is the integer $P(d) = \sum_{d_1 | d} \mu(d_1) N(g^{d | d_1})$.

The proof of the following lemma is based on Thurston's theory of homeomorphisms of surfaces [18].

LEMMA 2. [7] The Nielsen zeta function of a homeomorphism f of a compact surface F is either a rational function or the radical of a rational function.

PROOF The case of a surface with $\chi(F) > 0$ and the case of a torus were considered in [7]. If a surface has $\chi(F) = 0$ and F is not a torus, then any homeomorphism is isotopic to a periodic one (see [11]) and the Nielsen zeta function is a radical of a rational function by Lemma 1. In the case of a hyperbolic ($\chi(F) < 0$) surface, according to Thurston's classification theorem, the homeomorphism f is isotopic either to a periodic or a pseudo-Anosov, or a reducible homeomorphism. In the first case, the assertion of the lemma follows from Lemma 1. If f is a pseudo-Anosov homeomorphism of a compact surface, then for each $n > 0, N(f^n) = F(f^n)$ [20]. Consequently, in this case the Nielsen zeta function coincides with the Artin-Mazur zeta function: $N_f(z) = F_f(z)$. Since in [3] Markov partitions are constructed for a pseudo-Anosov homeomorphism, Manning's proof [13] of the rationality of the Artin-Mazur zeta function for diffeomorphisms satisfying Smale's axiom A carries over to the case of pseudo-Anosov homeomorphisms. Thus, the Nielsen zeta function $N_f(z)$ is also rational. Now if f is isotopic to a reduced homeomorphism ϕ, then there exists a reducing system S of disjoint circles $S_1, S_2, \ldots, S_m$ on $int F$ such that

1) each circle S_i does not bound a disk in F;

2) S_i is not isotopic to $S_j, i \neq j$;

3) the system of circles S is invariant with respect to ϕ;

4) the system S has an open ϕ-invariant tubular neighborhood $\eta(S)$ such that each ϕ-component Γ_j of the set $F - \eta(S)$ is mapped into itself by some iterate $\phi^{n_j}, n_j > 0$, of the map ϕ; here ϕ^{n_j} on Γ_j is either a pseudo-Anosov or a periodic homeomorphism;

5) each band $\eta(S_i)$ is mapped into itself by some iterate $\phi^{m_i}, m_i > 0$; here ϕ^{m_i} on $\eta(S_i)$ is a generalized twist (possibly trivial).

Since the band $\eta(S_i)$ is homotopically equivalent to the circle S^1, the Nielsen zeta function $N_{\phi_i^m}(z)$ is rational (see [7]). The zeta functions $N_\phi(z)$ and $N_{\phi_i^m}(z)$ are connected on the ϕ - component Γ_j by the formula $N_\phi(z) = \sqrt[n_j]{N_{\phi_j^n}(z^{n_j})}$; analogously, on the band $\eta(S_i), N_\phi(z) = \sqrt[m_j]{N_{\phi_j^m}(z^{m_j})}$. The fixed points of ϕ^n belonging to different components Γ_j and bands $\eta(S_i)$ are nonequivalent [9], so the Nielsen number $N(\phi^n)$ is equal to the sum of the Nielsen numbers $N(\phi^n/\Gamma_j)$ and $N(\phi^n/\eta(S_i))$ of ϕ-components and bands. Consequently, by the properties of the exponential, the Nielsen zeta function $N_\phi(z) = N_f(z)$ is equal to the product of the Nielsen zeta functions of the ϕ-components Γ_j and the bands $\eta(S_i)$, i.e. is the radical of a rational function.

1.2. Homeomorphisms of Seifert fibre spaces. Let M be a compact 3-dimensional Seifert fibre space. That is, a space which is foliated by simple closed curves, called fibres, such that a fibre L has a neighborhood which is either a solid Klein bottle or a fibred solid torus T_r, where r denotes the number of times a fibre near L wraps around L. A fibre is regular if it has neighborhoods fibre equivalent to the solid torus $S^1 \times D^2$, otherwise it is called a critical fibre. There is a natural quotient map $p : M \to F$ where F is a 2-dimensional orbifold, and hence, topologically a compact surface. The projection of the critical fibres, which we denote by S, consists of a finite set of points in the interior of F together with a finite subcollection of the boundary components. The reader is referred to [17] for details, for definitions about 2-dimensional orbifolds, their Euler characteristics and other properties of Seifert fibre spaces. An orbifold is hyperbolic if it has negative Euler characteristic. A hyperbolic orbifold admits a hyperbolic structure with totally geodesic boundary.

Any fibre-preserving homeomorphism $f : M \to M$ naturally induces a relative surface homeomorphism of the pair (F, S), which we will denote by $\hat{f}$. Recall from [17] that there is a unique orientable Seifert fibre space with orbifold $P(2, 2)$ - projective plane with cone singular points of order 2, 2. We denote this manifold by $M_{P(2,2)}$.

LEMMA 3. [17, 11] Suppose M is a compact orientable Seifert fiber space which is not $T^3, S^1 \times D^2, T^2 \times I$ or $M_{P(2,2)}$. Then there is a Seifert fibration $p : M \to F$ so that any orientation preserving homeomorphism on M is isotopic to a fiber preserving homeomorphism with respect to this fibration.

Observe that if M and F are both orientable, then M admits a coherent orientation of all of its fibres and the homeomorphism f either preserves fibre orientation of all the fibres or it reverses fibre orientation.

LEMMA 4. [**12**] Let M be a compact, orientable aspherical, 3-dimensional Seifert fibre space such that the quotient orbifold F is orientable and all fibres have neighborhoods of type T_1. Let $f : M \to M$ be a fibre-preserving homeomorphism inducing $\hat{f} : F \to F$. If f preserves fibre orientation, then the Nielsen number $N(f) = 0$. If f reverses fibre orientation, then $N(f) = 2N(\hat{f})$.

PROOF First, by small isotopy, arrange that f has a finite number of fibres which are mapped to themselves. If f preserves fibre orientation, a further isotopy, which leaves $\hat{f}$ unchanged, ensures that none of these fibres contains a fixed point. Thus, Fix $(f) = \emptyset$ and so, $N(f) = 0$. If f reverses fibre orientation there are exactly two fixed points on each invariant fibre. Since M is aspherical, Lemma 3.2 in [**17**] ensures that for any invariant fibre, the two fixed points on that fibre are in distinct fixed point classes. On the other hand, the restriction on the fibre types in the hypothesis implies that a Nielsen path in F can always be lifted to a Nielsen path in M. In fact, there will be two distinct lifts of each path in F. As a result, f has two fixed point classes covering each fixed point class of $\hat{f}$. Since the index of a fixed point class of f is the same as that of its projection under p, the result follows.

REMARK 1. As one can see, Lemma 4 deals with a restricted class of Seifert fibre spaces. M. Kelly gave some examples in [**12**] which indicate how the critical fibres of a Seifert fibre space effect Nielsen classes in the case when $r > 1$ in T_r and which difficulties arise in this case.

DEFINITION 1. A special Seifert fibre space is a Seifert fiber space such that the quotient orbifold F is orientable and all fibres have neighborhoods of type T_1.

THEOREM 1. Suppose M is a closed orientable aspherical manifold which is a special Seifert fibre space and $f : M \to M$ is an fibre-preserving homeomorphism. Then the Nielsen zeta function $N_f(z)$ is a rational function or a radical of a rational function.

PROOF If f preserves fibre orientation, then f^n also preserves fibre orientation and so, by Lemma 4, the Nielsen numbers $N(f^n) = 0$ for all n and the Nielsen zeta function $N_f(z) = 1$. If f reverses fibre orientation, then f^2 preserves fibre orientation, f^3 reverses fibre orientation and so on. Thus, we have $N(f^{2k+1}) = 2N(\hat{f}^{2k+1})$ and $N(f^{2k}) = 0$ for $k = 0, 1, 2, \ldots$. As a result,

$$\sum_{n=1}^{\infty} \frac{N(f^n)}{n} z^n = \sum_{k=0}^{\infty} \frac{N(f^{2k+1})}{2k+1} z^{2k+1} = \sum_{k=0}^{\infty} \frac{2N(\hat{f}^{2k+1})}{2k+1} z^{2k+1},$$

so that the Nielsen zeta function $N_f(z)$ equals

$$
\begin{aligned}
N_f(z) &= \exp\left(\sum_{k=0}^{\infty} \frac{2N(\hat{f}^{2k+1})}{2k+1} z^{2k+1}\right) \\
&= \exp\left(2 \cdot \sum_{n=1}^{\infty} \frac{N(\hat{f}^n)}{n} z^n - 2 \cdot \sum_{k=1}^{\infty} \frac{N(\hat{f}^{2k})}{2k} z^{2k}\right) \\
&= (N_{\hat{f}}(z))^2 \cdot \exp\left(-\sum_{k=1}^{\infty} \frac{N((\hat{f}^2)^k)}{k} (z^2)^k\right) \\
&= (N_{\hat{f}}(z))^2 / N_{\hat{f}^2}(z^2).
\end{aligned}
$$

From Lemma 2 it follows that the Nielsen zeta functions $N_{\hat{f}}(z)$ and $N_{\hat{f}^2}(z^2)$ are either rational functions or the radicals of rational functions. Consequently, the Nielsen zeta function $N_f(z)$ is a rational function or a radical of a rational function.

1.3. Main theorem. Basic concepts about 3-manifold topology can be found in [**10**], in particular the Jaco-Shalen-Johannson decomposition of Haken manifolds and Seifert fiber spaces. See [**18**] for discussions on hyperbolic 3-manifolds. We recall some basic facts about compact connected orientable 3-manifolds. A 3-manifold M is irreducible if every embedded 2-sphere bounds an embedded 3-disk. By the sphere theorem [**8**], an irreducible 3-manifold is a $K(\pi,1)$ Eilenberg-MacLane space if and only if it is a 3-disk or has infinite fundamental group.

A properly-embedded orientable connected surface in a 3-manifold is incompressible if it is not a 2-sphere and the inclusion induces an injection on the fundamental groups. An irreducible 3-manifold is Haken if it contains an embedded orientable incompressible surface. We use the notation $N(X)$ to denote a regular neighborhood of a set X. A Haken 3-manifold M can be decomposed along a canonical set $\mathcal{T}$ of incompressible tori into pieces such that each component of $M - N(\mathcal{T})$ is either a hyperbolic manifold, or a twisted I-bundle over a Klein bottle, or a Seifert fiber space with hyperbolic orbifold. This decomposition is called the Jaco-Shalen-Johannson (JSJ) decomposition.

DEFINITION 2. A special Haken manifold is a Haken manifold M such that each component of $M - N(\mathcal{T})$ in the JSJ decomposition of M is either a hyperbolic manifold, or a twisted I-bundle over a Klein bottle, or an aspherical special Seifert fiber space with hyperbolic orbifold.

We need some definitions from the paper [**11**].

DEFINITION 3. Suppose $f : M \to M$ is a map, and A, B are f-invariant sets of M. If there is a path γ from A to B such that $\gamma \sim f \circ \gamma$ rel (A, B), then we say that A, B are f-related.

DEFINITION 4. Suppose M is a compact 3-manifold with torus boundaries. A map $f : M \to M$ is standard on the boundary if for any component T of ∂M, the map f/T is one of the following types: (1) a fixed point free map; (2) a periodic map with isolated fixed points; (3) a fiber preserving, fiber orientation reversing map with respect to some S^1 fibration of T.

DEFINITION 5. A map f on a compact 3-manifold M is said to have the FR-property (fixed-point relating property) if the following is true: if $A \in \mathrm{Fix}(f)$ and B is either a fixed point of f or an f-invariant component of ∂M, and A, B are f-related by a path γ, then γ is A, B homotopic to a path in $\mathrm{Fix}(f)$.

DEFINITION 6. A map f is a type I standard map if (1) f has the FR-property, (2) f is standard on the boundary, (3) $\mathrm{Fix}(f)$ consists of isolated points, and (4) f is of flipped pseudo-Anosov type at each fixed point. A map f is a type II standard map if it satisfies (1), (2) above, as well as (3) $\mathrm{Fix}(f)$ is a properly embedded 1-dimensional submanifold and (4) f preserves a normal structure on $\mathrm{Fix}(f)$.

The main result of this section is the following theorem.

THEOREM 2. Suppose M is a closed orientable manifold which is a special Haken manifold, and $f : M \to M$ is an orientation preserving homeomorphism.

Then the Nielsen zeta function $N_f(z)$ is a rational function or a radical of a rational function.

PROOF Let $\mathcal{T}$ be a (possibly empty) set of invariant tori of the JSJ decomposition of M. Then each component of $M - IntN(\mathcal{T})$ is either a hyperbolic manifold, or a twisted I-bundle over a Klein bottle, or an aspherical special Seifert fiber space with hyperbolic orbifold. Isotop f so that it maps $N(\mathcal{T})$ homeomorphically to itself. Suppose that P is a Seifert fibered component of $M - IntN(\mathcal{T})$ such that $f(P) = P$. By [17], f/P is isotopic to a fiber preserving map. Recall that a torus T in M is a vertical torus if it is a union of fibers in M. By [11], Lemma 1.10, we can find a set of vertical tori $\mathcal{T}^*$ in P, cutting P into pieces which are either a twisted I-bundle over the Klein bottle, or have hyperbolic orbifold, and a fiber-preserving isotopy of f, so that after isotopy, the restriction of f on each invariant piece is a periodic or a pseudo-Anosov orbifold map. Adding all such $\mathcal{T}^*$ to $\mathcal{T}$, we get a collection of tori $\mathcal{T}'$, such that : (1) $f(N(\mathcal{T}') = \mathcal{N}(\mathcal{T}')$; (2) each component M_i of $M - IntN(\mathcal{T}')$ either is a twisted I-bundle over a Klein bottle or has hyperbolic orbifold; (3) If f maps a Seifert fibered component M_i to itself, and if M_i has hyperbolic orbifold, then f is fiber preserving, and the orbifold map $\hat{f}$ on F_i is either periodic or pseudo-Anosov. By [11], we can isotop f so that the restriction f/M_i is a standard map for all M_i and after that we can further isotop f, rel ∂N_j, on each component N_j of $N(\mathcal{T}')$, so that it is a standard map on N_j. By the definition of standard maps, Fix(f) intersects each of M_i and N_j in points and 1-manifolds, so Fix(f) is a disjoint union of points, arcs, and circles. Jiang, Wang and Wu proved [11], Theorem 9.1, that different components of Fix(f) are not equivalent in the Nielsen sense. We repeat their proof here for completeness. If there exist two different components C_0, C_1 of Fix(f) which are equivalent, then there is a path α connecting C_0, C_1 such that $f \circ \alpha \sim \alpha$ rel ∂. Denote by T'' the set of tori $\partial N(\mathcal{T}')$. Among all such α, choose one such that the number of components $\sharp(\alpha - T'')$ of the set $\alpha - T''$ is minimal. In the argument below, we will find another such curve α'' with $\sharp(\alpha'' - T'') < \sharp(\alpha - T'')$, which would contradict the choice of α. Let D be a disk, and let $h : D \to M$ be a homotopy $f \circ \alpha \sim \alpha$ rel ∂. We may assume that h is transverse to T'', and $\sharp h^{-1}(T'')$ is minimal among all such h. Then $h^{-1}(T'')$ consists of a properly embedded 1-manifold on D, together with possibly one or two isolated points mapped to the ends of α. T'' is π_1-injective in M, so one can modify h to remove all circles in $h^{-1}(T'')$. Note that $h^{-1}(T'')$ must contain some arcs, otherwise α would lie in some M_i or $N(T_j)$, which is impossible because the restriction of f in each piece has the FR-property. Now consider an outermost arc b in $h^{-1}(T'')$. Let $\beta = h(b)$. The ends of β can not both be on α, otherwise we can use the outermost disk to homotope α and reduce $\sharp(\alpha - T'')$, contradicting the choice of α. Since f is a homeomorphism, the same thing is true for $f \circ \alpha$. Therefore, β has one end on each of α and $f \cdot \alpha$. The arc b cuts off a disk Δ on D whose interior is disjoint from $h^{-1}(T'')$. The boundary of Δ gives rise to a loop $h(\partial \Delta) = \alpha_1 \cup \beta \cup (f \circ \alpha_1)^{-1}$, where α_1 is the subpath of α starting from an end point x of α. let T be the torus in T'' which contains β. Then the restriction of h on Δ gives a homotopy $\alpha_1 \sim f \cdot \alpha_1$ rel(x, T). Since f has the FR-property on each component of $M - IntN(\mathcal{T})$ and $N(\mathcal{T})$, by definition there is a path γ in Fix(f) such that $\gamma \sim \alpha_1$ rel (x, T). Since γ is in Fix(f), the path $\alpha' = \gamma^{-1} \cdot \alpha$ has the property that $f \circ \alpha' = (f \circ \gamma^{-1}) \cdot (f \circ \alpha) = \gamma^{-1} \cdot (f \circ \alpha) \sim \gamma^{-1} \cdot \alpha = \alpha'$ rel ∂. Since $\alpha_1 \sim \gamma$ rel (x, T), the path $\gamma^{-1} \cdot \alpha_1$ is rel ∂ homotopic to a path δ on T.

Write $\alpha = \alpha_1 \cdot \alpha_2$. Then $\alpha' = \gamma^{-1} \cdot \alpha = (\gamma^{-1} \cdot \alpha_1) \cdot \alpha_2 \sim \delta \cdot \alpha_2$ rel ∂. By a small perturbation on δ, we get a path $\alpha'' \sim \alpha'$ rel ∂ such that $\sharp(\alpha'' - \mathcal{T}'') < \sharp(\alpha - \mathcal{T}'')$. Since $f \circ \alpha'' \sim f \circ \alpha' \sim \alpha' \sim \alpha''$ rel ∂, this contradicts the minimality of $\sharp(\alpha - \mathcal{T}'')$.

We can prove in the same way that the fixed points of each iteration f^n, belonging to different components M_i and N_j are nonequivalent, so the Nielsen number $N(f^n)$ is equal to the sum of the Nielsen numbers $N(f^n/M_i)$ and $N(f^n/N_j)$ of the components. Consequently, by the properties of the exponential, the Nielsen zeta function $N_f(z)$ is equal to the product of the Nielsen zeta functions for the induced homeomorphisms of the components M_i and N_j. By Corollary 1, the Nielsen zeta function of the hyperbolic component M_i is a radical of a rational function. By Theorem 1 the Nielsen zeta function is a radical of a rational function for the component M_i which is an aspherical special Seifert fibre space. The Nielsen zeta function is a rational function or a radical of a rational function for a homeomorphism of a torus or a Klein bottle, by Lemma 2. This implies that the Nielsen zeta function is a rational function or a radical of a rational function for the induced homeomorphism of a component N_j which is an I-bundle over a torus and for the induced homeomorphism of a component M_i which is a twisted I-bundle over a Klein bottle. So, the Nielsen zeta function $N_f(z)$ of the homeomorphism f of the whole manifold M is a rational function or a radical of a rational function as a product of the Nielsen zeta functions for the induced homeomorphisms of the components M_i and N_j.

2. Asymptotic expansions for fixed point classes and twisted conjugacy classes of pseudo-Anosov homeomorphisms

2.1. Twisted conjugacy and Reidemeister Numbers. Let Γ be a group and $\phi : \Gamma \to \Gamma$ an endomorphism. Two elements $\alpha, \alpha' \in \Gamma$ are said to be ϕ-*conjugate* iff there exists $\gamma \in \Gamma$ with

$$\alpha' = \gamma \alpha \phi(\gamma)^{-1}.$$

We shall write $\{x\}_\phi$ for the ϕ-conjugacy class of the element $x \in \Gamma$. The number of ϕ-conjugacy classes is called the *Reidemeister number* of ϕ, denoted by $R(\phi)$. If ϕ is the identity map then the ϕ-conjugacy classes are the usual conjugacy classes in the group Γ.

In [6] we have conjectured that the Reidemeister number is infinite as long as the endomorphism is injective and the group has exponential growth. Below, we prove this conjecture for surface groups and pseudo-Anosov maps and, in fact, we obtain an asymptotic expansion for the number of twisted conjugacy classes whose norm is at most x.

Let $f : X \to X$ be given, and let a specific lifting $\tilde{f} : \tilde{X} \to \tilde{X}$ be chosen as reference. Let Γ be the group of covering translations of $\tilde{X}$ over X. Then every lifting of f can be written uniquely as $\gamma \circ \tilde{f}$, with $\gamma \in \Gamma$. So the elements of Γ serve as coordinates of liftings with respect to the reference $\tilde{f}$. Now, for every $\gamma \in \Gamma$, the composition $\tilde{f} \circ \gamma$ is a lifting of f; so there is a unique $\gamma' \in \Gamma$ such that $\gamma' \circ \tilde{f} = \tilde{f} \circ \gamma$. This correspondence $\gamma \to \gamma'$ is determined by the reference $\tilde{f}$, and is obviously a homomorphism.

DEFINITION 7. The endomorphism $\tilde{f}_* : \Gamma \to \Gamma$ determined by the lifting $\tilde{f}$ of f is defined by

$$\tilde{f}_*(\gamma) \circ \tilde{f} = \tilde{f} \circ \gamma.$$

It is well known that $\Gamma \cong \pi_1(X)$. We shall identify $\pi = \pi_1(X, x_0)$ and Γ in the following way. Pick base points $x_0 \in X$ and $\tilde{x}_0 \in p^{-1}(x_0) \subset \tilde{X}$ once and for all. Now points of $\tilde{X}$ are in 1-1 correspondence with homotopy classes of paths in X which start at x_0: for $\tilde{x} \in \tilde{X}$ take any path in $\tilde{X}$ from $\tilde{x}_0$ to $\tilde{x}$ and project it onto X; conversely, for a path c starting at x_0, lift it to a path in $\tilde{X}$ which starts at $\tilde{x}_0$, and then take its endpoint. In this way, we identify a point of $\tilde{X}$ with a path class $\langle c \rangle$ in X starting from x_0. Under this identification, $\tilde{x}_0 = \langle e \rangle$ is the unit element in $\pi_1(X, x_0)$. The action of the loop class $\alpha = \langle a \rangle \in \pi_1(X, x_0)$ on $\tilde{X}$ is then given by

$$\alpha = \langle a \rangle : \langle c \rangle \to \alpha \cdot c = \langle a \cdot c \rangle.$$

Now we have the following relationship between $\tilde{f}_* : \pi \to \pi$ and

$$f_* : \pi_1(X, x_0) \longrightarrow \pi_1(X, f(x_0)).$$

LEMMA 5. Suppose $\tilde{f}(\tilde{x}_0) = \langle w \rangle$. Then the following diagram commutes:

$$\begin{array}{ccc} \pi_1(X, x_0) & \xrightarrow{\ f_*\ } & \pi_1(X, f(x_0)) \\ {\scriptstyle \tilde{f}_*} \searrow & & \downarrow w_* \\ & \pi_1(X, x_0) & \end{array}$$

where w_* is the isomorphism induced by the path w.

In other words, for every $\alpha = \langle a \rangle \in \pi_1(X, x_0)$, we have

$$\tilde{f}_*(\langle a \rangle) = \langle w(f \circ a)w^{-1} \rangle.$$

REMARK 2. In particular, if $x_0 \in p(\mathrm{Fix}(\tilde{f}))$ and $\tilde{x}_0 \in \mathrm{Fix}(\tilde{f})$, then $\tilde{f}_* = f_*$.

LEMMA 6. Lifting classes of f (and hence fixed point classes, empty or not) are in 1-1 correspondence with $\tilde{f}_*$-conjugacy classes in π, the lifting class $[\gamma \circ \tilde{f}]$ corresponding to the $\tilde{f}_*$-conjugacy class of γ. We therefore have $R(f) = R(\tilde{f}_*)$.

We shall say that the fixed point class $p(\mathrm{Fix}(\gamma \circ \tilde{f}))$, which is labeled with the lifting class $[\gamma \circ \tilde{f}]$, *corresponds* to the $\tilde{f}_*$-conjugacy class of γ. Thus $\tilde{f}_*$-conjugacy classes in π serve as coordinates for fixed point classes of f, once a reference lifting $\tilde{f}$ is chosen.

2.2. Asymptotic expansions.

We assume that X is a compact surface of negative Euler characteristic and $f : X \to X$ is a pseudo-Anosov homeomorphism, i.e. there is a number $\lambda > 1$ and a pair of transverse measured foliations (F^s, μ^s) and (F^u, μ^u) such that $f(F^s, \mu^s) = (F^s, \frac{1}{\lambda}\mu^s)$ and $f(F^u, \mu^u) = (F^u, \lambda\mu^u)$. The mapping torus T_f of $f : X \to X$ is the space obtained from $X \times [0, 1]$ by identifying $(x, 1)$ with $(f(x), 0)$ for all $x \in X$. It is often more convenient to regard T_f as the space obtained from $X \times [0, \infty)$ by identifying $(x, s + 1)$ with $(f(x), s)$ for all $x \in X, s \in [0, \infty)$. On T_f there is a natural semi-flow $\phi : T_f \times [0, \infty) \to T_f, \phi_t(x, s) = (x, s + t)$ for all $t \geq 0$. Then the map $f : X \to X$ is the return map of the semi-flow ϕ. A point $x \in X$ and a positive number $\tau > 0$ determine the orbit curve $\phi_{(x, \tau)} := \phi_t(x)_{0 \leq t \leq \tau}$ in T_f. The fixed points and periodic points of f then correspond to closed orbits of various periods. Take the base point x_0 of X as the base point of T_f. According to van Kampen's Theorem, the fundamental group $G := \pi_1(T_f, x_0)$ is obtained from π by adding a new generator z and adding the relations $z^{-1}gz = \tilde{f}_*(g)$ for all $g \in \pi = \pi_1(X, x_0)$, where z is the generator of $\pi_1(S^1, x_0)$. This means that G is a semi-direct product $G = \pi \rtimes Z$ of π with Z.

We now describe some known results:

LEMMA 7. If Γ is a group and ϕ is an endomorphism of Γ then an element $x \in \Gamma$ is always ϕ-conjugate to its image $\phi(x)$.

PROOF. If $\gamma = x^{-1}$, then one has immediately $\gamma x = \phi(x)\phi(\gamma)$. The existence of a γ satisfying this equation implies that x and $\phi(x)$ are ϕ-conjugate.

LEMMA 8. Two elements x, y of π are $\tilde{f}_*$-conjugate if and only if xz and yz are conjugate in the usual sense in G. Therefore $R(f) = R(\tilde{f}_*)$ is the number of usual conjugacy classes in the coset $\pi \cdot z$ of π in G.

PROOF. If x and y are $\tilde{f}_*$-conjugate, then there is a $\gamma \in \pi$ such that $\gamma x = y\tilde{f}_*(\gamma)$. This implies $\gamma x = yz\gamma z^{-1}$ and therefore $\gamma(xz) = (yz)\gamma$. So xz and yz are conjugate in the usual sense in G. Conversely suppose xz and yz are conjugate in G. Then there is a $\gamma z^n \in G$ with $\gamma z^n xz = yz\gamma z^n$. From the relation $zxz^{-1} = \tilde{f}_*(x)$, we obtain $\gamma \tilde{f}_*^n(x)z^{n+1} = y\tilde{f}_*(\gamma)z^{n+1}$ and therefore $\gamma \tilde{f}_*^n(x) = y\tilde{f}_*(\gamma)$. This shows that $\tilde{f}_*^n(x)$ and y are $\tilde{f}_*$-conjugate. However, by Lemma 7, x and $\tilde{f}_*^n(x)$ are $\tilde{f}_*$-conjugate, so x and y must be $\tilde{f}_*$-conjugate.

There is a canonical projection $\tau : T_f \to R/Z$ given by $(x, s) \mapsto s$. This induces a map $\pi_1(\tau) : G = \pi_1(T_f, x_0) \to Z$.

We see that the Reidemeister number $R(f)$ is equal to the number of homotopy classes of closed paths γ in T_f whose projections onto R/Z are homotopic to the path

$$\sigma : \quad [0,1] \quad \to \quad R/Z$$
$$s \quad \mapsto \quad s.$$

Corresponding to this, there is a group-theoretical interpretation of $R(f)$ as the number of usual conjugacy classes of elements $\gamma \in \pi_1(T_f)$ satisfying $\pi_1(\tau)(\gamma) = z$.

LEMMA 9. [19, 14] The interior of the mapping torus $Int(T_f)$ admits a hyperbolic structure of finite volume if and only if f is isotopic to a pseudo-Anosov homeomorphism.

So, if the surface X is closed and f is isotopic to a pseudo-Anosov homeomorphism, the mapping torus T_f can be realised as a hyperbolic 3-manifold, H^3/G, where H^3 is the Poincare upper half space $\{(x, y, z) : z > 0, (x, y) \in R^2\}$ with the metric $ds^2 = (dx^2 + dy^2 + dz^2)/z^2$. The closed geodesics on a hyperbolic manifold are in one-to-one correspondence with the free homotopy classes of loops. These classes of loops are in one-to-one correspondence with the conjugacy classes of loxodromic elements in the fundamental group of the hyperbolic manifold. This correspondence allowed Ch. Epstein (see [2], p. 127) to study the asymptotics of such functions as $p_n(x)=$ #{primitive closed geodesics of length less than x represented by an element of the form gz^n} using the Selberg trace formula. A primitive closed geodesic is one which is not an iterate of another closed geodesic. Later, Phillips and Sarnak [15] generalised the results of Epstein and obtained for n-dimensional hyperbolic manifolds the asymptotics of the number of primitive closed geodesics of length at most x lying in a fixed homology class. The proof of this result makes routine use of the Selberg trace formula. In the more general case of variable negative curvature, such asymptotics were obtained by Pollicott and Sharp [16]. They used a dynamical approach based on the geodesic flow. We will only need an asymptotic expansion for $p_1(x)$. Note that closed geodesics represented by an element of the form gz are automatically primitive, because they wrap exactly once around the mapping torus (once around the generator z). We have the following asymptotic expansion [**2**, **15**, **16**]:

$$p_1(x) = \frac{e^{hx}}{x^{3/2}} \left(\sum_{n=0}^{N} \frac{C_n}{x^{n/2}} + o\left(\frac{1}{x^{N/2}} \right) \right), \tag{1}$$

for any $N > 0$, where $h = \dim T_f - 1 = 2$ is the topological entropy of the geodesic flow on the unit-tangent bundle ST_f, and the constant $C_0 > 0$ depends on the volume of the hyperbolic 3-manifold T_f. N. Anantharaman [**1**] has shown that the constants C_n vanish if n is odd. So, we have the following leading asymptotic behaviour:

$$p_1(x) \sim C_0 \frac{e^{hx}}{x^{3/2}}, \text{ as } x \to \infty \tag{2}$$

Notation: We write $f(x) \sim g(x)$ if $\frac{f(x)}{g(x)} \to 1$ as $x \to \infty$.

Now, using the one-to-one correspondences in Lemma 6 and Lemma 8, we define the norm of a fixed point class, or of the corresponding lifting class, or of the corresponding twisted conjugacy class $\{g\}_{\tilde{f}_*}$ in the fundamental group of the surface $\pi = \pi_1(X, x_0)$, as the length of the primitive closed geodesic γ on T_f, which is represented by an element of the form gz. So, for example, the norm function l^* on the set of twisted conjugacy classes equals $l^* = l \circ B$, where l is the length function on geodesics ($l(\gamma)$ is the length of the primitive closed geodesic γ) and B is a bijection between the set of twisted conjugacy classes $\{g\}_{\tilde{f}_*}$ in the fundamental group of the surface $\pi = \pi_1(X, x_0)$ and the set of closed geodesics represented by an element of the form gz in the fundamental group $G := \pi_1(T_f, x_0)$. We introduce the following counting functions:

FPC(x)= # {fixed point classes of f of norm less than x},
L(x)= # {lifting classes of f of norm less than x},
Tw(x)= #{twisted conjugacy classes for $\tilde{f}_*$ in the fundamental group of the surface of norm less than x}.

THEOREM 3. Let X be a closed surface of negative Euler characteristic and let $f : X \to X$ be a pseudo-Anosov homeomorphism. Then

$$FPC(x) = L(x) = Tw(x) = \frac{e^{2x}}{x^{3/2}} \left(\sum_{n=0}^{N} \frac{C_n}{x^{n/2}} + o\left(\frac{1}{x^{N/2}} \right) \right),$$

where the constant $C_0 > 0$ depends on the volume of the hyperbolic 3-manifold T_f, and the constants C_n vanish if n is odd.

PROOF The proof follows from Lemmas 8 and 9 and the asymptotic expansion (1).

COROLLARY 2. For pseudo-Anosov homeomorphisms of closed surfaces, the Reidemeister number is infinite.

We can generalise Theorem 3 in the following way:

THEOREM 4. Let M be a compact manifold and $f : M \to M$ be a homeomorphism. Suppose that the mapping torus T_f admits a Riemannian metric of negative sectional curvature. Then there exist constants $C_0, C_1, C_2, \ldots$ with $C_0 > 0$ such that

$$FPC(x) = L(x) = Tw(x) = \frac{e^{hx}}{x^{3/2}} \left(\sum_{n=0}^{N} \frac{C_n}{x^{n/2}} + o\left(\frac{1}{x^{N/2}}\right) \right)$$

for any $N > 0$, where $h > 0$ is the topological entropy of the geodesic flow on the unit-tangent bundle ST_f and the constants C_n vanish if n is odd.

QUESTION 1. For which compact manifolds M and homeomorphisms $f : M \to M$ does the mapping torus T_f admit a Riemannian metric of negative curvature or negative sectional curvatures?

QUESTION 2. How can one define the norm of a fixed point class or of a twisted conjugacy class in the general case?

References

[1] N. Anantharaman, Precise counting results for closed orbits of Anosov flows. Preprint, 1998.

[2] C. Epstein, The spectral theory of geometrically periodic hyperbolic 3-manifolds. Memoirs of the AMS, vol. 58, number 335, 1985.

[3] A. Fathi and M. Shub, Some dynamics of pseudo-Anosov diffeomorphisms. Astérisque 66-67 (1979), 181-207.

[4] A.L. Fel'shtyn, New zeta functions for dynamical systems and Nielsen fixed point theory. Lecture Notes in Math., vol. 1346, Springer, 1988, 33-55.

[5] A. L. Fel'shtyn, Dynamical zeta functions, Nielsen theory and Reidemeister torsion. Memoirs of the AMS, 150 pages, to appear.

[6] Fel'shtyn A. L., Hill R., Trace formulae, zeta functions, congruences and Reidemeister torsion in Nielsen theory. Forum Mathematicum, vol. 10, n. 6, 1998, 641-663.

[7] Fel'shtyn A. L., Pilyugina V. B., The Nielsen zeta function. Funct. Anal. Appl., vol. 19, n. 4, 1985, 61 -67.

[8] J. Hempel, 3-manifolds. Annals of Math. Studies 86, Princeton, 1976.

[9] N. V. Ivanov, Nielsen numbers of maps of surfaces. Journal Sov. Math., 26 (1984), 1636-1641.

[10] W. Jaco, Lectures on three-manifold topology. Regional Conference Series in Mathematics, AMS, vol. 43, 1977.

[11] B. Jiang, S. Wang, Y. Wu, Homeomorphisms of 3-manifolds and realisation of Nielsen number. Preprint n. 18 of Institute of Mathematics, Peking University, 1997.

[12] M. Kelly, Nielsen numbers and homeomorphisms of geometric 3-manifolds. Topology Proceedings, vol. 19, 1994.

[13] A. Manning, Axiom A diffeomorphisms have rational zeta function. Bull. London Math. Soc. 3 (1971), 215-220.

[14] J.-P. Otal, Le theoreme d'hyperbolisation pour les varietes fibrées de dimension 3. Astérisque vol. 235, 1996.

[15] R. Phillips, P. Sarnak, Geodesics in homology classes. Duke Math. Journal, vol. 55, n.2, 1987, 287-297.

[16] M. Pollicott, R. Sharp, Asymptotic expansions for closed orbits in homology classes. Preprint ESI 594, 1998, Vienna.

[17] P. Scott, The geometry of 3-manifolds. Bull. London Math. Soc., vol. 15, n. 56, 401-487.

[18] W. Thurston, The geometry and topology of 3-manifolds. Princeton University Press, 1978.

[19] W. Thurston, Hyperbolic structures on 3-manifolds, II: surface groups and 3-manifolds which fiber over the circle. Preprint.

[20] W. Thurston, On the geometry and dynamics of diffeomorphisms of surfaces. Bull. AMS, 19 (1988), 417-431.

Current address: Institut fur Mathematik, Ernst-Moritz-Arndt-Universitat Greifswald, Jahnstrasse 15a, D-17489 Greifswald, Germany.

E-mail address: `felshtyn@mail.uni-greifswald.de`

Contemporary Mathematics
Volume **290**, 2001

Computing the Riemann Zeta Function by Numerical Quadrature

William F. Galway

ABSTRACT. The Riemann zeta function, $\zeta(s)$, where $s = \sigma + it$, $\sigma, t \in \mathbb{R}$, may be computed to moderate accuracy, using $O(t^{1/2+\epsilon})$ arithmetic operations, by use of the Riemann-Siegel formula, which for fixed σ gives an asymptotic expansion of $\zeta(s)$ as $t \to \infty$. In contrast, the Euler-Maclaurin summation formula may be used to compute $\zeta(s)$ to arbitrarily high accuracy, requiring $O(t^{1+\epsilon} + d^{1+\epsilon})$ operations to compute d digits [**CO92**].

We analyze an integral underlying the Riemann-Siegel formula and show that numerical quadrature may be used to compute $\zeta(s)$ to arbitrarily high accuracy. This method requires $O(t^{1/2+\epsilon})$ operations to find d digits of $\zeta(s)$, for fixed d as $t \to \infty$. For large t, this quadrature method should allow computation of $\zeta(\sigma + it)$ to high precision in much less time than by the Euler-Maclaurin formula. Further, precise error bounds for the Riemann-Siegel formula are known only in the case $\sigma = 1/2$, while the quadrature method should allow computation of $\zeta(\sigma + it)$ to a prescribed accuracy, for arbitrary σ, with efficiency comparable to that of the Riemann-Siegel formula. Although we do not find precise error bounds here, we do give heuristics for choosing parameters, such as step size, used in this method.

1. Introduction

We begin by relating $\zeta(s)$ to an integral used by Wolfgang Gabcke, to which he applied the saddle point method to find precise error bounds for the Riemann-Siegel formula when $\sigma = 1/2$ [**Gab79**]. (This representation was discovered by Riemann. However, in his development of the Riemann-Siegel formula, Riemann applied the saddle point method to a different integral representation.)

Throughout this paper, $\mathbb{N}$ denotes the set of non-negative integers. Given $N \in \mathbb{N}$, define the integral $I_N(s)$ by

$$(1.1) \qquad I_N(s) := \int_{N \diagdown N+1} f(z; s) \, dz,$$

where $f(z; s) := e^{i\pi z^2} z^{-s} / (e^{i\pi z} - e^{-i\pi z})$ and where for $z_0, z_1 \in \mathbb{C}$ we write $z_0 \diagup z_1$ for the path $z = (z_0 + z_1)/2 + e^{-3\pi i/4}\alpha$, $-\infty < \alpha < \infty$. As usual, we define z^{-s} as $\exp(-s \ln(z))$, with a branch cut along the negative real axis of the z-plane.

2000 *Mathematics Subject Classification.* Primary 11M06, 11Y35.

The integral (1.1) converges to give an entire function of s, since for any $\epsilon > 0$ the numerator of the integrand satisfies

$$(1.2) \qquad \left| e^{i\pi z^2} z^{-s} \right| = |z|^{-\sigma} e^{t \arg(z) - 2\pi xy} = O(e^{(-\pi + \epsilon)\alpha^2}) \quad \text{as } \alpha \to \pm\infty,$$

where $z = N + 1/2 + e^{-3i\pi/4}\alpha = x + iy$. Division by the denominator multiplies this by a factor of $O(e^{-\pi|y|})$ as $y \to \pm\infty$, and the path avoids the poles at $z \in \mathbb{Z}$.

Recall that $\zeta(s)$ satisfies the functional equation $\zeta(s) = \chi(s)\zeta(1 - s)$, where

$$\chi(s) := \pi^{s-1/2} \frac{\Gamma((1-s)/2)}{\Gamma(s/2)} = 2^{s-1}\pi^s \sec(\pi s/2)/\Gamma(s).$$

Riemann showed ([**Sie66**, §3], [**Edw74**, §7.9]) that

$$(1.3) \qquad\qquad\qquad \zeta(s) = I_0(s) + \chi(s)\,\overline{I_0(1 - \bar{s})}.$$

This representation converges for $s \in \mathbb{C}$ with the exception of the pole at $s = 1$ and removable singularities at $s = 2k + 3$, $k \in \mathbb{N}$, arising from poles of $\chi(s)$ at these points. Outside these singular points, an algorithm for computing $I_0(s)$ yields an algorithm for computing $\zeta(s)$.

By the Cauchy residue formula, shifting the path for $I_0(s)$ to $N \diagup N+1$ gives

$$(1.4) \qquad\qquad\qquad I_0(s) = \sum_{n=1}^{N} n^{-s} + I_N(s).$$

To yield a numerically tractable integral, we choose N so the path passes near the saddle point of $f(z; s)$ (Figure 1). (For fixed σ, we will have $N \sim \sqrt{t/(2\pi)}$ as $t \to \infty$.) Gabcke restricts t to be real, and applies the saddle point method to $I_N(s)$ to derive and analyze the Riemann-Siegel expansion, asymptotic as $t \to \infty$, for $Z(t) := \chi(1/2 + it)^{-1/2}\zeta(1/2 + it)$. For $t \in \mathbb{R}$, $Z(t)$ satisfies the properties $Z(t) \in \mathbb{R}$, $|Z(t)| = |\zeta(1/2 + it)|$. The function $Z(t)$ is preferred over $\zeta(s)$ in studies of zeros of $\zeta(s)$ along the critical line, $s = 1/2 + it$, since the location of zeros can be reduced to the task of locating sign changes of $Z(t)$. However, in this paper we will treat only the computation of $\zeta(s)$.

2. A quadrature formula for $I_0(s)$

Instead of applying the saddle point method to give an asymptotic expansion, we note that the integral (1.1) defining $I_N(s)$ is well suited to numerical quadrature when N lies near the saddle point of $f(z; s)$. The results of this section follow easily from the Cauchy residue formula. (Other examples of applying quadrature to saddle point integrals are given in [**Tem77**]. See [**Hen88**, §4.9] for examples of sums expressed as integrals of the form used below.)

LEMMA 2.1. *Let $\mathcal{L}$ be a path between N and $N + 1/2$, parallel to and traveling in the same direction as $N \diagup N+1$. Let $\mathcal{R}$ be a similar path between $N + 1/2$ and $N + 1$. Let $H(w) = 1/(1 - e^{2i\pi w})$. Given $h \in \mathbb{C}$ with $\arg(h) = -3\pi/4$, and given z_1 lying on the path $N \diagup N+1$, we have*

$$(2.1) \qquad I_N(s) = h \sum_{m \in \mathbb{Z}} f(z_1 + mh; s)$$

$$+ \int_{\mathcal{L}} f(z; s) H\left(\frac{z - z_1}{h}\right) dz + \int_{\mathcal{R}} f(z; s) H\left(\frac{z_1 - z}{h}\right) dz.$$

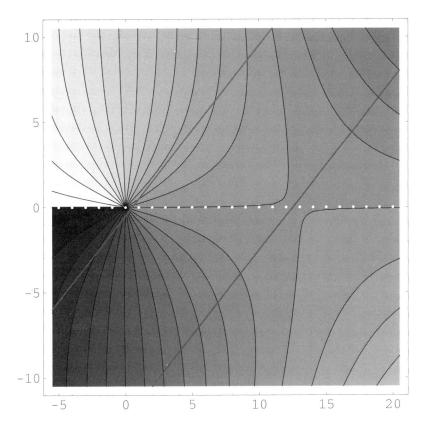

FIGURE 1. Contour plot of $\log_{10}|f(z;s)|$, where $f(z;s)$ is the integrand for $I_N(s)$, $s = 1.5 + 1000i$. The paths $0 \diagup 1$ and $12 \diagup 13$ are shown. The contour interval is 100, larger values are lighter. White dots indicate $z \in \mathbb{Z}$; the poles of $f(z;s)$ are not apparent at the scale used. For reference, $|f(-5.5 + 0.1i; 1.5 + 1000i)| \approx 10^{1356.55}$ while $|f(-5.5 - 0.1i; 1.5 + 1000i)| \approx 10^{-1359.41}$. On the path $12 \diagup 13$, $f(z;s)$ takes a maximum value of ≈ 0.0117.

We think of the "side-integrals" along the "side-paths", $\mathcal{L}$ and $\mathcal{R}$ as error terms. We may bound these terms by noting $|H(w)| \leq e^{2\pi \operatorname{Im}(w)}/(1 - e^{2\pi \operatorname{Im}(w)})$ when $\operatorname{Im}(w) < 0$. Writing Δ for the distance between the paths $\mathcal{L}$ and $N \diagup N+1$, this gives

$$(2.2) \qquad \left| \int_{\mathcal{L}} f(z;s) H\left(\frac{z - z_1}{h}\right) dz \right| \leq \frac{e^{-2\pi\Delta/|h|}}{1 - e^{-2\pi\Delta/|h|}} \int_{\mathcal{L}} |f(z;s)| \, |dz|$$

$$= O_\Delta(e^{-2\pi\Delta/|h|}) \quad \text{as } |h| \to 0.$$

Similarly, writing Δ for the distance between the paths $\mathcal{R}$ and $N \diagup N+1$, we have

$$(2.3) \qquad \left| \int_{\mathcal{R}} f(z;s) H\left(\frac{z_1 - z}{h}\right) dz \right| = O_\Delta(e^{-2\pi\Delta/|h|}) \quad \text{as } |h| \to 0.$$

For fixed Δ, the side-integrals decrease exponentially in $1/|h|$. Moving the side-paths further from the central path increases Δ, and may further decrease these error terms. The optimal choices for $\mathcal{L}$ and $\mathcal{R}$ should lie near saddle points of the integrands in (2.2) and (2.3), respectively. We illustrate the new paths in Figure 2. In Section 3 we give heuristics for finding points $z_{\mathcal{L}}$ and $z_{\mathcal{R}}$ which are near the desired saddle points.

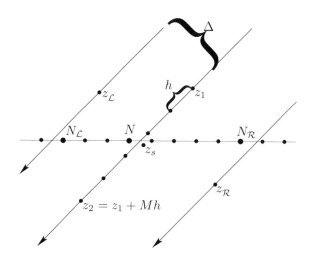

FIGURE 2. Paths and parameters related to Formula (2.4).

Moving the side-paths introduces "side-sums", arising from residues picked up from the poles of $f(z; s)$. Corollary 2.2 gives the resulting formula.

COROLLARY 2.2. *Given $N_{\mathcal{L}}$, N, $N_{\mathcal{R}} \in \mathbb{N}$, $0 < N_{\mathcal{L}} \leq N < N_{\mathcal{R}}$, let $\mathcal{L}$ be a path between $N_{\mathcal{L}} - 1$ and $N_{\mathcal{L}}$, parallel to and traveling in the same direction as $N \nearrow N + 1$, and let $\mathcal{R}$ be a similar path between $N_{\mathcal{R}}$ and $N_{\mathcal{R}} + 1$. Then*

$$I_N(s) = h \sum_{m \in \mathbb{Z}} f(z_1 + mh; s) - \sum_{n=N_{\mathcal{L}}}^{N} n^{-s} H\left(\frac{n - z_1}{h}\right) + \sum_{n=N+1}^{N_{\mathcal{R}}} n^{-s} H\left(\frac{z_1 - n}{h}\right)$$

$$+ \int_{\mathcal{L}} f(z; s) H\left(\frac{z - z_1}{h}\right) dz + \int_{\mathcal{R}} f(z; s) H\left(\frac{z_1 - z}{h}\right) dz.$$

Combining this result with Equation (1.4) gives a quadrature formula for $I_0(s)$:

(2.4) $$I_0(s) = \sum_{n=1}^{N} n^{-s} + h \sum_{m=0}^{M} f(z_1 + mh; s)$$

$$- \sum_{n=N_{\mathcal{L}}}^{N} n^{-s} H\left(\frac{n - z_1}{h}\right) + \sum_{n=N+1}^{N_{\mathcal{R}}} n^{-s} H\left(\frac{z_1 - n}{h}\right)$$

$$+ \mathcal{E}_{\Sigma} + \mathcal{E}_{\int}.$$

In (2.4), $\mathcal{E}_{\Sigma}$ is the error introduced by truncating $\sum_m f(z_1 + mh; s)$ to a finite sum,

$$\mathcal{E}_{\Sigma} = h \sum_{m<0} f(z_1 + mh; s) + h \sum_{m>M} f(z_1 + mh; s),$$

and $\mathcal{E}_f$ is given in terms of our side-integrals,

$$(2.5) \qquad \mathcal{E}_f = \int_{\mathcal{L}} f(z;s) H\left(\frac{z - z_1}{h}\right) dz + \int_{\mathcal{R}} f(z;s) H\left(\frac{z_1 - z}{h}\right) dz.$$

Note that the side-sums in Formula (2.4) serve to "smooth" the truncated Dirichlet series, $\sum_{n=1}^{N} n^{-s}$, giving a more accurate result, as discussed below. This is similar to a parameterized family of expansions studied by Berry and Keating [**BK92**], which are analogous to the classical Riemann-Siegel formula but more accurate, and which use a sum $\sum_n a_n n^{-s}$ where a_n drops "smoothly" to zero as $n \to \infty$.

3. Heuristics for choosing parameters

Using the functional equation for $\zeta(s)$ and the fact that $\zeta(\bar{s}) = \overline{\zeta(s)}$, we can reduce the problem of computing $\zeta(s)$ to the case where $\sigma \geq 1/2$, $t \geq 0$. Referring to Equation (1.3), we see that to compute $\zeta(s)$ with an error bounded by $\varepsilon > 0$, it suffices to compute $I_0(s)$ with an error bounded by $\varepsilon/2$ and to compute $I_0(1 - \bar{s})$ with an error bounded by $|\chi(s)|^{-1} \varepsilon/2$. With an obvious change of variables we need only consider the problem of approximating $I_0(\sigma + it)$, $\sigma \in \mathbb{R}$, $t > 0$, with an error bounded by $\varepsilon > 0$.

Here we give heuristics for making choices for the parameters in Formula (2.4). These choices are suitable for "moderate" σ, "large" t, and an error bound ε which is not "too small". Since we have not yet done a complete analysis of this algorithm, we attempt to ensure only that our error is $O(\varepsilon)$, uniformly under restrictions spelled out below. We begin by sketching a computational procedure for determining our parameters, rather than giving analytic expressions for them. We write $\lfloor x \rfloor$ for the greatest integer function, and $\lceil x \rceil$ for the "ceiling" function, i.e., $\lceil x \rceil = -\lfloor -x \rfloor$.

Figure 2 illustrates the paths and most of the parameters used in our analysis. Let $g(z;s) := \exp(i\pi z^2) z^{-s}$, and note that

$$(3.1) \qquad f(z;s) = O_\delta\left(g(z;s) e^{-\pi|\operatorname{Im} z|}\right) = O(g(z;s)),$$

where δ measures the distance from z to the nearest integer. Let $z_s = \sqrt{s/(2i\pi)}$, so that z_s is a saddle point of $g(z;s)$, where we take the branch with $\operatorname{Re} z_s > 0$. We set $N = \lfloor \operatorname{Re} z_s - \operatorname{Im} z_s \rfloor$ so that $N \nearrow N+1$ passes near z_s.

In part, our goal is to bound $|\mathcal{E}_\Sigma| + |\mathcal{E}_f|$, where $\mathcal{E}_\Sigma$ and $\mathcal{E}_f$ are the error terms in Formula (2.4). We also want to choose our parameters to minimize the number of operations. To simplify analysis, we assume that this is equivalent to minimizing the number of terms in Formula (2.4), i.e., to minimizing $N + M + N_{\mathcal{L}} - N_{\mathcal{R}}$. Since N must lie within a very limited range to yield a tractable integrand, and to further simplify our analysis, we then assume that minimizing the number of terms is equivalent to minimizing M. Near the end of this section, we will see that M typically dominates $N_{\mathcal{L}} - N_{\mathcal{R}}$, so this assumption is reasonable.

To bound the error, we focus on the magnitude of $f(z;s)$ and $g(z;s)$ at four key points. Two of these points, z_1 and z_2, lie on $N \nearrow N+1$ and denote the endpoints of the sum $\sum_m f(z_1 + mh; s)$. The other two points, $z_{\mathcal{L}}$, $z_{\mathcal{R}}$, lie on the paths $\mathcal{L}$ and $\mathcal{R}$. They are chosen to lie near the maximum along these paths of the integrands appearing in the integrals $\int_{\mathcal{L}} \cdots$ and $\int_{\mathcal{R}} \cdots$, respectively, of Equation (2.5).

Our paths may be parameterized as $z = e^{-i\pi/4}R + e^{-3\pi i/4}\alpha$, $-\infty < \alpha < \infty$, where R measures the signed distance from the path to the origin. (We have $R < 0$ if the path passes to the left of the origin.) With this parameterization, we have

$$(3.2) \qquad \frac{d}{d\alpha}\ln|g(z;s)| = -\frac{2\pi\alpha^3 + (2\pi R^2 + \sigma)\alpha + Rt}{\alpha^2 + R^2}.$$

A careful analysis of (3.2) shows that taking $R > \sqrt{|\sigma|/\pi}$ suffices to ensure that $|g(z;s)|$ is both unimodal and has a "sharp" maximum as a function of α.

With this in mind, assume that $N \nearrow N+1$ is sufficiently far from the origin, i.e., that $(N+1/2) > \sqrt{2|\sigma|/\pi}$. We then choose z_1, z_2, to lie on the path $N \nearrow N+1$, subject to the restrictions that they lie on opposite sides of the point where $|g(z;s)|$ reaches its maximum along this path, and to satisfy the four additional conditions

$$(3.3) \qquad\qquad |f(z_1;s)| \le \varepsilon,$$

$$(3.4) \qquad\qquad |f(z_2;s)| \le \varepsilon,$$

$\operatorname{Im} z_1 > 0$, $\operatorname{Im} z_2 < 0$. An analysis of (3.2) and the bound (3.1) can be used to show that these conditions ensure that $|\mathcal{E}_\Sigma| \ll |f(z_1;s)| + |f(z_2;s)| \ll \varepsilon$. Of course, to avoid an unnecessarily long interval of summation, we should choose z_1 and z_2 to satisfy (3.3) and (3.4) with equality "nearly" holding.

To analyze $\mathcal{E}_f$, we begin by examining the "left" side-integral appearing in Equation (2.5):

$$\int_{\mathcal{L}} f(z;s)H\left(\frac{z-z_1}{h}\right)dz.$$

As outlined in Section 2, if $\mathcal{L}$ lies a distance Δ to the left of z_1, and if z lies on $\mathcal{L}$, we have

$$H((z-z_1)/h) = e^{-2i\pi(z-z_1)/h} + O_\Delta(e^{-4\pi\Delta/|h|}),$$

where the O-term is uniform provided Δ is bounded away from zero. In other words, if z lies well to the left of $N \nearrow N+1$ then $H((z-z_1)/h)$ is well approximated by $e^{-2i\pi(z-z_1)/h}$. For this reason, for fixed $h > 0$, we set $z_{\mathcal{L}}$ to be the saddle point of $g(z;s)e^{-2i\pi(z-z_1)/h}$ having positive real part, and let $\mathcal{L}$ pass through $z_{\mathcal{L}}$. Again, if R denotes the distance from $\mathcal{L}$ to the origin, and if $R > \sqrt{|\sigma|/\pi}$, then $|g(z;s)|$ assumes a single maximum along $\mathcal{L}$, at $z = z_{\mathcal{L}}$. Under the additional condition that $\mathcal{L}$ is bounded a distance δ from the singularities of $f(z;s)$, we can show that

$$(3.5) \qquad \int_{\mathcal{L}} f(z;s)H\left(\frac{z-z_1}{h}\right)dz \ll \left|g(z_{\mathcal{L}};s)e^{-2i\pi(z_{\mathcal{L}}-z_1)/h}\right|,$$

with an O-constant depending on δ and Δ. By deforming $\mathcal{L}$ to a region where the integrand is well behaved, we believe that (3.5) can be made to hold uniformly in δ, even when the original path passes near or through a pole of $f(z;s)$. Since $z_{\mathcal{L}}$ is a saddle point, we expect our choice of $\mathcal{L}$ to be nearly optimal, again provided Δ is bounded away from zero, and that R is sufficiently large.

A similar analysis applies to the "right" side-integral along $\mathcal{R}$, where we set $z_{\mathcal{R}}$ to be the saddle point of $g(z;s)e^{2i\pi(z-z_1)/h}$ with positive real part, and then let $\mathcal{R}$ pass through $z_{\mathcal{R}}$. We conclude, given fixed $h > 0$, that if $\mathcal{L}$ is sufficiently far to the right of the origin, and if $\mathcal{L}$ and $\mathcal{R}$ are bounded away from $N \nearrow N+1$, then this choice of $\mathcal{L}$, $\mathcal{R}$ should be nearly optimal and that

$$\left|\mathcal{E}_f\right| \ll \left|g(z_{\mathcal{L}};s)e^{-2i\pi(z_{\mathcal{L}}-z_1)/h}\right| + \left|g(z_{\mathcal{R}};s)e^{2i\pi(z_{\mathcal{R}}-z_1)/h}\right|.$$

Based on this analysis for fixed h, we determine M and h by choosing M to be the least integer subject to the following six conditions: $M > 0$, $h = (z_2 - z_1)/M$,

$$(3.6) \qquad z_\mathcal{L} = (2h)^{-1} + \sqrt{(2h)^{-2} + s/(2i\pi)},$$

$$(3.7) \qquad z_\mathcal{R} = -(2h)^{-1} + \sqrt{(2h)^{-2} + s/(2i\pi)},$$

$$(3.8) \qquad \varepsilon \geq \left| g(z_\mathcal{L}; s) e^{-2i\pi(z_\mathcal{L} - z_1)/h} \right|,$$

$$(3.9) \qquad \varepsilon \geq \left| g(z_\mathcal{R}; s) e^{2i\pi(z_\mathcal{R} - z_1)/h} \right|.$$

In (3.6) and (3.7) we use the branch of the square root with positive real part. Finally, we set $N_\mathcal{L} = \lceil \operatorname{Re} z_\mathcal{L} - \operatorname{Im} z_\mathcal{L} \rceil$, $N_\mathcal{R} = \lfloor \operatorname{Re} z_\mathcal{R} - \operatorname{Im} z_\mathcal{R} \rfloor$, so that $\mathcal{L}$ passes between $N_\mathcal{L} - 1$ and $N_\mathcal{L}$, while $\mathcal{R}$ passes between $N_\mathcal{R}$ and $N_\mathcal{R} + 1$.

To summarize, provided t is sufficiently large and ε is not too small, the above conditions on our parameters should ensure an error that is uniformly $O(\varepsilon)$. In contrast, when t or ε are small enough, our paths may pass too near the origin to ensure the validity of our analysis. (It is even possible that $N_\mathcal{L}$ will be negative.)

s	ε	M	$|z_2 - z_1|$	$N_\mathcal{L}$	N	$N_\mathcal{R}$
	10^{-25}	36	5.64	9	12	17
$1/2 + 10^3 i$	10^{-50}	75	8.18	7	12	19
	10^{-100}	155	11.80	6	12	23
	10^{-25}	37	5.58	395	398	403
$1/2 + 10^6 i$	10^{-50}	75	8.12	393	398	405
	10^{-100}	149	11.69	390	398	408
	10^{-25}	33	5.46	12612	12615	12619
$1/2 + 10^9 i$	10^{-50}	70	8.04	12610	12615	12621
	10^{-100}	143	11.64	12607	12615	12624
	10^{-25}	7	2.25	12614	12615	12617
$5 + 10^9 i$	10^{-50}	43	6.24	12611	12615	12620
	10^{-100}	116	10.45	12608	12615	12623
$-4 + 10^9 i$	$\dfrac{10^{-100}}{|\chi(5 + 10^9 i)|}$	116	10.45	12608	12615	12623

TABLE 1. Parameters for the quadrature method, as functions of s and ε.

For fixed σ and ε, as $t \to \infty$, we can give more analytically tractable estimates of the parameters found above. The reader may wish to compare these estimates against the values in Table 1, which were found using the more complicated procedure outlined in the previous paragraphs. We begin by noting that $z_s = \sqrt{t/(2\pi)} + O(\sigma/\sqrt{t})$, so $N \sim \sqrt{t/(2\pi)}$ as $t \to \infty$. We then approximate $g(z; s)$ by taking the Taylor expansion of $\ln g(z; s)$ about z_s. This gives

$$\ln g(z; s) = i\pi z_s^2 - s \ln(z_s) + 2i\pi(z - z_s)^2 + O\left(\left| s(z - z_s)/z_s \right|^3 \right)$$

uniformly for z "near" z_s, e.g., for $|z - z_s| \leq |z_s|/2$. Dropping the O-term and exponentiating gives a sufficiently accurate approximation to both $g(z; s)$ and to $f(z; s)$ to estimate our parameters. Setting $\Delta = \sqrt{\max(0, \ln(|z_s|^{-\sigma}/\varepsilon)/(2\pi))}$, we find that z_1 and z_2 lie near $N + 1/2 \pm e^{i\pi/4}\Delta$ respectively; $N_\mathcal{L}$ and $N_\mathcal{R}$ lie near

$N + 1/2 \mp \sqrt{2}\Delta$ respectively; and $M \approx 4\Delta^2$. (The case $\Delta = 0$ corresponds to the situation where $I_0(s)$ is well approximated by $\sum_{n=1}^{N} n^{-s}$.) Nearly identical parameters suffice to compute $I_0(1 - \bar{s})$ with an error bounded by $|\chi(s)|^{-1} \varepsilon$, since $|\chi(\sigma + it)| \sim (t/(2\pi))^{1/2-\sigma}$ as $t \to \infty$ [**Tit86**, Eq. 4.12.3].

There are several remaining issues which should be dealt with to complete the analysis given in this section. To begin, we need a more careful analysis to find explicit O-constants. There should be constants C_1 and C_2 such that

$$|\mathcal{E}_\Sigma| \leq C_1(|f(z_1; s)| + |f(z_1; s)|)$$

$$\left|\mathcal{E}_f\right| \leq C_2\left(|g(z_\mathcal{L}; s)e^{-2i\pi(z_\mathcal{L}-z_1)/h}| + |g(z_\mathcal{R}; s)e^{2i\pi(z_\mathcal{R}-z_1)/h}|\right).$$

Given C_1, C_2, we could then modify the conditions (3.3), (3.4), (3.8) and (3.9) to ensure $|\mathcal{E}_\Sigma| + |\mathcal{E}_f| \leq \varepsilon$. As it is, our parameter ε serves more as an "error goal" instead of an error bound, and we only have experimental evidence that our cruder analysis gives satisfactory accuracy (see Table 2). Secondly, we have ignored the computational cost of determining the parameters used in Formula (2.4). This cost will depend largely on the method used to ensure conditions such as (3.3). Finally, our analysis assumes that $\chi(s)$ in Equation (1.3) and all values in Formula (2.4) are computed exactly. A complete analysis would require determining error bounds required of $\chi(s)$, and of the subexpressions in Formula (2.4), needed to ensure a total error bounded by ε. In the following section we discuss some computations and briefly outline how we have dealt with these remaining issues.

4. Examples

Table 1 illustrates the parameters found by using the heuristics of the previous section. These heuristics were implemented using the GP/PARI calculator, version 2.1.0 using the 32-bit MicroSparc kernel.

In rough outline, we used the following procedure to find z_1 satisfying the condition (3.3). With the parameterization $z = N + 1/2 - e^{-3\pi i/4}\alpha$ we began with an initial guess for α and then repeatedly doubled α until $|f(z; s)| \leq \varepsilon$. Similarly, we repeatedly halved α until $|f(z; s)| \geq \varepsilon$. Starting with these bracketing values, we used PARI's `solve` function to find the solution to $|f(z; s)| = \varepsilon$. Only limited accuracy is needed for this solution, so we implemented this procedure using 19 digits of precision. Similar techniques were used to satisfy the conditions (3.4), (3.8), and (3.9). Although the use of PARI's `solve` function is somewhat computationally expensive, we have found this method to be quite practical, and as $t \to \infty$ or $\varepsilon \to 0$ the cost of determining parameters appears to be dominated by the cost of evaluating Formula (2.4).

Table 2 illustrates the accuracy achieved when using our heuristics to approximate $\zeta(s)$. Beginning at $s_0 = \sigma + it_0$, we computed 100 values of $\zeta(s)$ at equally spaced steps. When t_0 is sufficiently large compared to σ, our analysis of Section 3 suggests that N should increase by unity as t increases from t_0 to $t_0 + 2\sqrt{2\pi t_0}$. For this reason, we used a step value of $i\sqrt{2\pi \operatorname{Im}(s_0)}/5$, which ensured that N increased by a few units, and which made it likely that, for some values of s, our saddle point z_s lay near a singularity of $f(z; s)$. For a given "error goal", ε, and for each value of s, we found $I_0(s)$ using parameters selected for an error goal of $\varepsilon/2$, while we found $I_0(1 - \bar{s})$ using parameters selected for an error goal of $|\chi(s)|^{-1}\varepsilon/2$, and finally $\zeta(s)$ was computed using Equation (1.3). The values of $\zeta(s)$ found were compared

s_0	ε	min	max	mean
	10^{-25}	$2.87 \cdot 10^{-32}$	$1.45 \cdot 10^{-28}$	$3.03 \cdot 10^{-29}$
$1/2 + 10^3 i$	10^{-50}	$1.49 \cdot 10^{-56}$	$7.89 \cdot 10^{-54}$	$2.30 \cdot 10^{-54}$
	10^{-100}	$3.37 \cdot 10^{-106}$	$5.36 \cdot 10^{-104}$	$1.95 \cdot 10^{-104}$
	10^{-25}	$6.91 \cdot 10^{-30}$	$2.02 \cdot 10^{-28}$	$5.07 \cdot 10^{-29}$
$5 + 10^3 i$	10^{-50}	$4.02 \cdot 10^{-55}$	$6.12 \cdot 10^{-54}$	$2.95 \cdot 10^{-54}$
	10^{-100}	$1.93 \cdot 10^{-105}$	$4.71 \cdot 10^{-104}$	$2.00 \cdot 10^{-104}$

TABLE 2. Minimum error, maximum error, and mean error in computations of $\zeta(s_0 + 2ik\sqrt{2\pi \, \mathrm{Im}(s_0)}/5)$, $0 \le k < 100$.

against the values returned by PARI's `zeta` function, which uses Euler-Maclaurin summation to find $\zeta(s)$. All computations were performed using 10 extra digits of precision. For example, when $\varepsilon = 10^{-25}$ the computations were done using at least 35 digits of precision, using the GP/PARI construct `default(realprecision,35)`.

5. Remarks and conclusions

As noted following Equation (1.3), the quadrature method described here is not appropriate for computing $\zeta(2k + 3)$, $k \in \mathbb{N}$. More generally, for large σ and t near 0 the analysis of the quadrature method is difficult, largely because in this situation a path passing near z_s also passes near the branch cut along $z < 0$.

The quadrature method has several strengths. These include the ability to find $\zeta(s)$ to arbitrary accuracy (for most values of s), and the simplicity of error analysis for this method. This analysis, outlined in Section 3, is relatively insensitive to our choice of σ and to our error bound ε. In contrast, the Riemann-Siegel formula provides only limited accuracy. Good error bounds for the Riemann-Siegel formula are available only when $\sigma = 1/2$, and the error analysis becomes significantly more complicated with each additional term included in the expansion [Gab79].

However, these advantages are unlikely to be of interest when performing numerical tests of the Riemann Hypothesis, where a few terms of the Riemann-Siegel formula provide sufficient accuracy. For example, two terms of the Riemann-Siegel expansion sufficed to verify the Riemann Hypothesis up to $t \approx 545 \cdot 10^6$ [vdLtRW86]. Other problems have required the location of zeros to greater accuracy. For example, as a small part of a computation performed in order to disprove Merten's conjecture, Odlyzko and teRiele found the first 2000 zeros of $\zeta(s)$ to 100 digit accuracy [OtR85]. Although the quadrature method may be of some use when zeros of $\zeta(s)$ are needed to high accuracy, we believe this method is more likely to be useful in applications such as the analytic computation of $\pi(x)$, where $\zeta(\sigma + it)$, $\sigma > 1$ must be computed to high accuracy, with rigorous error bounds [LO87].

For large t, when finding a single value of $\zeta(\sigma + it)$, the time spent computing $\sum_{n=1}^N n^{-s}$ will dominate the time spent in computing the other terms in Formula (2.4). However, when computing $\zeta(\sigma + it)$ with fixed σ and many closely and equally spaced t, the cost of computing the truncated Dirichlet series can be greatly reduced, e.g., by using the algorithm of Odlyzko and Schönhage [OS88]. In this situation it may be worth noting that many of the other quantities in Formula (2.4) may be calculated infrequently, and then be reused many times. For large t_0, with t in a relatively short interval $[t_0, t_1]$, the parameters z_1, N, M, $N_{\mathcal{L}}$,

$N_{\mathcal{R}}$ can stay fixed as t varies, and the quantities $H(\dots)$ in the side-sums need be calculated only once. Many of the subexpressions—such as $\ln(z_1 + mh)$, implicitly calculated while evaluating $h \sum_{m=0}^{M} f(z_1 + mh; s)$—may be reused so long as N stays constant. However, in this case a more complicated error analysis than given in Section 3 would be required to ensure a given error bound.

It should be noted that the integral representation for $\zeta(s)$ used here may be generalized to Dirichlet L-functions. Deuring has used these representations to develop analogues of the Riemann-Siegel formula for L-functions [**Deu67**]. In a similar manner, it should be possible to generalize the quadrature method to the computation of L-functions.

Several other promising methods for computing $\zeta(s)$ have been proposed recently. These include the work of Berry and Keating mentioned above, methods discussed in the survey article by Borwein, Bradley, and Crandall [**BBC00**], and a method given in the Ph.D. thesis of Michael Rubinstein [**Rub98**]. Rubinstein's thesis considers the problem of computing values of a very general family of L-functions—his results apply to $\zeta(s)$ and to Dirichlet L-functions as special cases.

6. Acknowledgments

I would like to thank Harold G. Diamond and the anonymous referee of this paper for their useful and insightful comments.

References

[BBC00]   Jonathan M. Borwein, David M. Bradley, and Richard E. Crandall, *Computational strategies for the Riemann zeta function*, J. Comput. Appl. Math. **121** (2000), no. 1-2, 247–296, Numerical analysis in the 20th century, Vol. I, Approximation theory. MR 1 780 051

[BK92]   M. V. Berry and J. P. Keating, *A new asymptotic representation for $\zeta(\frac{1}{2} + it)$ and quantum spectral determinants*, Proc. Roy. Soc. London Ser. A **437** (1992), no. 1899, 151–173. MR **93j**:11057

[CO92]   Henri Cohen and Michel Olivier, *Calcul des valeurs de la fonction zêta de Riemann en multiprécision*, C. R. Acad. Sci. Paris Sér. I Math. **314** (1992), no. 6, 427–430. MR **93g**:11134

[Deu67]   Max Deuring, *Asymptotische Entwicklungen der Dirichletschen L-Reihen*, Math. Ann. **168** (1967), 1–30. MR 35 #4173

[Edw74]   H. M. Edwards, *Riemann's zeta function*, Pure and Applied Mathematics, Vol. 58, Academic Press, New York, 1974. MR 57 #5922

[Gab79]   Wolfgang Gabcke, *Neue Herleitung und Explizite Restabschätzung der Riemann-Siegel-Formel*, Ph.D. thesis, Georg-August-Universität zu Göttingen, 1979.

[Hen88]   Peter Henrici, *Applied and computational complex analysis. Vol. 1*, John Wiley & Sons Inc., New York, 1988, Power series—integration—conformal mapping—location of zeros, Reprint of the 1974 original. MR **90d**:30002

[LO87]   J. C. Lagarias and A. M. Odlyzko, *Computing $\pi(x)$: an analytic method*, Journal of Algorithms **8** (1987), no. 2, 173–191. MR **88k**:11095

[OS88]   Andrew M. Odlyzko and A. Schönhage, *Fast algorithms for multiple evaluations of the Riemann zeta function*, Trans. Amer. Math. Soc. **309** (1988), no. 2, 797–809. MR **89j**:11083

[OtR85]   A. M. Odlyzko and H. J. J. te Riele, *Disproof of the Mertens conjecture*, J. Reine Angew. Math. **357** (1985), 138–160. MR **86m**:11070

[Rub98]   Michael Oded Rubinstein, *Evidence for a spectral interpretation of the zeros of L-functions*, Ph.D. thesis, Princeton University, June 1998.

[Sie66]   C. L. Siegel, *Über Riemanns Nachlaß zur analytischen Zahlentheorie*, Gesammelte Abhandlungen, vol. 1, Springer-Verlag, New York, 1966.

[Tem77]   Nico M. Temme, *The numerical computation of special functions by use of quadrature rules for saddle point integrals. I. Trapezoidal integration rules*, Tech. Report TW 164/77, Mathematisch Centrum, Afdeling Toegepaste Wiskunde, Amsterdam, 1977. MR 57 #4483

[Tit86]   E. C. Titchmarsh, *The theory of the Riemann zeta-function*, second ed., The Clarendon Press Oxford University Press, New York, 1986, Edited and with a preface by D. R. Heath-Brown. MR **88c:**11049

[vdLtRW86] J. van de Lune, H. J. J. te Riele, and D. T. Winter, *On the zeros of the Riemann zeta function in the critical strip. IV*, Math. Comp. **46** (1986), no. 174, 667–681. MR **87e:**11102

DEPARTMENT OF MATHEMATICS, UNIVERSITY OF ILLINOIS AT URBANA-CHAMPAIGN, 1409 W. GREEN STREET, URBANA, IL 61801

E-mail address: galway@math.uiuc.edu

URL: http://www.math.uiuc.edu/~galway

Contemporary Mathematics
Volume **290**, 2001

On Riemann's Zeta function

Shai Haran

We describe in §1 the Riemann Zeta function, the explicit sums, and the equivalence of Weil's positivity with the Riemann hypothesis. In §2 we reformulate the explicit sum in a suggestive form involving the trace of the action of the multiplicative group on the additive group. In §3 we describe the results of Landau-Pollack-Slepian. In §4 we describe the recent reformulation by Alain Connes. In §5 we describe the q-analogue interpolating between the p-adic numbers and the reals. In §6 we give a heuristic proof of the Riemann hypothesis. The emphasis is on the simplest formulations and our attention is focused on the Riemann Zeta function.

1. The Explicit Sums

The (completed) Riemann Zeta function is defined for $\mathrm{Re}\,(s) > 1$ by the Euler product

$$(1.1) \qquad \zeta_{\mathbb{A}}(s) = \prod_{p \geq \eta} \zeta_p(s)$$

where for a finite prime $p \neq \eta$,

$$(1.2) \qquad \zeta_p(s) = (1 - p^{-s})^{-1} = \sum_{n \geq 0} p^{-ns}$$

and for the "real prime" $p = \eta$,

$$(1.3) \qquad \zeta_\eta(s) = \pi^{-s/2}\Gamma(s/2) = \int_{-\infty}^{\infty} e^{-\pi a^2} |a|^s \, d^*a.$$

Expanding the product over the finite primes using the unique factorization of integers into products of prime powers,

$$(1.4) \qquad \zeta_{\mathbb{A}}(s) = \sum_{n \geq 1} n^{-s} \int_{-\infty}^{\infty} e^{-\pi a^2} |a|^s \, d^*a = \int_{0}^{\infty} \left[\sum_{\substack{n \in \mathbb{Z} \\ n \neq 0}} e^{-\pi a^2 n^2} \right] |a|^s \, d^*a.$$

Splitting the integral into two integrals on $[0, 1]$ and on $[1, \infty]$, and using the Poisson summation formula

$$(1.5) \qquad \sum_{n \in \mathbb{Z}} e^{-\pi a^2 n^2} = a^{-1} \sum_{n \in \mathbb{Z}} e^{-\pi a^{-2} n^2}$$

2000 *Mathematics Subject Classification.* Primary 11M26.

to transform the integral on $[0,1]$ to another integral on $[1,\infty]$, taking care of the terms corresponding to $n = 0$, we get

$$(1.6) \qquad \zeta_{\mathbb{A}}(s) = \int\limits_{1}^{\infty} \sum_{n\neq 0} \left[e^{-\pi a^2 n^2} \right] (a^s + a^{1-s})\, d^*a - \left[\frac{1}{s} + \frac{1}{1-s} \right].$$

This shows $\zeta_{\mathbb{A}}(s)$ has meromorphic continuation to all $s \in \mathbb{C}$, with simple poles at $s = 0, 1$ (and residues $-1, 1$ respectively), and the functional equation

$$(1.7) \qquad\qquad\qquad\qquad \zeta_{\mathbb{A}}(s) = \zeta_{\mathbb{A}}(1-s).$$

From (1.1), $\zeta_{\mathbb{A}}(s)$ has no zeros in $\mathrm{Re}\,(s) > 1$, and by (1.7) also in $\mathrm{Re}\,(s) < 0$, and the Riemann Hypothesis "RH" is the assertion that all zeros of $\zeta_{\mathbb{A}}(s)$ satisfy $\mathrm{Re}\,(s) = 1/2$.

For $f \in C_c^\infty(\mathbb{R}^+)$, we have the Mellin transform

$$(1.8) \qquad\qquad\qquad\qquad \hat{f}(s) = \int\limits_{0}^{\infty} f(a) a^s \, d^*a.$$

$\hat{f}(s)$ is an entire function of $s \in \mathbb{C}$, fast decreasing in vertical strips $a \leq \mathrm{Re}\,(s) \leq b$, and Fourier inversion reads

$$(1.9) \qquad\qquad\qquad\qquad f(a) = \frac{1}{2\pi} \int\limits_{\sigma-i\infty}^{\sigma+i\infty} \hat{f}(s) a^{-s} \, ds.$$

By the Residue theorem,

$$\sum_{\zeta_{\mathbb{A}}(1/2+s)=0} \hat{f}(s) - \hat{f}(1/2) - \hat{f}(-1/2) = \frac{1}{2\pi} \oint \hat{f}(s)\, d\log \zeta_{\mathbb{A}}(1/2+s)$$

$$= \frac{1}{2\pi} \left[\int\limits_{c-i\infty}^{c+i\infty} - \int\limits_{-c-i\infty}^{-c+i\infty} \right] \hat{f}(s)\, d\log \zeta_{\mathbb{A}}(1/2+s)$$

for any $c > 1/2$. The explicit sum is the equality

(1.10)
$$\sum_{\zeta_{\mathbb{A}}(1/2+s)=0} \hat{f}(s) - \hat{f}(1/2) - \hat{f}(-1/2)$$

$$= \frac{1}{2\pi} \left[\int_{c-i\infty}^{c+i\infty} - \int_{-c-i\infty}^{-c+i\infty} \right] \hat{f}(s) \, d\log\zeta_{\mathbb{A}}(1/2+s) \qquad \text{for any } c > 1/2,$$

$$= \frac{1}{2\pi} \int_{c-i\infty}^{c+i\infty} [\hat{f}(s) - \hat{f}(-s)] \, d\log\zeta_{\mathbb{A}}(1/2+s), \qquad \text{by (1.7)}$$

$$= \sum_{p\geq\eta} \frac{1}{2\pi} \int_{c-i\infty}^{c+i\infty} [\hat{f}(s) - \hat{f}(-s)] \, d\log\zeta_p(1/2+s), \quad \text{by (1.1)}$$

$$= \sum_{p\geq\eta} \frac{1}{2\pi} \int_{c-i\infty}^{c+i\infty} \hat{f}(s) \, d\log\frac{\zeta_p(1/2+s)}{\zeta_p(1/2-s)} \qquad \text{for any } c \in (-1/2, 1/2),$$

$$\stackrel{\text{def}}{=} -\sum_{p\geq\eta} \mathcal{W}_p(f)$$

where the Weil distribution $\mathcal{W}_p(f)$ is the inverse Mellin transform of $d\log\frac{\zeta_p(1/2-s)}{\zeta_p(1/2+s)}$ (note the minus sign at the end of (1.10)).

For a finite prime $p \neq \eta$, we have

(1.11)
$$\mathcal{W}_p(f) = \frac{1}{2\pi} \int_{-i\infty}^{i\infty} \hat{f}(s) \, d\log\frac{\zeta_p(1/2-s)}{\zeta_p(1/2+s)}$$

$$= \frac{1}{2\pi} \int_{-i\infty}^{i\infty} \hat{f}(s) \log p \left[\sum_{n\geq1} p^{-n(1/2+s)} + \sum_{n\geq1} p^{-n(1/2-s)} \right] ds$$

$$= \log p \cdot \sum_{\substack{n\in\mathbb{Z}\\n\neq0}} p^{-|n|/2} \cdot \frac{1}{2\pi} \int_{-i\infty}^{i\infty} \hat{f}(s) p^{ns} \, ds$$

$$= \log p \cdot \sum_{n\neq0} p^{-|n|/2} f(p^n) \qquad\qquad \text{by (1.9)}$$

Note that we have for $p \neq \eta$,

(1.12)
$$f \geq 0 \Rightarrow \mathcal{W}_p(f) \geq 0.$$

Ignoring the contribution of the real prime $\mathcal{W}_\eta(f)$, and the contribution of the poles $\hat{f}(\pm1/2)$, we see that (1.10) says that the sum of $\hat{f}(s)$ over the zeros of $\zeta_{\mathbb{A}}\left(\frac{1}{2}+s\right)$ is "equal" to the (weighted) sum of f over the prime powers. But the contribution of the real prime is very important. It can be written in a finite closed form in various ways [H90], for example (see [H01]):

(1.13) $\mathcal{W}_\eta(f) = \displaystyle\int_0^\infty \frac{f(a) - f(1)}{1 - \min\{a, a^{-1}\}^2} \cdot \min\{a, a^{-1}\}^{1/2} \, d^*a + (\gamma + \pi/2 + \log 8\pi) f(1).$

Note that we have only

(1.14) $f \geq 0$ and $f(1) = 0 \Rightarrow \mathcal{W}_\eta(f) \geq 0.$

If in the explicit sum (1.10) we replace the function $f(a)$ by the function

(1.15) $f * f^*(a) = \int d^*a_0 f(a_0) \cdot \overline{f(a_0 \cdot a^{-1})}$

whose Mellin transform is

(1.16) $(\widehat{f * f^*})(s) = \hat{f}(s) \cdot \overline{\hat{f}(-\bar{s})},$

we get

(1.17) $\sum_{\zeta_\mathbb{A}(\frac{1}{2}+s)=0} \hat{f}(s)\overline{\hat{f}(-\bar{s})} = 2\mathrm{Re}\,\hat{f}\left(\frac{1}{2}\right)\overline{\hat{f}\left(-\frac{1}{2}\right)} - \sum_{p \geq \eta} \mathcal{W}_p(f * f^*).$

Note that if RH holds, the sum (1.17) is positive. Conversely, since we have enough functions $\hat{f}(s)$ to localize the zeros of $\zeta_\mathbb{A}\left(\frac{1}{2} + s\right)$, the RH is equivalent to the positivity of (1.17). We can even assume without loss of generality that $\hat{f}\left(\pm\frac{1}{2}\right) = 0$, and then RH is equivalent to the positivity of

(1.18) $-\sum_{p \geq \eta} \mathcal{W}_p(f * f^*) \geq 0.$

Note that we cannot ignore any prime in the sum (1.18): for each $p_0 \geq \eta$ there is a function $f = f_{p_0}$ such that the sum in (1.18) with $\mathcal{W}_{p_0}(f * f^*)$ omitted is negative ($\zeta_\mathbb{A}(s)/\zeta_{p_0}(s)$ has "trivial" zeros off the line). On the other hand, for a given function f of compact support$\subseteq (c^{-1}, c)$, the sum (1.18) is finite, containing only primes $p < c^2$ (and the real prime η, which always appears).

Note that taking $f \geq 0$, $f * f^* \geq 0$ and (1.18) seems to contradict (1.12), (1.14), but in this case $\hat{f}\left(\pm\frac{1}{2}\right) > 0$ and $f * f^*(1) = \|f\|^2 > 0$.

2. Reformulation as normalized trace

For each finite prime $p \neq \eta$ we have the completion of the field of rational numbers $\mathbb{Q}$ with respect to the p-adic absolute value, the p-adic numbers $\mathbb{Q}_p$, and for $p = \eta$ we have the reals $\mathbb{R} = \mathbb{Q}_\eta$. We are interested in the absolute values

(2.1)
$$| \,|_p \colon \mathbb{Q}_p^*/\mathbb{Z}_p^* \xrightarrow{\sim} p^\mathbb{Z}, \quad p \neq \eta,$$

$$| \,|_\eta \colon \mathbb{Q}_\eta^*/\mathbb{Z}_\eta^* = \mathbb{R}^*/\pm 1 \xrightarrow{\sim} \mathbb{R}^+,$$

and with $f \in C_c^\infty(\mathbb{R}^+)$ we associate a function on $\mathbb{Q}_p$, $p \geq \eta$, by the rule $f|_p(x) = f(|x|_p)$. The contribution to the explicit sum $\mathcal{W}_p(f)$, for $p \geq \eta$, can be written as a *local formula* [H90],

(2.2)
$$\mathcal{W}_p(f) = \mathcal{F}_p(-\log|x|_p)\mathcal{F}_p^{-1}(a^{-1/2} \cdot f)|_p(1)$$
$$= \left.\frac{\partial}{\partial s}\right|_{s=0} \mathcal{F}_p|x|_p^{-s}\mathcal{F}_p^{-1}(a^{-1/2} \cdot f)|_p(1).$$

Here $\mathcal{F}_p$ is the Fourier transform on $\mathbb{Q}_p$,

$$\mathcal{F}_p f(y) = \int f(x)\psi_p(xy)\,dx,$$

$\psi_p(x)$ the "canonical" additive character of $\mathbb{Q}_p$. For $\mathrm{Re}\,(s) > 0$ we have

$$(2.3) \qquad \mathcal{F}_p |x|_p^{-s} \mathcal{F}_p^{-1}(a^{-1/2} \cdot f)|_p(1) = \frac{\zeta_p(1-s)}{\zeta_p(s)} \int_{\mathbb{Q}_p} dx |1-x|_p^{s-1} |x|_p^{-1/2} f(|x|_p)$$

the Riesz potential.

Formula (2.2) is not natural: we apply the additive Fourier transform $\mathcal{F}_p$ to the multiplicative function $f|_p$ (alternatively, we apply additive convolution in (2.3) to $f|_p$). The advantage of (2.2) is that one can easily translate it into a natural formula involving a trace, indeed in more than one way. We go on to describe one such translation.

We begin with the case of a finite prime $p \neq \eta$, where we have the additive Haar measure dx on $\mathbb{Q}_p$, normalized by $dx(\mathbb{Z}_p) = 1$, and the multiplicative Haar measure $d^*x = \zeta_p(1)\frac{dx}{|x|_p}$. We shall actually be interested only in the space $\mathbb{Q}_p / \mathbb{Z}_p^* \simeq p^{\mathbb{Z}} \cup \{0\}$, with the induced measures

$$(2.4)_p \qquad\qquad dx(p^n) = p^{-n}(1 - p^{-1}), \quad d^*x(p^n) = 1,$$

and the associated Hilbert space

$$\mathcal{H}_p = L_2(\mathbb{Q}_p, dx)^{\mathbb{Z}_p^*} = \ell_2(p^{\mathbb{Z}}, dx).$$

We shall also be interested in the deformation given by the Tate measure for $\beta > 0$:

$$(2.5)_p \qquad\qquad \tau_p^\beta(d^*x) = |x|_p^\beta \frac{d^*x}{\zeta_p(\beta)}, \quad \tau_p^\beta(p^n) = p^{-n\beta}(1 - p^{-\beta}),$$

$$\mathcal{H}_p^\beta = L_2(\mathbb{Q}_p, \tau_p^\beta)^{\mathbb{Z}_p^*} = \ell_2(p^{\mathbb{Z}}, \tau_p^\beta).$$

At $\beta = 1$, $\tau_p^1 = dx$ is the additive measure.

We have a unitary action π_p^β of $p^{\mathbb{Z}} = \mathbb{Q}_p^* / \mathbb{Z}_p^*$ on $\mathcal{H}_p^\beta$, and the characteristic function $\phi_{\mathbb{Z}_p}$ of $p^{\mathbb{N}} \cup \{0\}$ is a cyclic vector, giving rise to an isomorphism

$$(2.6)_p \qquad \tau_p \colon \mathcal{H}_p^\beta \xrightarrow{\sim} \hat{\mathcal{H}}_p^\beta \overset{\mathrm{def}}{=} L_2 \left(\mathbb{R} \Big/ \frac{2\pi}{\log p} \mathbb{Z}, \left| \zeta_p\left(\frac{\beta}{2} + it\right) \right|^2 dt \cdot c_p^\beta \right),$$

$$c_p^\beta = \frac{\log p}{2\pi}(1 - p^{-\beta}),$$

$$\varphi \mapsto \tau_p^{\beta/2+it} \varphi = \int \varphi(x) |x|_p^{\beta/2+it} \frac{d^*x}{\zeta_p(\beta/2+it)}$$

$$\pi_p^\beta(p^n)\varphi(x) = p^{n\beta/2}\varphi(p^{-n}x) \longleftrightarrow \pi_p^\beta(p^n)\tau_p^{\beta/2+it}\varphi = p^{-nit} \cdot \tau_p^{\beta/2+it}\varphi.$$

We get the following $2N + 1$-dimensional subspace of $\mathcal{H}_p^\beta$, that can be characterized in the following equivalent ways,

$(2.7)_p$

$$\begin{aligned} \mathcal{H}_p^\beta(N) &= \left\{ \varphi \in \mathcal{H}_p^\beta, \mathrm{supp}\,\varphi,\ \mathrm{supp}\,\mathcal{F}_p^\beta\varphi \subseteq \{x,\, |x|_p \leq p^N\} \right\} && \text{``additive''} \\ &= \left\{ \varphi \in \mathcal{H}_p^\beta,\ \tau_p^{\beta/2+it}\varphi \in \oplus_{|n|\leq N}\, \mathbb{C}\, p^{nit} \right\} && \text{``analytic''} \\ &= \oplus_{|n|\leq N}\, \mathbb{C}\, \pi_p^\beta(p^n)\phi_{\mathbb{Z}_p} && \text{``multiplicative''}. \end{aligned}$$

We can apply Gram-Schmidt to $\{1, p^{nit}+p^{-nit},\ \ n \geq 1\}$ and $\{p^{nit}-p^{-nit},\ \ n \geq 1\}$ to obtain an orthogonal basis for $\hat{\mathcal{H}}_p^\beta$, and then translate this to a basis of $\mathcal{H}_p^\beta$, getting $\{\phi_n^\beta\}_{n\in\mathbb{Z}}$, with $\phi_0^\beta = \phi_{\mathbb{Z}_p}$ the cyclic vector, and for $n \geq 1$,

$$(2.8)_p \qquad\qquad \phi_n^\beta = (1 + \mathcal{F}_p^\beta)\delta_{p^{-n}}, \quad \phi_{-n}^\beta = (1 - \mathcal{F}_p^\beta)\delta_{p^{-n}},$$

where the p-adic Fourier-Bessel transform $\mathcal{F}_p^\beta$ is given explicitly by

$$\mathcal{F}_p^\beta \delta_{p^{-n}}(p^k) = \begin{cases} p^{n\beta}(1 - p^{-\beta}) & k \geq n \\ -p^{(n-1)\beta} & k = n - 1 \\ 0 & k < n - 1 \end{cases}$$

We have now the orthogonal direct sums

$$(2.9)_p \qquad \mathcal{H}_p^\beta = \underset{n \in \mathbb{Z}}{\oplus} \mathbb{C}\,\phi_n^\beta, \quad \mathcal{H}_p^\beta(N) = \underset{|n| \leq N}{\oplus} \mathbb{C}\,\phi_n^\beta.$$

At $\beta = 1$, $\mathcal{F}_p^1$ is the Fourier transform, and we have a natural isomorphism $\mathcal{H}^1(N) \simeq \ell_2\left(\frac{1}{p^N}\mathbb{Z}/p^N\,\mathbb{Z}\right)^{(\mathbb{Z}/p^{2n}\,\mathbb{Z})^*}$.

We define the number operator at $\beta = 1$ by

$$(2.10)_p \qquad \mathbb{N}\phi_n = (\log p)|n| \cdot \phi_n,$$

and we have the associated Heat semigroup acting on $\mathcal{H}_p$,

$$(2.11)_p \qquad e^{-t\mathbb{N}_p}\,\phi_n = p^{-t|n|}\phi_n.$$

We have now the local formula via Heat kernel, [H93], $p \neq \eta$:

$$(2.12)_p \qquad \frac{\zeta(2t)}{\zeta(t)^2}\mathrm{tr}\left(e^{-t\mathbb{N}_p}\,\pi_p(f|_p)\right) = f(1) + \frac{1 - p^{-t}}{1 + p^{-t}}\sum_{n \neq 0} f(p^n)p^{-|n|/2}$$

$$= f(1) + \frac{t}{2}\mathcal{W}_p(f) + O(t^2) \quad \text{as } t \to 0.$$

Let us turn to the description of the real analogue. We let dx denote the usual Haar measure on $\mathbb{Q}_\eta = \mathbb{R}$, d^*x the multiplicative measure, and we consider the Hilbert space

$$(2.4)_\eta \qquad \mathcal{H}_\eta = L_2(\mathbb{R}, dx)^{\pm 1} \quad \text{(even functions)}.$$

We also have the β-deformation given by

$$(2.5)_\eta \qquad \begin{aligned} \tau_\eta^\beta(dx) &= |x|_\eta^\beta \frac{d^*x}{\zeta_\eta(\beta)} \\ \mathcal{H}_\eta^\beta &= L_2(\mathbb{R}, \tau_\eta^\beta(dx))^{\pm 1} \end{aligned}$$

We have a unitary action π_η^β of $\mathbb{R}^+ = \mathbb{R}^*/\pm 1$ on $\mathcal{H}_\eta^\beta$, and $\phi_{\mathbb{Z}_\eta}(x) = e^{-\pi x^2}$ is a cyclic vector, giving rise to an isomorphism

$$(2.6)_\eta \qquad \begin{aligned} \tau_\eta \colon \mathcal{H}_\eta^\beta &\overset{\sim}{\to} \hat{\mathcal{H}}_\eta^\beta = L_2\left(\mathbb{R}, \left|\zeta_\eta\left(\frac{\beta}{2} + it\right)\right|^2 dt \cdot c_\eta^\beta\right), \quad c_\eta^\beta = \frac{1}{4\pi}\frac{1}{\zeta_\eta(\beta)}, \\ \varphi &\mapsto \tau_\eta^{\beta/2+it}\varphi = \int \varphi(x)|x|_\eta^{\beta/2+it}\frac{d^*x}{\zeta_\eta(\beta/2 + it)}, \\ \pi_\eta^\beta(a)\varphi(x) &= a^{-\beta/2}\varphi(a^{-1}x) \longleftrightarrow \pi_\eta^\beta(a)\tau_\eta^{\beta/2+it}\varphi = a^{it} \cdot \tau_\eta^{\beta/2+it}\varphi. \end{aligned}$$

We have the following finite dimensional subspace of $\mathcal{H}_\eta^\beta$,

$$(2.7)_\eta \qquad \begin{aligned} \mathcal{H}_\eta^\beta(N) &= \{\varphi \in \mathcal{H}_\eta^\beta(N), \ \tau_\eta^{\beta/2+it}\varphi \in \underset{0 \leq n \leq N}{\oplus} \mathbb{C} \cdot t^n\} \quad \text{``analytic''} \\ &= \oplus_{0 \leq n \leq N} \mathbb{C}\left(\frac{\beta}{2} + x\frac{\partial}{\partial x}\right)^n \phi_{\mathbb{Z}_\eta}(x) \qquad \text{``multiplicative''}. \end{aligned}$$

Note that $\frac{\beta}{2} + x\frac{\partial}{\partial x}$ it the infinitesimal generator of the unitary action π_η^β of $\mathbb{R}^+$, and note that we do not have an additive characterization as in $(2.7)_p$. We can orthogonalize the polynomials $\{s^n\}$, and then translate it to an orthogonal basis $\{\phi_n^\beta\}_{n \geq 0}$ of $\mathcal{H}_\eta^\beta$. At $\beta = 1$ we get the (even) Hermite functions

$(2.8)_\eta$ $$\phi_n^1(x) = \frac{(2n)!}{n!4^n} e^{-\pi x^2} \sum_{0 \leq j \leq n} \frac{(-8\pi x^2)^j}{(2j)!(n-j)!}, \quad \mathcal{F}_\eta^1 \phi_n = (-1)^n \phi_n^1.$$

For general $\beta > 0$ we get $\phi_n^\beta = L_n^{\beta/2 - 1}(2\pi x^2)e^{-\pi x^2}$, $L_n^{\beta/2 - 1}(x)$ the Laguerre polynomials, see H[01]. We have the orthogonal direct sums

$(2.9)_\eta$ $$\mathcal{H}_\eta^\beta = \bigoplus_{n \geq 0} \mathbb{C}\,\phi_n^\beta, \quad \mathcal{H}_\eta^\beta(N) = \bigoplus_{0 \leq n \leq N} \mathbb{C}\,\phi_n^\beta.$$

The number operator is defined at $\beta = 1$ by

$(2.10)_\eta$ $$\mathbb{N}_\eta \phi_n = \frac{1}{2}\log(2n+1) \cdot \phi_n$$

and the associated Heat semigroup acts on $\mathcal{H}_\eta = \mathcal{H}_\eta^1$

$(2.11)_\eta$ $$e^{-t\mathbb{N}_\eta}\,\phi_n = (2n+1)^{-t/2}\phi_n.$$

We can write the local formula via the Heat kernel, [H93],
$(2.12)_\eta$
$$\frac{\zeta_\eta(2t)}{\zeta_\eta(t)^2}\operatorname{tr}\left(e^{-t\mathbb{N}_\eta}\,\pi_\eta(f|_\eta)\right) = f(1) + \frac{t}{2}(\mathcal{W}_\eta(f) - f(1)\log\pi) + O(t^2), \quad \text{as } t \to 0.$$

We can put together the local formulas $(2.12)_p$, $p \geq \eta$, to obtain a global formula as will be shown. We consider the ring of adeles

(2.13) $$\mathbb{A} = \mathbb{R} \times \prod_{p \neq \eta} \mathbb{Q}_p : \mathbb{Z}_p \quad \text{(restricted direct product)},$$

it contains $\mathbb{Q}$ as a discrete cocompact subring, and its group of units is the ideles

(2.14) $$\mathbb{A}^* = \mathbb{R}^+ \times \mathbb{Q}^* \times \prod_{p \neq \eta} \mathbb{Z}_p^* = \mathbb{R}^* \times \prod_{p \neq \eta} \mathbb{Q}^* : \mathbb{Z}_p^*$$

We shall actually be interested in the simpler space

(2.15) $$A = \mathbb{A}/(\pm 1)\prod_p \mathbb{Z}_p^* = \mathbb{R}/\pm 1 \times \prod_{p \neq \eta} p^{\mathbb{Z}} \cup \{0\} : p^{\mathbb{N}} \cup \{0\},$$

and the associated Hilbert space

(2.16) $$\mathcal{H}_A = L_2(A, dx) = L_2(\mathbb{A}, dx)^{(\pm 1)\prod_p \mathbb{Z}_p^*} = \bigotimes_{p \geq \eta} \mathcal{H}_p \quad \text{w.r.t } \phi_{\mathbb{Z}_p}.$$

We also have for $\beta > 0$ the Tate measure $\tau_\mathbb{A}^\beta = \otimes_{p \geq \eta} \tau_p^\beta$, and the associated Hilbert space

$$\mathcal{H}_A^\beta = L_2(A, \tau_A^\beta) = L_2(\mathbb{A}, \tau_\mathbb{A}^\beta)^{(\pm 1)\prod_p \mathbb{Z}_p^*} = \bigotimes_{p \geq \eta} \mathcal{H}_p^\beta \quad \text{w.r.t. } \phi_{\mathbb{Z}_p}.$$

On $\mathcal{H}_A^\beta$ we have the unitary action π_η^β of $\mathbb{R}^+$, as well as the unitary action π^β of $\mathbb{Q}^+ = \oplus_p p^{\mathbb{Z}}$; but note that $\mathbb{Q}^+$ acts on both the p-adic and real variables preserving the product measure $\tau_\mathbb{A}^\beta = \otimes_{p \geq \eta}\tau_p^\beta$. We have an orthogonal basis for $\mathcal{H}_A$ given by

(2.17) $$\phi_{\vec{n}} = \bigotimes_{p \geq \eta} \phi_{n_p}, \quad \vec{n} = \{n_p\}, \quad n_p \in \mathbb{Z}, \quad n_\eta \in \mathbb{N},$$

$$n_p = 0 \text{ for all but finitely many } p\text{'s},$$

the associated number operator and Heat semigroup are

(2.18)
$$\mathbb{N}_A = \sum_{p \geq \eta} \mathbb{N}_p, \quad e^{-t\mathbb{N}_A} \underset{p \geq \eta}{\otimes} e^{-t\mathbb{N}_p},$$

$$e^{-t\mathbb{N}_A} \phi_{\vec{n}} = (2n_\eta + 1)^{-t/2} \prod_{p \neq \eta} p^{-t|n|_p} \cdot \phi_{\vec{n}}.$$

We can put together the loal formulas $(2.12)_p$, $p \geq \eta$, into a global formula [H91],

(2.19)
$$-\sum_{p \geq \eta} \mathcal{W}_p(f) = \sum_{q \in \mathbb{Q}^+} \underset{t=0}{\mathrm{CT}} \, 2\pi^{t/2} \cdot \mathrm{tr}\left(e^{-t\mathbb{N}_A} \, \pi_\eta(f)\pi(q)\right).$$

Here $\underset{t=0}{\mathrm{CT}}$ is the constant term in the expansion around $t = 0$, it actually vanishes unless q is a prime power; the sum over $q = p^n$, $n \neq 0$, gives the p-adic contribution $-\mathcal{W}_p(f)$, and for $q = 1$ we get the real contribution $-\mathcal{W}_\eta(f)$. But note that the derivative $-\frac{\partial}{\partial t}\big|_{t=0}$ of the local formulas $(2.12)_p$ now become the constant term in (2.19), and the sign in (2.19), both of which come out of

$$\frac{\zeta_A(2t)}{\zeta_A(t)^2} = -\frac{t}{2}(1 + O(t^2)), \quad \text{as } t \to 0.$$

The RH is thus equivalent to the positivity, [H91]
(2.20)
$$\sum_{q \in \mathbb{Q}^+} \underset{t=0}{\mathrm{CT}} \, 2\pi^{t/2} \cdot \mathrm{tr}\left(e^{-t\mathbb{N}_A} \, \pi_\eta(f)\pi_\eta(f)^*\pi(q)\right) \geq 0, \quad f \in C_c^\infty(\mathbb{R}^+), \quad \hat{f}\left(\pm\frac{1}{2}\right) = 0.$$

The above global formula already suggests that the zeros of $\zeta_A(s)$ are connected to the action of $\mathbb{R}^+$ (via $\pi_\eta(f)$) on the "noncommutative" space $A/\mathbb{Q}^+ = \mathbb{A}/\mathbb{A}^{(1)}$. This space has a beautiful structure. For a set of primes S, let $x_S \in A$ be given by $x_{S,p} = 1$ (resp. 0) for $p \notin S$ (resp. $p \in S$) and let $\bar{x}_S = x_S \cdot \mathbb{Q}^+$ denote the associated orbit. The space $A/\mathbb{Q}^+$ contains the dense $\mathbb{R}^+$ orbit, $\mathbb{R}^+ \cdot \bar{x}_\phi = \mathbb{R}^+ \cdot \mathbf{1}$, and as $t \in \mathbb{R}^+$ approached 0, $t \cdot \mathbf{1} \in \mathbb{R}^+ \cdot \mathbf{1}$ approaches simultaneously the circles $t \cdot \bar{x}_{\{p\}} \in \mathbb{R}^+ \cdot \bar{x}_{\{p\}} \simeq \mathbb{R}^+/p^\mathbb{Z}$, $p \neq \eta$; these circles are "glued together" by the noncommutative spaces $\mathbb{R}^+ \cdot \bar{x}_{\{p_1,\ldots,p_\ell\}} \simeq \mathbb{R}^+/p_1^\mathbb{Z} \cdots p_\ell^\mathbb{Z}$. But all these constitute a subset of $A/\mathbb{Q}^+$ of measure zero — there is still the "dust" consisting of $x \cdot \mathbb{Q}^+$ with $|x_p| < 1$ for infinitely many p's, and this dust has full measure! When we consider A with the Tate measure τ_A^β, for $\beta \in (0,1]$ the action of $\mathbb{Q}^+$ is ergodic. But for $\beta > 1$ the measure τ_A^β is supported on the ideles, $\tau_A^\beta(dx) = |x|_A^\beta \cdot \frac{d^*x}{\zeta_A(\beta)}$, the dense orbit has full measure, the "dust" gets measure zero.

3. The prolate spheroidal wave functions

Let B_c denote the function

(3.1)
$$B_c(x) = \begin{cases} 1 & |x|_p \leq c \\ 0 & |x|_p > c \end{cases}$$

and view B_c as a projection acting on $\mathcal{H}_p$ by multiplication, and let

(3.2)
$$\hat{B}_c = \mathcal{F}_p B_c \mathcal{F}_p^{-1}$$

denote the dual projection on $\mathcal{H}_p$. For $p \neq \eta$, and $c = p^n$, $n \geq 0$, the projections B_c and $\hat{B}_c$ commute, and by the additive characterization $(2.7)_p$ we have

(3.3)
$$B_{p^N} \hat{B}_{p^N} \mathcal{H}_p = B_{p^N} \mathcal{H}_p \cap \hat{B}_{p^N} \mathcal{H}_p = \mathcal{H}_p(N).$$

For the real prime η, the projections B_c and $\hat{B}_c$ never commute, and we do not have an additive characterization in $(2.7)_p$: for $\varphi \in B_c \mathcal{H}_\eta$, $\mathcal{F}_\eta \varphi$ is an entire analytic function, and so $\mathcal{F}_\eta \varphi$ is never supported in $[-c, c]$, thus

$$(3.4) \qquad\qquad B_c \mathcal{H}_\eta \cap \hat{B}_c \mathcal{H}_\eta = \{0\}.$$

This problem was analyzed by Landau, Pollak, Slepian [LPS], enabling the definition of a subspace $B^{\langle c \rangle} \mathcal{H}_\eta \subseteq B_c \mathcal{H}_\eta$ consisting of functions φ supported in $[-c, c]$, and such that $\mathcal{F}_\eta \varphi$ is "essentially" supported in $[-c, c]$, thus $B^{\langle c \rangle} \mathcal{H}_\eta$ replaces the empty intersection in (3.4).

Consider the operator $B_c \hat{B}_c$ acting on the space $B_c \tilde{\mathcal{H}}_\eta = L_2([-c, c], dx)$ (even and odd functions). It is an integral operator given by the kernel

$$(3.5) \qquad\qquad p_c(x, y) = \frac{\sin 2\pi c(x - y)}{\pi(x - y)}.$$

We get an orthogonal basis consisting of eigenfunctions of $B_c \hat{B}_c$,

$$(3.6) \qquad\qquad B_c \tilde{\mathcal{H}}_\eta = \bigoplus_{n \geq 0} \mathbb{C} \cdot \varphi_n^c, \quad B_c \hat{B}_c \varphi_n^c = \lambda_n^c \cdot \varphi_n^c,$$

where $1 > \lambda_0^c > \lambda_1^c > \lambda_2^c > \ldots > 0$.

It follows that we get an orthogonal basis for $\hat{B}_c \tilde{\mathcal{H}}_\eta$ by applying $\hat{B}_c$ to the base φ_n^c,

$$(3.7) \qquad\qquad \hat{B}_c \tilde{\mathcal{H}}_\eta = \bigoplus_{n \geq 0} \mathbb{C} \cdot \hat{B}_c \varphi_n^c,$$

and

$$\|B_c \mathcal{F}_\eta \varphi_n^c\|^2 = \|\hat{B}_c \varphi_n^c\|^2 = (\hat{B}_c \varphi_n^c, \hat{B}_c \varphi_n^c) = (\hat{B}_c \varphi_n^c, B_c \varphi_n^c) = \lambda_n^c \cdot \|\varphi_n^c\|^2.$$

Thus for λ_n^c near 1, most of the energy of $\mathcal{F}_\eta \varphi_n^c$ is concentrated in $[-c, c]$, and for λ_n^c near 0, most of the energy of $\mathcal{F}_\eta \varphi_n^c$ is concentrated outside $[-c, c]$. The remarkable discovery of [LPS] is that the λ_n^c are near 1 for $n \leq 4c^2$, then they decrease sharply to near 0 for $n \geq 4c^2 + O(\log c)$. The easiest way of seeing this is to note that

$$(3.8) \qquad\qquad \sum_{n \geq 0} \lambda_n^c = \mathrm{tr}\,(B_c \hat{B}_c) = \int_{-c}^{c} dx\, p_c(x, x) = 4c^2,$$

and

$$(3.9) \qquad \sum_{n \geq 0} (\lambda_n^c)^2 = \mathrm{tr}\,((B_c \hat{B}_c)^2) = \iint_{-c}^{c} dx\, dy\, p_c(x, y)^2 \geq 4c^2 - \frac{\log \pi c^2}{\pi^2} - 1.$$

It follows from the equality

$$(3.10) \qquad\qquad \lambda_N \leq \frac{\sum_n \lambda_n - \sum_n \lambda_n^2}{N - \sum_n \lambda_n}$$

(valid for $N > \sum_n \lambda_n$, and any $1 \geq \lambda_0 \geq \lambda_1 \geq \ldots \geq 0$), that

$$(3.11) \qquad\qquad \lambda_N^c \leq \left(\frac{\log \pi c^2}{\pi^2} + 1 \right) \Big/ (N - 4c^2) \quad \text{for } N > 4c^2.$$

Thus for $N \geq 4c^2 + \frac{1}{\epsilon} \left(\frac{\log \pi c^2}{\pi^2} + 1 \right)$ we have $\lambda_N^c \leq \epsilon$.

An important observation of [LPS] is that the differential operator

$$(3.12) \qquad H_c = -\frac{\partial}{\partial x}(c^2 - x^2)\frac{\partial}{\partial x} + (2\pi c x)^2$$

$$= 2\pi c^2\left(2\pi x^2 + \frac{1}{2\pi}\frac{\partial^2}{\partial x^2}\right) + \frac{\partial}{\partial x}x^2\frac{\partial}{\partial x}$$

$$= 2\pi c^2 H_\infty + H_0$$

satisfies $\mathcal{F}_\eta H_c = H_c \mathcal{F}_\eta$ (indeed, this is true for $c = 0, \infty$) and $B_c H_c = H_c B_c$ (indeed, B_c commutes with the multiplication operator $(2\pi c x)^2$, and it commutes with $\frac{\partial}{\partial x}(c^2 - x^2)\frac{\partial}{\partial x}$ by integration by parts using the vanishing of $c^2 - x^2$ at the end points $x = \pm c$). It follows that H_c commutes with $\hat{B}_c = \mathcal{F}_\eta B_c \mathcal{F}_\eta^{-1}$, and with the integral operator $B_c \hat{B}_c$, hence the φ_n^c are also eigenfunctions of H_c,

$$(3.13) \qquad H_c \varphi_n^c = \chi_n^c \cdot \varphi_n^c$$

and actually $0 < \chi_0^c < \chi_1^c < \chi_2^c < \ldots$, thus the φ_n^c are the prolate spheroidal wave functions, [F].

It follows that φ_n^c has n zeros in the interval $[-c, c]$, and is an even (resp. odd) function when n is even (resp. odd). We write $\phi_n^{1,c} = \varphi_{2n}^c$ for the even functions. Let $B^{\langle c \rangle}$ denote the projection on the $2c^2 + 1$ dimensional subspace of $\mathcal{H}_\eta$

$$(3.14) \qquad B^{\langle c \rangle}\mathcal{H}_\eta = \bigoplus_{0 \le n \le 2c^2} \mathbb{C}\,\phi_n^{1,c}.$$

This is the substitute for the empty intersection (3.4).

Note that for c large, H_c is essentially $2\pi c^2 H_\infty$, and the eigenfunctions of H_∞ are just the Hermite functions ϕ_n of $(2.8)_\eta$,

$$(3.15) \qquad H_\infty \phi_n = (4n + 1)\phi_n.$$

It follows that $\phi_n^{1,c}$ converge for $c \to \infty$ to a scalar multiple of ϕ_n, and we can normalize the $\phi_n^{1,c}$ by

$$(3.16) \qquad (\phi_n^{1,c}, \phi_n) = \|\phi_n\|^2, \quad \text{so } \phi_n^{1,c} \xrightarrow[c \to \infty]{} \phi_n.$$

4. Reformulation as asymptotic trace

The maps $|\ |_p \colon \mathbb{Q}_p^* \to \mathbb{R}^+$ have compact kernel, and a natural normalization of the Haar measure on $\mathbb{Q}_p^*$, $p \ge \eta$, is the measure $d^\circ x$ characterized by

$$(4.1) \qquad \int_{1 \le |x|_p \le c} d^\circ x \sim \log c, \quad \text{as } c \to \infty.$$

The measure $d^\circ x$ is different from $d^* x = \zeta_p(1)\frac{dx}{|x|_p}$, we have

$$(4.2) \qquad d^\circ x = n_p \cdot \frac{dx}{|x|_p}, \quad \text{with } n_p = \frac{\partial}{\partial s}\bigg|_{s=0} \frac{\zeta_p(1 - s)}{\zeta_p(s)} = \begin{cases} \dfrac{\log p}{1 - p^{-1}} & p \ne \eta \\[2mm] \dfrac{1}{2} & p = \eta. \end{cases}$$

For a function $f \in C_c^\infty(\mathbb{R}^+)$, such that $f(1) = 0$, we can write the local formula (2.2), (2.3), as

$$(4.3) \qquad \begin{aligned} \mathcal{W}_p(f) &= \left(\frac{\partial}{\partial s}\bigg|_{s=0} \frac{\zeta_p(1 - s)}{\zeta_p(s)}\right) \cdot \int (a^{-1/2} \cdot f)|_p(x)\frac{dx}{|1 - x|_p} \\ &= \int f(|x|_p)|x|_p^{1/2}\frac{d^\circ x}{|1 - x|_p}. \end{aligned}$$

Let $\pi_p^0(f)\varphi(x) = \int d^\circ a\, f(|a|_p)|a|_p^{1/2}\varphi(ax)$ denote the integrated unitary action π_p with respect to the measure $d^\circ a$, and consider the operator $\hat{B}_c B_c \pi_p^0(f)$ acting on $\mathcal{H}_p$: it is an integral operator with kernel

$$(4.4) \qquad \int d^\circ a\, f(|a|_p)|a|_p^{-1/2} B_c(a^{-1}x)\mathcal{F}_p B_c(y - a^{-1}x).$$

Its trace is given by setting $y = x$ and integrating with respect to x,

(4.5)

$$\mathrm{tr}\left(\hat{B}_c B_c \pi_p^0(f)\right) = \int dx \int d^\circ a\, f(|a|_p)|a|_p^{-1/2} B_c(a^{-1}x) \int dz\, B_c(z)\psi_p((1 - a^{-1})xz)$$

$$(x := a \cdot x) = \int dx\, B_c(x) \int dz\, B_c(z) \int d^\circ a\, f(|a|_p)|a|_p^{1/2}\psi_p((a - 1)xz)$$

$$\left(z := \frac{z}{x}\right) = \int dz \int d^\circ x\, B_c(x)B_c\left(\frac{z}{x}\right) \int da\, f(|a|_p)|a|_p^{-1/2}\psi_p((a - 1)z)$$

$$= \int dz\, (B_{c^2}(z)\log c^2/|z|_p) \int da\, f(|a|_p)|a|_p^{-1/2}\psi_p((a - 1)z)$$

$$= \mathcal{F}_p^{-1}(B_{c^2}(z)\log c^2/|z|_p)\mathcal{F}_p(a^{-1/2} \cdot f)|_p(1).$$

Using the local formula (2.2) we have

$$\mathrm{tr}\left(\hat{B}_c B_c \pi_p^0(f)\right) = \mathcal{F}_p^{-1}(\log c^2/|z|)\mathcal{F}_p(a^{-1/2} \cdot f)|_p(1)$$

(4.6)

$$- \mathcal{F}_p^{-1}(1 - B_{c^2}(z))\log\frac{c^2}{|z|} \cdot \mathcal{F}_p(a^{-1/2} \cdot f)|_p(1)$$

$$= 2\log c \cdot f(1) + W_p(f) - \mathcal{R}_p^c$$

where

(4.7)

$$\mathcal{R}_p^c = \mathcal{F}_p^{-1}(1 - B_{c^2}(z))\log\frac{c^2}{|z|} \cdot \mathcal{F}_p(a^{-1/2} \cdot f)|_p(1)$$

$$= \int_{|z|_p > c^2} dz\, \log(c^2/|z|_p) \cdot \psi(-z) \cdot \mathcal{F}_p(a^{-1/2} \cdot f)|_p(z)$$

decreases to 0 as $c \to \infty$; in fact for $p \neq \eta$ it vanishes for c^2 greater than the support of $\mathcal{F}_p(a^{-1/2} \cdot f)|_p$. Thus we get the local formula in asymptotic form,

$$(4.8) \qquad \mathrm{tr}\,(B_c\hat{B}_c \pi_p^0(f)) = 2\log c \cdot f(1) + \mathcal{W}_p(f) + o(1), \quad \text{as } c \to \infty, \quad p \geq \eta.$$

Let us turn next to the global theory of $A/\mathbb{Q}^+$. Consider the Schwartz space

$$(4.9) \qquad \boldsymbol{S}(A) = \boldsymbol{S}(\mathbb{A})^{(\pm 1)\prod_p \mathbb{Z}_p^*} = \bigotimes_{p \geq \eta} \boldsymbol{S}(\mathbb{Q}_p)^{\mathbb{Z}_p^*} \quad (\text{algebraic } \otimes \text{ w.r.t. } \phi_{\mathbb{Z}_p})$$

and its subspace of codimension 2

$$(4.10) \qquad \boldsymbol{S}(A)_0 = \{\varphi \in \boldsymbol{S}(A), \quad \varphi(0) = 0 = \mathcal{F}\varphi(0)\}.$$

With $\varphi \in \boldsymbol{S}(A)_0$ we associate the function on $\mathbb{R}^+$,

$$(4.11) \qquad E\varphi(a) = a^{1/2} \cdot \sum_{q \in \mathbb{Q}^+} \varphi(a \cdot q).$$

It is fast decreasing as $a \to \infty$, and by Poisson summation

$$(4.12) \qquad E\varphi(a^{-1}) = E\mathcal{F}\varphi(a)$$

so $E\varphi(a)$ is also fast decreasing as $a \to 0$.

Note that

(4.13) $$E\pi(q)\varphi = E\varphi, \quad \pi(q)\varphi(x) = \varphi(x \cdot q), \qquad q \in \mathbb{Q}^+,$$

so the $E\varphi$'s can be thought of as functions on $A/\mathbb{Q}^+$ (restricted to the dense orbit $\mathbb{R}^+ \cdot \bar{1}$). In fact the situation is very simple, since every $\varphi \in \boldsymbol{S}(A)$ can be written uniquely as finite linear combination

(4.14) $$\varphi = \sum_{1 \leq j \leq \ell} \pi(q_j)\phi_{\hat{\mathbb{Z}}} \otimes \varphi_j, \quad q_j \in \mathbb{Q}^+,$$

where $\varphi_j \in \boldsymbol{S}(\mathbb{R})^{\pm 1}$, and $\phi_{\hat{\mathbb{Z}}}$ the characteristic function of $\hat{\mathbb{Z}} = \prod_{p \neq \eta} \mathbb{Z}_p$. Thus we can restrict ourselves to real functions φ in

(4.15) $$\boldsymbol{S}(\mathbb{R})_0^{\pm 1} = \{\varphi \in \boldsymbol{S}(\mathbb{R}), \quad \varphi(-x) = \varphi(x), \quad \varphi(0) = 0 = \mathcal{F}\varphi(0)\},$$

and $E\varphi$ is given simply by

(4.16) $$E\varphi(a) = a^{1/2} \sum_{n \geq 1} \varphi(a \cdot n), \quad a \in \mathbb{R}^+.$$

We have the following picture:

(4.17)
$$\boldsymbol{S}(\mathbb{R})_0^{\pm 1} \overset{E}{\hookrightarrow} \quad L_2(\mathbb{R}^+, d^*a) \quad \simeq L_2\left(\mathbb{R}, \frac{dt}{2\pi}\right)$$
$$f(a) \qquad \longleftrightarrow \hat{f}(it),$$
$$\varphi \longmapsto \quad E\varphi(a) \qquad \longleftrightarrow \hat{E}\varphi(t) = \zeta_\mathbb{A}\left(\frac{1}{2} + it\right) \cdot \tau_\eta^{1/2+it}\varphi.$$

Note that E is injective on $\boldsymbol{S}(\mathbb{R})_0^{\pm 1}$, and has dense image (a function in $L_2\left(\mathbb{R}^+, \frac{dt}{2\pi}\right)$, orthogonal to the image will have to be a distribution supported on the zeros of $\zeta_\mathbb{A}\left(\frac{1}{2} + it\right)$, hence not an L_2 function).

A "basis" for the Schwartz space $\boldsymbol{S}(A)$ is given by $\{\phi_{\bar{n}}\}$ of (2.17), and we have

(4.18) $$\hat{E}\phi_{\bar{n}}(t) = \zeta_\mathbb{A}\left(\frac{1}{2} + it\right) \cdot \tau_\mathbb{A}^{\frac{1}{2}+it}\phi_{\bar{n}},$$

where

$$\tau_\mathbb{A}^{1/2+it}\phi_{\bar{n}} \in \mathbb{C}[t, p^{it}, p^{-it}] = \bigoplus_{q \in \mathbb{Q}^+} \mathbb{C}[t]q^{it}$$

are polynomials satisfying a functional equation

$$\tau_\mathbb{A}^{1/2-it}\phi_{\bar{n}} = \pm\tau_\mathbb{A}^{1/2+it}\phi_{\bar{n}},$$

and moreover (cf. [H93], the case $p = \eta$ is in [BN]),

(4.19) $$\tau_\mathbb{A}^s\phi_{\bar{n}} = 0 \Rightarrow \text{Re}\,(s) = \frac{1}{2}.$$

Let $B_0^{\langle c \rangle}$ denote the projection onto the codimensional 2-subspace of $B^{\langle c \rangle}\mathcal{H}_\eta$

(4.20) $$B_0^{\langle c \rangle}\mathcal{H}_\eta = \{\varphi \in \bigoplus_{0 \leq n \leq 2c^2} \mathbb{C} \cdot \phi_n^c, \quad \varphi(0) = 0 = \mathcal{F}_\eta\varphi(0)\},$$

and let $B_E^{\langle c \rangle}$ denote the projection on $L_2(\mathbb{R}^+, d^*a)$ with image the $2c^2 - 1$ dimensional subspace

(4.21) $$E[B^{\langle c \rangle}\mathcal{H}_\eta] = \{E\varphi, \quad \varphi \in B_0^{\langle c \rangle}\mathcal{H}_\eta\}.$$

For $\varphi \in B_0^{\langle c \rangle} \mathcal{H}_\eta$ we have supp $\varphi \subseteq [-c, c]$, hence supp $E\varphi \subseteq (0, c]$. Also, the Fourier transform $\mathcal{F}_\eta \varphi$ is (essentially) supported in $[-c, c]$, hence $E\varphi(a) = E\mathcal{F}_\eta \varphi(a^{-1})$ is (essentially) supported in $[c^{-1}, \infty)$, and all in all $E\varphi$ is (essentially) supported in $[c^{-1}, c]$. Letting $\mathbf{1}_{[c^{-1}, c]}$ denote the characteristic function of $[c^{-1}, c]$ viewed as a projection on $L_2(\mathbb{R}^+, d^*a)$, we have (essentially) the inequlity of projections

$$(4.22) \qquad B_E^{\langle c \rangle} \leq \mathbf{1}_{[c^{-1}, c]}.$$

Connes [98] conjecture that for $f \in C_c^\infty(\mathbb{R}^+)$, with $\hat{f}\left(\pm \frac{1}{2}\right) = 0$,

$$(4.23) \qquad \operatorname{tr}\left((\mathbf{1}_{[c^{-1}, c]} - B_E^{\langle c \rangle})\pi_\eta(f)\pi_\eta(f)^*\right) \xrightarrow[c \to \infty]{} -\sum_{p \geq \eta} \mathcal{W}_p(f * f^*).$$

Since the operator in (4.23) is (essentially) positive, the limit $c \to \infty$ of its trace is positive, and we get Weil's positivity, hence RH. Since

$$(4.24) \qquad \operatorname{tr}\left(\mathbf{1}_{[c^{-1}, c]}\pi_\eta(f)\right) = 2 \log c \cdot f(1)$$

we can formulate (4.23) also in the equivalent form

$$(4.25) \qquad \operatorname{tr}\left(B_E^{\langle c \rangle}\pi_\eta(f)\right) = 2 \log c \cdot f(1) + \sum_{p \geq \eta} \mathcal{W}_p(f) + o(1), \quad \text{as } c \to \infty.$$

5. q-series interpolating the p-adic and the reals

There is a world that interpolates between the p-adic and the reals: the world of basic hypergeometric series or q-series, quantum groups, etc. In this world we approximate $\mathbb{R}^+ = \mathbb{Q}_\eta^*/(\pm 1)$ by $q^{\mathbb{Z}} \subseteq \mathbb{R}^+$, $q \to 1$, resembling the situation over the p-adic numbers where $\mathbb{Q}_p^*/\mathbb{Z}_p^* = p^{\mathbb{Z}}$. We shall give a glimpse into this world, see [H01] for more. The q-zeta function is the inverse q-factorial

$$(5.1) \qquad \zeta_{(q)}(\alpha) = \prod_{n \geq 0}(1 - q^{\alpha + n})^{-1} = \sum_{n \geq 0} \frac{q^{\alpha n}}{(1 - q) \cdots (1 - q^n)}, \quad q \in (0, 1),$$

and the q-Beta function is

$$(5.2) \qquad \zeta_{(q)}(\alpha, \beta) = \frac{\zeta_{(q)}(\alpha)\zeta_{(q)}(\beta)}{\zeta_{(q)}(\alpha + \beta)\zeta_{(q)}(1)}.$$

We have the p-adic limit $\textcircled{p}$: $q := p^{-d}, \quad \alpha := \alpha/d, \quad \beta := \beta/d, \quad d \to \infty$

$$(5.3) \qquad \begin{aligned} \zeta_{(p^{-d})}(\alpha/d) &\xrightarrow[d \to \infty]{} \zeta_p(\alpha), \\ \zeta_{(p^{-d})}(\alpha/d, \beta/d) &\xrightarrow[d \to \infty]{} \zeta_p(\alpha, \beta) = \frac{\zeta_p(\alpha)\zeta_p(\beta)}{\zeta_p(\alpha + \beta)}, \end{aligned}$$

and we have the real limit $\textcircled{$\eta$}$: $q := q_0^{2/d}, \quad \alpha := \alpha/2, \quad \beta := \beta/2, \quad d \to \infty$,

$$(5.3)_\eta \qquad \begin{aligned} \frac{\zeta_{(q^{2/d})}(\alpha/2)}{\zeta_{(q^{2/d})}(1/2)^\alpha}\left(\frac{\zeta_{(q^{2/d})}(1)}{1 - q^{2/d}}\right)^{\alpha - 1} &\xrightarrow[d \to \infty]{} \zeta_\eta(\alpha), \\ (1 - q^{2/d})\zeta_{(q^{2/d})}(\alpha/2, \beta/2) &\xrightarrow[d \to \infty]{} \zeta_\eta(\alpha, \beta) = \frac{\zeta_\eta(\alpha)\zeta_\eta(\beta)}{\zeta_\eta(\alpha + \beta)}. \end{aligned}$$

Letting

(5.4)
$$\mathcal{W}_{(q/d)}(f) = \frac{1}{2\pi} \int_{-i\infty}^{i\infty} \hat{f}(s) \, d \log \frac{\zeta_{(q^d)}\left(\dfrac{1/2-s}{d}\right)}{\zeta_{(q^d)}\left(\dfrac{1/2+s}{d}\right)} (1-q^d)^s$$

$$= (\log q)^{-1} \sum_{n \neq 0} \frac{q^{|n|/2}}{1-q^{d|n|}} f(q^n) + f(1) \log(1-q^d),$$

we have

(5.5)
$$\lim_{\substack{\longrightarrow \\ (p)}} \mathcal{W}_{(q/d)}(f) = \mathcal{W}_{(p^{-1}/\infty)}(f) = \mathcal{W}_p(f),$$

$$\lim_{\substack{\longrightarrow \\ (\eta)}} \mathcal{W}_{(q/d)}(f) = \mathcal{W}_{(1/2)}(f) = \mathcal{W}_\eta(f) - f(1) \log \pi.$$

The limits (5.5) lead to approximations of Weil's positivity. For example, taking f with $\hat{f}\left(\pm\frac{1}{2}\right) = 0$, and approximating $\mathcal{W}_\eta(f * f^*)$ by (5.5), exponentiating the result we get

(5.6)
$$(2\pi(1-q))^{-\sum_{m\in\mathbb{Z}}|f(q^m)|^2} \geq \prod_{\substack{m_1, m_2 \in \mathbb{Z} \\ m_1 \neq m_2}} (q^{-1})^{\frac{q^{(|m_1-m_2|)/2} f(q^{m_1})\bar{f}(q^{m_2})}{1-q^{(|m_1-m_2|)/2}}}$$

$$\cdot \prod_{p \neq \eta} p^{p^{-(|m_1-m_2|)/2} f(q^{m_1})\bar{f}(q^{m_1} p^{m_1-m_2})}$$

The validity of (5.6) for a sequence of $q \uparrow \mathbf{1}$ implies the RH.

Consider the Hilbert space

(5.7)
$$\mathcal{H}_{(q)}^\beta = \ell_2(g^{\mathbb{Z}}, \tau_{(q)}^\beta), \quad \tau_{(q)}^\beta(g^j) = \frac{\zeta_{(q)}(1)}{\zeta_{(q)}(\beta)} q^{j\beta}, \qquad \beta > 0,$$

and the unitary action of $g^{\mathbb{Z}}$ on $\mathcal{H}_{(q)}^\beta$ given by

(5.8)
$$\pi_q^\beta(g^j)\varphi(g^i) = q^{-j\beta/2}\varphi(g^{i-j}).$$

The vector $\phi_{\mathbb{Z}_q} \in \mathcal{H}_{(q)}^\beta$, given by

(5.9)
$$\phi_{\mathbb{Z}_q}(g^j) = \frac{1}{\zeta_{(q)}(1+j)},$$

is a cyclic vector, giving rise to an isomorphism

(5.10)
$$\tau_q^{\beta/2+it} : \mathcal{H}_{(q)}^\beta \xrightarrow{\sim} \hat{\mathcal{H}}_{(q)}^\beta = L_2\left(\mathbb{R} \Big/ \frac{2\pi}{\log q}\mathbb{Z}, |\zeta_{(q)}(\beta/2+it)|^2 \, dt \cdot c_q^\beta\right),$$

$$\varphi \longrightarrow \tau_q^{\beta/2+it}\varphi = \sum_j \varphi(g^j) q^{j(\beta/2+it)} \frac{\zeta_{(q)}(1)}{\zeta_{(q)}(\beta/2+it)},$$

$$\tau_q^{\beta/2+it}(\pi_q^\beta(g^j)\varphi) = q^{jit} \cdot \tau_q^{\beta/2+it}\varphi.$$

On the space $\hat{\mathcal{H}}^{\beta}_{(q)}$, we have the operator $\mathcal{F}^{\beta}_q f(t) = f(-t)$, and it corresponds under the isomorphism (5.10) to the q-Fourier-Bessel transform $\mathcal{F}^{\beta}_q$ on $\mathcal{H}^{\beta}_{(q)}$,

(5.11)
$$\mathcal{F}^{\beta}_q \varphi(g^j) = \sum_i \varphi(g^i) \mathcal{F}^{\beta}_q(g^{i+j}) \tau^{\beta}_{(q)}(g^i),$$

$$\mathcal{F}^{\beta}_q(g^j) = \sum_{k \geq 0}(-1)^k q^{\frac{k(k-1)}{2}} \frac{\zeta_{(q)}(1)}{\zeta_{(q)}(1+k)} \frac{\zeta_{(q)}(\beta)}{\zeta_{(q)}(\beta+k)} q^{(1+j)k}.$$

In the p-adic limit $\textcircled{p}$: $g^j := p^j$, $q := p^{-d}$, $\beta := \frac{\beta}{d}$, $d \to \infty$, we get the theory described in $(2.5)_p$

$$\lim_{\textcircled{p}} \tau^{\beta}_{(q)} = \tau^{\beta}_p(dx) = |x|^{\beta}_p \frac{d^* x}{\zeta_p(\beta)},$$

$$\lim_{\textcircled{p}} \mathcal{H}^{\beta}_{(q)} = L_2\left(\mathbb{Q}_p, |x|^{\beta}_p \frac{d^* x}{\zeta_p(\beta)}\right)^{\mathbb{Z}^*_p},$$

$(5.12)_p$ $\quad \lim_{\textcircled{p}} \phi_{\mathbb{Z}_q} = \phi_{\mathbb{Z}_p}$ = characteristic function of $\mathbb{Z}_p \subseteq \mathbb{Q}_p$, or $p^{\mathbb{N}} \subseteq p^{\mathbb{Z}}$,

$$\lim_{\textcircled{p}} \mathcal{F}^{\beta}_q = \mathcal{F}^{\beta}_p \text{ the } p\text{-adic Fourier Bessel transform given by the kernel}$$

$$\mathcal{F}^{\beta}_p(p^j) = \frac{\phi_{\mathbb{Z}_p}(p^j) - p^{\beta} \phi_{\mathbb{Z}_p}(p^{j+1})}{1 - p^{\beta}}.$$

In the real limit $\textcircled{$\eta$}$: $g = q \to 1$, $\beta := \frac{\beta}{2}$, and $j \to \infty$, such that $\frac{q^j}{1-q} \to \pi x^2$, we get the theory of $(2.5)_\eta$,

$$\lim_{\textcircled{η}} \tau^{\beta}_{(q)} = \tau^{\beta}_\eta(dx) = |x|^{\beta}_\eta \frac{d^* x}{\zeta_\eta(\beta)},$$

$$\lim_{\textcircled{η}} \mathcal{H}^{\beta}_{(q)} = L_2\left(\mathbb{R}, |x|^{\beta}_\eta \frac{d^* x}{\zeta_\eta(\beta)}\right)^{\pm 1},$$

$(5.12)_\eta$ $\quad \lim_{\textcircled{η}} \phi_{\mathbb{Z}_q}(x) = \phi_{\mathbb{Z}_\eta}(x) = e^{-\pi x^2},$

$$\lim_{\textcircled{η}} \mathcal{F}^{\beta}_q = \mathcal{F}^{\beta}_\eta \text{ the Fourier Bessel transform given by the kernel}$$

$$\mathcal{F}^{\beta}_\eta(x) = \sum_{k \geq 0}(-1)^k \frac{\zeta_\eta(\beta)}{\zeta_\eta(\beta+2k)} \frac{(\pi x^2)^k}{k!}.$$

For $\beta = n$ an integer,

(5.13) $\qquad \mathcal{H}^n_p = L_2(\mathbb{Q}^{\oplus n}_p, dx)^{\mathrm{GL}(\mathbb{Z}_p)}, \qquad \mathcal{H}^n_\eta = L_2(\mathbb{R}^{\oplus n}, dx)^{O_n},$

and $\mathcal{F}^n_p$, $p \geq \eta$, is the n-dimensional Fourier transform applied to $\mathrm{GL}_n(\mathbb{Z}_p)$ (resp. O_n, $p = \eta$) symmetric functions.

Consider the function

(5.14) $\qquad B_c(g^j) = \begin{cases} 1 & j \geq c \\ 0 & j < c \end{cases}, \qquad c \in \mathbb{Z},$

as projection on $\mathcal{H}^{\beta}_{(q)}$, and consider the "dual" projection

$$(5.15) \qquad \hat{B}_c = \mathcal{F}^{\beta}_q B_c \mathcal{F}^{\beta}_q.$$

In the model $\hat{\mathcal{H}}^{\beta}_{(q)}$, B_c (resp. $\hat{B}_c$) is the projection onto

$$(5.16) \qquad \underset{n \geq c}{\oplus} \mathbb{C} \cdot \frac{q^{ns}}{\zeta_{(q)}(\beta/2 + s)} \quad \left(\text{resp. } \underset{n \geq c}{\oplus} \mathbb{C} \cdot \frac{q^{-ns}}{\zeta_{(q)}(\beta/2 - s)} \right).$$

For $c \leq 0$, $B_c \mathcal{H}^{\beta}_{(q)} \cap \hat{B}_c \mathcal{H}^{\beta}_{(q)}$ is a $2|c| + 1$ dimensional space whose p-adic limit gives $\mathcal{H}^{\beta}_p(|c|)$, $(2.7)_p$. On the other hand, for $c > 0$, the intersection is trivial

$$(5.17) \qquad B_c \mathcal{H}^{\beta}_{(q)} \cap \hat{B}_c \mathcal{H}^{\beta}_{(q)} = \{0\}.$$

We can imitate the Landau-Pollak-Slepian theory, producing a subspace $B^{\langle c \rangle} \mathcal{H}^{\beta}_{(q)} \subseteq B_c \mathcal{H}^{\beta}_{(q)}$ of functions φ supported in $\{g^j, \quad j \geq c\}$, and such that $\mathcal{F}^{\beta}_q \varphi$ is essentially supported also in $\{g^j, \quad j \geq p\}$. The operator $B_c \hat{B}_c$ on $B_c \mathcal{H}^{\beta}_{(q)}$ is an integral operator with kernel

$$(5.18) \qquad p^{\beta}_c(g^{j_1}, g^{j_2}) = \sum_{i \geq c} \mathcal{F}^{\beta}_q(g^{i+j_1}) \mathcal{F}^{\beta}_q(g^{i+j_2}) \tau^{\beta}_{(q)}(g^i),$$

and we have the orthogonal decomposition associated with its eigenfunctions,

$$(5.19) \qquad \begin{aligned} & B_c \mathcal{H}^{\beta}_{(q)} = \underset{n \geq 0}{\oplus} \mathbb{C} \cdot \phi^{\beta,c}_n, \quad B_c \hat{B}_c \phi^{\beta,c}_n = \lambda_n \cdot \phi^{\beta,c}_n, \\ & \lambda_n = \lambda^{\beta,c}_{q,n}, \quad 1 > \lambda_0 > \lambda_1 > \ldots > 0, \\ & \hat{B}_c \mathcal{H}^{\beta}_{(q)} = \underset{n \geq 0}{\oplus} \mathbb{C} \cdot \hat{B}_c \phi^{\beta,c}_n, \\ & \|B_c \mathcal{F}^{\beta}_q \phi^{\beta,c}_n\|^2 = \|\hat{B}_c \phi^{\beta,c}_n\|^2 = (\hat{B}_c \phi^{\beta,c}_n, \hat{B}_c \phi^{\beta,c}_n) = \lambda_n \|\phi^{\beta,c}_n\|^2. \end{aligned}$$

Consider the operators $\mathcal{X}, \mathcal{Y}$ on $\mathcal{H}^{\beta}_{(q)}$ given by

$$(5.20) \qquad \mathcal{X}\varphi(g^j) = q^j \varphi(g^j), \quad \mathcal{Y}\varphi(g^j) = \varphi(g^{j-1}).$$

They satisfy

$$(5.21) \qquad \begin{aligned} & \mathcal{X}\mathcal{Y} = q\mathcal{Y}\mathcal{X} \\ & \mathcal{X}^* = \mathcal{X}, \quad \mathcal{Y}^* = q^{\beta} \cdot \mathcal{Y}^{-1} = \mathcal{F}^{\beta}_q \mathcal{Y} \mathcal{F}^{\beta}_q \end{aligned}$$

and

$$(5.22) \qquad \mathcal{F}^{\beta}_q \mathcal{X} \mathcal{F}^{\beta}_q = (1 - \mathcal{Y}^*)\mathcal{X}^{-1}(1 - \mathcal{Y}).$$

We let $H = H^{\beta,c}_q$ be the operator on $\mathcal{H}^{\beta}_{(q)}$ given by

$$(5.23) \qquad \begin{aligned} H &= q^c \cdot [\mathcal{X} + \mathcal{F}^{\beta}_q \mathcal{X} F^{\beta}_q] - (1 - \mathcal{Y}^*)(1 - \mathcal{Y}) \\ &= q^c \cdot \mathcal{X} + (1 - \mathcal{Y}^*)(q^c \cdot \mathcal{X}^{-1} - 1)(1 - \mathcal{Y}), \quad \text{by } (5.22). \end{aligned}$$

The first line in (5.23) shows that H commutes with $\mathcal{F}^{\beta}_q$. The second line in (5.23) (and integration by parts, using the vanishing of $q^c \mathcal{X} - 1$ at the end point g^c) shows that H commutes with B_c. It follows that H commutes with $B_c \hat{B}_c$, hence the $\phi^{\beta,c}_n$ are eigenfunctions of H.

In the analytic model $\hat{\mathcal{H}}^\beta_{(q)}$, the operator H translates into the difference operator $\hat{H} = \hat{H}^{\beta,c}_q$, $\hat{H}\tau^{\beta/2+it}_q\varphi = \tau^{\beta/2+it}_q H\varphi$, where

(5.23)′
$$\hat{H}f(s) = q^c\cdot[(1-q^{\beta/2+s})f(s+1)+(1-q^{\beta/2-s})f(s-1)]+(1-q^{\beta/2+s})(1-q^{\beta/2-s})f(s).$$

In the real limit $\textcircled{$\eta$}$ of $(5.12)_\eta$, we have

$$\lim_{\textcircled{η}}\frac{1}{1-q}\mathcal{X} = \pi x^2,$$

$$\lim_{\textcircled{η}}\frac{1}{1-q}(1-\mathcal{Y}) = -\frac{x}{2}\frac{\partial}{\partial x},$$

$$\lim_{\textcircled{η}}\frac{1}{1-q}(1-\mathcal{Y}^*) = \frac{\beta}{2}+\frac{x}{2}\frac{\partial}{\partial x},$$

so that

(5.24)
$$\lim_{\textcircled{η}}\frac{1}{1-q}(\mathcal{Y}-\mathcal{Y}^*) = \frac{\beta}{2}+x\frac{\partial}{\partial x} = \pi^\beta_\eta\left(a\frac{\partial}{\partial a}\right)$$

is the infinitesimal generator of the unitary action π^β_η of $\mathbb{R}^+$ on $\mathcal{H}^\beta_\eta$.

If we also include in the real limit $\textcircled{$\eta$}$ the condition on $c \in \mathbb{Z}$,

(5.25)
$$c \to \infty, \text{ such that } \frac{q^c}{1-q} \to \pi c^2_\eta, \quad c_\eta \in \mathbb{R}^+ \text{ fixed,}$$

the projection $B_c, \hat{B}_c$ converge to their real counterpart $B_{c_\eta}, \hat{B}_{c_\eta} = \mathcal{F}^\beta_\eta B_c\mathcal{F}^\beta_\eta$, and so does the integral operator and its eigenfunctions: $\lim_{\textcircled{$\eta$}}\phi^{\beta,c}_{q,n} = \phi^{\beta,c}_{\eta,n}$. The limit of the operator H is given by

(5.26)
$$H^{\beta,c}_\eta = \lim_{\textcircled{η}}\frac{1}{(1-q)^2}H^{\beta,c}_q$$

$$= 4\pi c^2_\eta[\pi x^2 + \mathcal{F}^\beta_\eta\pi x^2\mathcal{F}^\beta_\eta] + \left(\frac{\beta}{2}+x\frac{\partial}{\partial x}\right)\left(x\frac{\partial}{\partial x}\right)$$

$$= (2\pi c_\eta x)^2 + (x^2 - c^2_\eta) + \frac{\partial^2}{\partial x^2} + (1+\beta)x\frac{\partial}{\partial x} + c^2_\eta(1-\beta)\frac{1}{x}\frac{\partial}{\partial x}$$

$$= (2\pi c_\eta x)^2 + x^{1-\beta}\frac{\partial}{\partial x}x^{\beta-1}(x^2-c^2_\eta)\frac{\partial}{\partial x}$$

At $\beta = 1$ this is the operator (3.12), and its eigenfunctions $\phi^{1,c}_{\eta,n}$ are the even prolate spheroidal function; at $\beta = 3$, the $\phi^{3,c}_{\eta,n}(x)\cdot x$ are the odd prolate spheroidal functions.

The generalized prolate spheroidal functions $\phi^{\beta,c}_{\eta,n}$ are continuous in the parameter β and $c \in [0,\infty]$. At $c = \infty$ the operator $H^{\beta,c}_\eta$ is the number operator of the Laguerre basis of $\mathcal{H}^\beta_\eta$, cf. H[01].

We have

(5.27)
$$B_c\hat{B}_c\phi^{\beta,c}_{\eta,n} = \lambda^{\beta,c}_{\eta,n}\cdot\phi^{\beta,c}_{\eta,n}, \quad 1 > \lambda_{\eta,0} > \lambda_{\eta,1} > \ldots > 0;$$

the $\phi^{\beta,c}_{\eta,n}$ are restrictions of entire analytic functions $\tilde{\phi}^{\beta,c}_{\eta,n}$, $\phi^{\beta,c}_{\eta,n} = B_c\tilde{\phi}^{\beta,c}_{\eta,n}$, with

(5.28)
$$\mathcal{F}^\beta_\eta\phi^{\beta,c}_{\eta,n} = \epsilon^{\beta,c}_{\eta,n}\tilde{\phi}^{\beta,c}_{\eta,n},$$

and

$$(5.29) \qquad (-1)^n \cdot \epsilon_{\eta,n}^{\beta,c} > 0, \quad (\epsilon_{\eta,n}^{\beta,c})^2 = \lambda_{\eta,n}^{\beta,c}.$$

When we let $c \to \infty$ and $|x| \to \infty$, with fixed n, we have the asymptotic behavior

(5.30)

$$\tilde{\phi}_{\eta,n}^{\beta,c}(x) \asymp \begin{cases} \phi_{\eta,n}^{\beta}(x) = L_n^{\beta/2-1}(2\pi x^2)e^{-\pi x^2} & |x| \le (2\pi)^{-1/4} \cdot c^{1/2} \\ k \cdot c^{(\beta+1)/2} \cdot \dfrac{e^{2\pi c\sqrt{c^2-x^2}}}{(c^2-x^2)^{1/4}} \dfrac{(c-\sqrt{c^2-x^2})^n}{(c+\sqrt{c^2-x^2})^{n+\beta/2}} & (2\pi)^{-1/4}c^{1/2} \le |x| \le c - \frac{1}{2\pi c} \\ k \cdot 2\pi c \cdot I_0(2\pi c\sqrt{c^2-x^2}) & c - \frac{1}{2\pi c} \le |x| \le c \\ k \cdot 2\pi c \cdot J_0(2\pi c\sqrt{x^2-c^2}) & c \le |x| \le c + \frac{1}{2\pi c} \\ k \cdot c \cdot 2\,\mathrm{Re}\,\dfrac{e^{2\pi i c\sqrt{x^2-c^2}}}{e^{i\pi/4}\sqrt{x^2-x^2}} \dfrac{(c-i\sqrt{x^2-c^2})^n}{(c+i\sqrt{x^2-c^2})^{n+\beta/2}} & c + \frac{1}{2\pi c} \le |x| \end{cases}$$

with $k = \frac{(-1)^n}{n!}(8\pi c^2)^n \cdot 2^{\beta/2} \cdot e^{-2\pi c^2}$. The eigenvalues $\lambda_{\eta,n}^{\beta,c}$ are close to 1 for $n \le 2c^2$, then they plunge to near zero in an interval of length $O(\log c)$ (independently of β!). We have as $c, n \to \infty$ with

$$(5.31) \qquad n = 2c^2 + \frac{\delta}{\pi}\log c + O(1)$$

the asymptotic behavior

$$(5.32) \qquad \lambda_{\eta,n}^{\beta,c} \approx \frac{1}{2\pi}\left|\Gamma\left(\frac{1+i\delta}{2}\right)\right|^2 e^{-\frac{\pi}{2}\delta} = (1+e^{\pi\delta})^{-1}.$$

6. Approximating a proof of RH using the $\beta > 1$ deformation

On the space

$$(6.1) \qquad \mathcal{H}_A^\beta = L_2(A, \tau_A^\beta) = \underset{p \ge \eta}{\otimes} \mathcal{H}_p^\beta \text{ w.r.t. } \phi_{\mathbb{Z}_p}$$

we have the unitary action π^β of $\mathbb{A}^*/(\pm 1)\prod_p \mathbb{Z}_p^* = \mathbb{R}^+\mathbb{Q}^+$, where $\mathbb{Q}^+$ preserves the measure τ_A^β, and we have the unitary self-adjoint operator $\mathcal{F}_A^\beta = \otimes_{p \ge \eta}\mathcal{F}_p^\beta$

$$(6.2) \qquad \begin{cases} \mathcal{F}_A^\beta = (\mathcal{F}_A^\beta)^* = (\mathcal{F}_A^\beta)^{-1} \\ \pi^\beta(a)\mathcal{F}_A^\beta = \mathcal{F}_A^\beta \pi^\beta(a^{-1}) \\ \mathcal{F}_A^\beta \phi_A = \phi_A, \quad \phi_A = \underset{p \ge \eta}{\otimes} \phi_{\mathbb{Z}_p}. \end{cases}$$

When $\beta > 1$, the measure τ_A^β is supported on $\mathbb{A}^*$, and letting for $x \in \mathbb{A}^*$

$$B_c(x) = \begin{cases} 1 & |x| \le c \\ 0 & |x| > c \end{cases}$$

we get a well defined projection operator on $\mathcal{H}_A^\beta$ (note: the operation of multiplication by $B_c(x)$ does not define a projection on $\mathcal{H}_A^1$, since at $\beta = 1$, $\tau_A^1 = dx$ is Haar measure, and $\mathbb{A}^*$ is a set of measure zero). We have for $\beta > 1$ also the dual projection $\hat{B}_c = \mathcal{F}_A^\beta B_c \mathcal{F}_A^\beta$. The operators B_c, $\hat{B}_c$ (resp. $\mathcal{F}_A^\beta$) commute (resp. anti-commute) with the $\mathbb{Q}^+$-action. We can identify the "$\mathbb{Q}^+$-quotient" of $\mathcal{H}_A^\beta$ with $L_2(\mathbb{R}^+, d^*a)$:

$$(6.3) \qquad \begin{aligned} \mathcal{S}_A^0 &= \{\varphi \in \mathcal{S}(A), \quad \mathcal{F}_A^1\varphi(0) = \mathcal{F}_A^1\mathcal{F}^\beta\varphi(0) = 0\} \searrow E^\beta \\ \mathcal{S}_\mathbb{R}^0 &= \{\varphi \in \mathcal{S}(\mathbb{R})^{\pm 1}, \quad \mathcal{F}_\eta^1\varphi(0) = \mathcal{F}_\eta^1\mathcal{F}_\eta^\beta\varphi(0) = 0\} \hookrightarrow L_2(\mathbb{R}^+, d^*a) \end{aligned}$$

with $E^{\beta}\varphi(a) = |a|^{\beta/2} \sum_{q \in \mathbb{Q}^+} \varphi(a \cdot q)$, for $\varphi \in \mathcal{S}_{\mathbb{A}}^0$. We have $E^{\beta}[\mathcal{S}_{\mathbb{A}}^0] = E^{\beta}[\mathcal{S}_{\mathbb{R}}^0]$, E^{β} is an injection on $\mathcal{S}_{\mathbb{R}}^0$ with dense image; for $\varphi \in \mathcal{S}_{\mathbb{R}}^0$

$$(6.4) \qquad E^{\beta}\varphi(a) = a^{\beta/2} \sum_{n \geq 1} \varphi(a \cdot n).$$

We get a unitary self-adjoint operator $\mathcal{F}_{\mathbb{Q}}^{\beta}$ on $L_2(\mathbb{R}^+, d^*a)$ such that

$$(6.5) \qquad \mathcal{F}_{\mathbb{Q}}^{\beta} E^{\beta} \varphi = E^{\beta} \mathcal{F}_{\mathbb{A}}^{\beta} \varphi,$$

and we have the projections $B_{\mathbb{Q}}^c = B_c$, $\hat{B}_{\mathbb{Q}}^c = \mathcal{F}_{\mathbb{Q}}^{\beta} B_c \mathcal{F}_{\mathbb{Q}}^{\beta}$. These projections do not commute. But for large c, uniformly in β, the operator $B_{\mathbb{Q}}^c \hat{B}_{\mathbb{Q}}^c$ approximates the projection $B_E^{\beta\langle c\rangle}$ onto the subspace of $E^{\beta}[\varphi]$ with $\varphi(x)$ supported in $|x| \leq c$ and $\mathcal{F}_{\eta}^{\beta}\varphi(x)$ (essentially) supported in $|x| \leq c$:

$$(6.6) \qquad B_E^{\beta\langle c\rangle} L_2(\mathbb{R}^+) = E^{\beta}\{\varphi \in \bigoplus_{n \leq 2c^2} \mathbb{C}\,\phi_{\eta,n}^{\beta,c}, \quad \mathcal{F}_{\eta}^1\varphi(0) = \mathcal{F}_{\eta}^1 \mathcal{F}^{\beta}\varphi(0) = 0\}.$$

More precisely, we have

$$(6.7) \qquad \operatorname{tr}(B_{\mathbb{Q}}^c \hat{B}_{\mathbb{Q}}^c \pi(f)) = \operatorname{tr}(B_E^{\beta\langle c\rangle}\pi(f)) + o(\beta, c)$$

where for a function $R(\beta, c)$ of the parameters β, c we write $R(\beta, c) = o(\beta, c)$ if for any $\epsilon > 0$ there exists $\beta_{\epsilon} > 1$, $c_{\epsilon} > 1$ such that $|R(\beta, c)| < \epsilon$ for all $(\beta, c) < (1, \beta_{\epsilon}) \times (c_{\epsilon}, \infty)$.

Since the $\phi_{\eta,n}^{\beta,c}$ are continuous in the parameters β, c we also have

$$(6.8) \qquad \operatorname{tr}(B_E^{\beta\langle c\rangle}\pi(f)) = \operatorname{tr}(B_E^{\langle c\rangle}\pi(f)) + o(\beta, c)$$

where $B_E^{\langle c\rangle} = B_E^{1\langle c\rangle}$ is the projection at $\beta = 1$.

Lift the function $f \in C_c^{\infty}(\mathbb{R}^+)$ to the function $\tilde{f} = f \otimes \phi_{\hat{\mathbb{Z}}^*}$ on the ideles. The operator $B_c \hat{B}_c \pi^{\beta}(\tilde{f})$ on $\mathcal{H}_{\mathbb{A}}^{\beta}$ is given by a kernel

$$(6.9) \qquad \tilde{K}(y, x) = B_c(y) \int_{\mathbb{A}^*} \frac{d^*z}{\zeta_{\mathbb{A}}(\beta)} |z|^{\beta} \mathcal{F}_{\mathbb{A}}^{\beta}(yz) B_c(z) \int_{\mathbb{A}^*} d^*a\, \tilde{f}(a)|a|^{\beta/2} \mathcal{F}_{\mathbb{A}}^{\beta}(zxa)$$

and the induced operator $B_{\mathbb{Q}}^c \hat{B}_{\mathbb{Q}}^c \pi(f)$ on $L_2(\mathbb{R}^+)$ is given by the kernel

$$(6.10) \qquad K(y, x) = \sum_{q \in \mathbb{Q}^+} \tilde{K}(y, x \cdot q).$$

We can calculate the trace by putting $x = y$ in (6.10) and integrating, so

$$(6.11) \qquad \operatorname{tr}(B_{\mathbb{Q}}^c \hat{B}_{\mathbb{Q}}^c \pi^{\beta}(f)) = \sum_{q \in \mathbb{Q}^+} \int_{\mathbb{R}^+} \frac{d^*x \cdot x^{\beta}}{\zeta_{\mathbb{A}}(\beta)} \tilde{K}(x, x \cdot q).$$

Substituting (6.9) in (6.11), then changing variables $z := \frac{z}{x}$, we get

$$(6.12) \qquad \begin{aligned} \operatorname{tr}(B_{\mathbb{Q}}^c \hat{B}_{\mathbb{Q}}^c \pi^{\beta}(f)) &= \sum_{q \in \mathbb{Q}^+} \mathcal{F}_{\mathbb{A}}^{\beta}(B_{c^2}(z) \log c^2/|z|) \mathcal{F}_{\mathbb{A}}^{\beta}||^{-\beta/2}\tilde{f}(q) \\ &= 2\log c \cdot f(1) + W^{\beta}(f) + R^{\beta,c}, \end{aligned}$$

with

$$(6.13) \qquad W^{\beta}(f) = \sum_{q \in \mathbb{Q}^+} \mathcal{F}_{\mathbb{A}}^{\beta}(-\log|z|) \mathcal{F}_{\mathbb{A}}^{\beta}||^{-\beta/2}\tilde{f}(q)$$

a continuous function of $\beta \geq 1$, whose value at $\beta = 1$ is the Weil distribution H[91]:

$$(6.14) \qquad W^1(f) = \sum_{q \in \mathbb{Q}^+} \sum_{p \geq \eta} \mathcal{F}_p(-\log |z|_p)\mathcal{F}_p||^{-1/2}\tilde{f}(q) = \sum_{p \geq \eta} W_p(f).$$

The remainder $\mathcal{R}^{\beta,c}$ in (6.12) is given by

$$(6.15) \qquad \mathcal{R}^{\beta,c} = \sum_{q \in \mathbb{Q}^+} \mathcal{F}_{\mathbb{A}}^\beta (1 - B_{c^2}(z)) \log(|z|/c^2) \, \mathcal{F}_{\mathbb{A}}^\beta ||^{-\beta/2}\tilde{f}(q),$$

and we expect that

$$(6.16) \qquad\qquad\qquad \mathcal{R}^{\beta,c} = o(\beta, c),$$

so that (6.12) reads

$$(6.17) \qquad \operatorname{tr}(B_{\mathbb{Q}}^c \hat{B}_{\mathbb{Q}}^c \pi(f)) = 2 \log c \cdot f(1) + \sum_{p \geq \eta} W_p(f) + o(\beta, c).$$

Combining this with (6.7), (6.8) we get

$$(6.18) \qquad \operatorname{tr}\left(B_E^{\langle c \rangle} \pi_\eta(f)\right) = 2 \log c \cdot f(1) + \sum_{p \geq \eta} W_p(f) + o(\beta, c)$$

and since both sides of (6.17) are independent of β we can replace $o(\beta, c)$ by $o(1)$ as $c \to \infty$, which is (4.25), implying RH.

Typically to the RH, closing the gap (6.16) in the above "proof" could be a formidable task. The final proof of RH will probably require a more *additive* and global construction of the finite dimensional spaces of "band and time limited" functions on $\mathbb{A}/\mathbb{Q}^*$.

References

[BN] D. Bump and E.K.S. Ng, On Riemann's Zeta function, *Math. Z.* **192** (1986), 195–204.

[C94] A. Connes, *Noncommutative Geometry*, Academic Press, 1994.

[C98] A. Connes, Trace formula in noncommutative geometry and the zeros of the Riemann's Zeta function, preprint (1998).

[F] C. Flammer, *Spheroidal wave functions*, Stanford University Press, 1957.

[H90] S. Haran, Riesz potentials and explicit sums in arithmetic, *Inventiones Math.* **101** (1990), 697–703.

[H91] S. Haran, Index theory, potential theory, and the Riemann hypothesis, in: L-functions and Arithmetic, Durham 1990, LMS Lecture Notes, Ser. 153 (1991), 257–270.

[H93] S. Haran, On Riemann's Zeta function and the mysteries of the prime at infinity, preprint 1993.

[H01] S. Haran, *The mysteries of the real prime*, Oxford University Press 2001.

[LPS] H.J. Landau, H.O. Pollak, D. Slepian, Prolate spheroidal wave functions, Fourier analysis and uncertainty, I, II, III, IV. *Bell System Tech. J.* **40** (1961–2), 43–63, 65–84, 1295–1336, (1964) 3009–3057.

[W] A. Weil, Sur le formules explicites de la théorie des nombres premiers, *Proc. R. Physiogr. Soc. Lund* **21** (1952), 252–265.

S. Haran
Dept. of Mathematics
Technion, Haifa, Israel
haran@tx.technion.ac.il

Contemporary Mathematics
Volume **290**, 2001

A Prime Orbit Theorem for Self-Similar Flows and Diophantine Approximation

MICHEL L. LAPIDUS AND MACHIEL VAN FRANKENHUYSEN

ABSTRACT. Assuming some regularity of the dynamical zeta function, we establish an explicit formula with an error term for the prime orbit counting function of a suspended flow. We define the subclass of self-similar flows, for which we give an extensive analysis of the error term in the corresponding prime orbit theorem.

1. Introduction

In [PP1], Parry and Pollicott obtain a Prime Orbit Theorem for certain dynamical systems—the so-called 'suspension flows'. (See also [PP2, Chapter 6].) The first results of this kind were obtained in special cases by Huber [Hu], Sinai [S], and Margulis [Mr], among others. See [PP1, 2] and the relevant references therein, as well as the historical note in [BKS, p. 154]. Parry and Pollicott derive the first term in the asymptotic expansion of the counting function of prime orbits, by applying the Wiener-Ikehara Tauberian Theorem to the logarithmic derivative of the dynamical zeta function. An alternate approach was taken by Lalley in [Lal1, 2], who considers, in particular, the (approximately) self-similar case. Using a nonlinear extension of the Renewal Theorem, he shows that in the nonlattice case, the leading asymptotics are nonoscillatory. In the lattice case, the leading asymptotics are periodic, and it becomes a natural question whether they are constant or nontrivially periodic.

In a recent book [LvF2], we have developed a theory of complex dimensions of fractal strings (one-dimensional drums with fractal boundary, see [LP,LM]). These (geometric) complex dimensions—defined as the poles of the associated geometric zeta function—enable us to describe the oscillations intrinsic to the geometry or

1991 *Mathematics Subject Classification.* Primary: 11N05, 28A80, 58F03, 58F20; Secondary: 11M41, 58F11, 58F15, 58G25.

Key words and phrases. Suspended flows, self-similar flows, self-similar fractal strings, lattice vs. nonlattice flows, dynamical systems, periodic orbits, dynamical zeta functions, geometric zeta functions, dynamical complex dimensions, Prime Orbit Theorem for suspended flows, explicit formulas, oscillatory terms, Diophantine approximation.

This work was partially supported by the National Science Foundation under the grants DMS-9623002 and DMS-0070497 (for M.L.L.).

the spectrum of fractal drums, via suitable 'explicit formulas', obtained in [LvF2, Chapter 4].

In this paper, we apply these explicit formulas to obtain an asymptotic expansion for the prime orbit counting function of suspension flows. The resulting formula involves a sum of oscillatory terms associated with the dynamical complex dimensions of the flow. We then focus on the special case of self-similar flows and deduce from our explicit formulas a Prime Orbit Theorem with error term. In the lattice case (to be defined below), the counting function of the prime orbits, $\psi_{\mathfrak{w}}(x)$, has oscillatory leading asymptotics and our explicit formula enables us to give a very precise expression for this function in terms of multiplicatively periodic functions. In the nonlattice case (which is the generic case), the leading term is nonoscillatory and we provide a detailed analysis of the error term. The precise order of the error term depends on the 'dimension free' region of the dynamical zeta function, as in the classical Prime Number Theorem. This region in turn depends on properties of Diophantine approximation of the weights of the flow.

For suspension flows, the dynamical complex dimensions are defined as the poles of the logarithmic derivative of the dynamical zeta function. On the other hand, the geometric complex dimensions of a fractal string are defined in [LvF1, 2] as the poles of the geometric zeta function, which coincides with the dynamical zeta function when the string and the flow are self-similar. Thus the geometric complex dimensions of a self-similar flow only depend on the poles of the corresponding zeta function, and they are counted with a multiplicity, whereas the dynamical complex dimensions of a flow depend on the zeros and the poles of the dynamical zeta function, and they usually have no multiplicity. Due to the fact that the dynamical zeta function of a self-similar flow has no zeros, the two sets of complex dimensions coincide in this case.

2. The Zeta Function of a Dynamical System

Let $N \geq 0$ be an integer and let $\Omega = \{1, \ldots, N\}^{\mathbb{N}}$ be the space of sequences over the alphabet $\{1, \ldots, N\}$. Let $\mathfrak{w}: \Omega \to (0, \infty]$ be a function, called the *weight*. On Ω, we have the left shift σ, given on a sequence (a_n) by $(\sigma a)_n = a_{n+1}$. We define the *suspended flow* $\mathcal{F}_{\mathfrak{w}}$ on the space $[0, \infty) \times \Omega$ as the following dynamical system (time evolution, see [PP2, Chapter 6]):

$$(2.1) \qquad \mathcal{F}_{\mathfrak{w}}(t, a) = \begin{cases} (t, a) & \text{if } 0 \leq t < \mathfrak{w}(a), \\ \mathcal{F}_{\mathfrak{w}}(t - \mathfrak{w}(a), \sigma a) & \text{if } t \geq \mathfrak{w}(a). \end{cases}$$

(Note that $\mathcal{F}_{\mathfrak{w}}(t, a)$ may not be defined. However, it is always defined on periodic sequences.) This formalism is seemingly less general than the one introduced in [PP2, Chapter 1]. However, defining $\mathfrak{w}(a) = \infty$ when the sequence a contains a prohibited word of length 2, and $e^{-s\infty} = 0$, allows us to deal with the general case.

Given a finite sequence $\mathfrak{x} = a_1, a_2, \ldots, a_l$ of length $l = l(\mathfrak{x})$, we let $a = a_1, a_2, \ldots, a_l, a_1, a_2, \ldots, a_l, \ldots$ be the corresponding periodic sequence, and we define $\sigma \mathfrak{x} = a_2, \ldots, a_l, a_1$. The *total weight* of the orbit of σ on $\mathfrak{x}$ is

$$(2.2) \qquad \mathfrak{w}_t(\mathfrak{x}) = \mathfrak{w}(a) + \mathfrak{w}(\sigma a) + \cdots + \mathfrak{w}(\sigma^{l-1} a).$$

We now define (see [Bo, R] and [PP2, Chapter 5]):

DEFINITION 2.1. The *dynamical zeta function* of $\mathcal{F}_\mathfrak{w}$ is defined as

$$(2.3) \qquad \zeta_\mathfrak{w}(s) = \exp\left(\sum_\mathfrak{x} \frac{1}{l(\mathfrak{x})} e^{-s\mathfrak{w}_t(\mathfrak{x})}\right),$$

where the sum extends over all finite sequences $\mathfrak{x}$ of positive length.

For $N = 0$, the alphabet is empty, and we interpret $\mathcal{F}_\mathfrak{w}$ as the static flow on a point, and $\zeta_\mathfrak{w}(s) = 1$. Further, for $N = 1$, we have the dynamical system of a point moving around a circle of length $\mathfrak{w}_t(1) = \mathfrak{w}(1, 1, \dots)$, and $\zeta_\mathfrak{w}(s) = (1 - e^{-s\mathfrak{w}_t(1)})^{-1}$.

We also introduce the logarithmic derivative

$$(2.4) \qquad -\frac{\zeta_\mathfrak{w}'}{\zeta_\mathfrak{w}}(s) = \sum_\mathfrak{x} \frac{\mathfrak{w}_t(\mathfrak{x})}{l(\mathfrak{x})} e^{-s\mathfrak{w}_t(\mathfrak{x})}.$$

For $N \geq 1$, this series does not converge for $s = 0$. We assume that (2.4) converges for some value of $s > 0$, and the abcissa of convergence of this series will be denoted by D, the *dimension* of $\mathcal{F}_\mathfrak{w}$.[1] Clearly, $D \geq 0$. Then (2.4) is absolutely convergent for $\operatorname{Re} s > D$. Moreover, as in [LvF1, 2], we assume that there exists a function $S: \mathbb{R} \to \mathbb{R}$, called the *screen*, satisfying $S(t) < D$ for every $t \in \mathbb{R}$, such that $-\zeta_\mathfrak{w}'/\zeta_\mathfrak{w}$ has a meromorphic extension to a neighborhood of the region

$$(2.5) \qquad W = \{s = \sigma + it \colon \sigma \geq S(t)\},$$

called the *window*. In Section 4, we will also assume that $-\zeta_\mathfrak{w}'/\zeta_\mathfrak{w}$ satisfies the growth conditions $(\mathbf{H}_1)$ and $(\mathbf{H}_2)$, to be introduced in Section 3. We will then say that $\mathcal{F}_\mathfrak{w}$ satisfies $(\mathbf{H}_1)$ and $(\mathbf{H}_2)$.

DEFINITION 2.2. The poles of $-\zeta_\mathfrak{w}'/\zeta_\mathfrak{w}(s)$ in W are called the *complex dimensions* of the flow $\mathcal{F}_\mathfrak{w}$. The *set of complex dimensions* of $\mathcal{F}_\mathfrak{w}$ in W is denoted by $\mathcal{D}_\mathfrak{w}(W)$ or $\mathcal{D}_\mathfrak{w}$ for short.

The nonreal complex dimensions of a flow come in complex conjugate pairs $\omega, \overline{\omega}$ (provided that W is symmetric about the real axis). If $\zeta_\mathfrak{w}$ has a meromorphic extension to W as well, then the complex dimensions of $\mathcal{F}_\mathfrak{w}$ are simple and they are located at the zeros and poles of $\zeta_\mathfrak{w}$,

$$\mathcal{D}_\mathfrak{w}(W) = \{\omega \in W \colon \zeta_\mathfrak{w}(\omega) = 0 \text{ or } \infty\},$$

and the residue at a complex dimension ω (i.e., $\operatorname{res}(-\zeta_\mathfrak{w}'/\zeta_\mathfrak{w}; \omega)$) is $-\operatorname{ord}(\zeta_\mathfrak{w}; \omega)$, where $\operatorname{ord}(\zeta_\mathfrak{w}; \omega) = n$ is the order of $\zeta_\mathfrak{w}$ at ω: $\zeta_\mathfrak{w}(s) = C(s - \omega)^n + O((s - \omega)^{n+1})$. In general, the complex dimensions of $\mathcal{F}_\mathfrak{w}$ in W are not simple, and the residues are not necessarily integers. By abuse of notation, we write $\operatorname{ord}(\zeta_\mathfrak{w}; \omega) = \operatorname{res}(\zeta_\mathfrak{w}'/\zeta_\mathfrak{w}; \omega)$ if $\zeta_\mathfrak{w}'/\zeta_\mathfrak{w}$ has a meromorphic extension with a simple pole at ω, even if the residue is not an integer (and consequently, $\zeta_\mathfrak{w}$ is not analytic at ω).

2.1. Periodic Orbits, Euler Product. A periodic sequence a in Ω with period l, $a = a_1, \dots, a_l, a_1, \dots, a_l, \dots$, gives rise to the finite orbit $\{a, \sigma a, \dots, \sigma^{l-1}a\}$ of σ. It is clear that l is a multiple of the cardinality $\#\{a, \sigma a, \dots, \sigma^{l-1}a\}$ of this orbit.

DEFINITION 2.3. A finite sequence $\mathfrak{x}$ is *primitive* if its length $l(\mathfrak{x})$ coincides with the length of the corresponding periodic orbit of σ.

---

[1] The dimension often coincides with the topological entropy of the flow; see [PP2, Chapter 5] and the references therein.

We denote by $\sigma\backslash\Omega$ the space of periodic orbits of σ. Thus

$$(2.6) \qquad \sigma\backslash\Omega = \left\{\{\sigma^k\mathfrak{x}\colon k \in \mathbb{N}\}\colon \mathfrak{x} \text{ is a finite sequence}\right\}.$$

We reserve the letter $\mathfrak{p}$ for elements of $\sigma\backslash\Omega$. So $\mathfrak{p}$ will denote a periodic orbit of σ, and we write $\#\mathfrak{p}$ for its length. The *total weight* of an orbit $\mathfrak{p}$ is

$$(2.7) \qquad \mathfrak{w}_t(\mathfrak{p}) = \sum_{a\in\mathfrak{p}} \mathfrak{w}(a).$$

THEOREM 2.4 (Euler sum). *For* $\operatorname{Re} s > D$, *we have the following expression for the logarithmic derivative of* $\zeta_\mathfrak{w}$:

$$(2.8) \qquad -\frac{\zeta'_\mathfrak{w}}{\zeta_\mathfrak{w}}(s) = \sum_{\mathfrak{p}\in\sigma\backslash\Omega}\sum_{k=1}^{\infty} \mathfrak{w}_t(\mathfrak{p})e^{-sk\mathfrak{w}_t(\mathfrak{p})},$$

where $\mathfrak{p}$ *runs through all periodic orbits of* $\mathcal{F}_\mathfrak{w}$.

PROOF. We write the sum in (2.4) over the finite sequences $\mathfrak{x}$ as a sum over the primitive sequences and repetitions of these. An orbit $\mathfrak{p}$ of σ contains $\#\mathfrak{p}$ different primitive sequences of length $\#\mathfrak{p}$, so we obtain

$$\sum_{\mathfrak{x}} \frac{\mathfrak{w}_t(\mathfrak{x})}{l(\mathfrak{x})}e^{-s\mathfrak{w}_t(\mathfrak{x})} = \sum_{\mathfrak{x}:\text{primitive}}\sum_{k=1}^{\infty} \frac{k\mathfrak{w}_t(\mathfrak{x})}{kl(\mathfrak{x})}e^{-ks\mathfrak{w}_t(\mathfrak{x})}$$

$$= \sum_{\mathfrak{p}\in\sigma\backslash\Omega} \#\mathfrak{p}\sum_{k=1}^{\infty} \frac{k\mathfrak{w}_t(\mathfrak{p})}{k\#\mathfrak{p}}e^{-ks\mathfrak{w}_t(\mathfrak{p})}.$$

The theorem follows. $\qquad\qquad\square$

DEFINITION 2.5. The following function counts the periodic orbits and their multiples by their total weight:

$$(2.9) \qquad \psi_\mathfrak{w}(x) = \sum_{k\mathfrak{w}_t(\mathfrak{p})\leq\log x} \mathfrak{w}_t(\mathfrak{p}).$$

The function $\psi_\mathfrak{w}(x)$ is the counterpart of $\psi(x) = \sum_{p^k\leq x}\log p$, which counts prime powers p^k with a weight $\log p$; see Example 3.6.

COROLLARY 2.6. *We have the following relation between* $\zeta'_\mathfrak{w}/\zeta_\mathfrak{w}$ *and* $\psi_\mathfrak{w}$:

$$(2.10) \qquad -\frac{\zeta'_\mathfrak{w}}{\zeta_\mathfrak{w}}(s) = \int_0^{\infty} x^{-s}d\psi_\mathfrak{w}(x),$$

for $\operatorname{Re} s > D$.

The integral on the right-hand side of (2.10) is a Riemann-Stieltjes integral associated with the monotonic function $\psi_\mathfrak{w}$.

COROLLARY 2.7 (Euler product). *The function* $\zeta_\mathfrak{w}(s)$ *has the following expansion as a product*:

$$(2.11) \qquad \zeta_\mathfrak{w}(s) = \prod_{\mathfrak{p}\in\sigma\backslash\Omega} \frac{1}{1 - e^{-s\mathfrak{w}_t(\mathfrak{p})}},$$

where $\mathfrak{p}$ *runs over all periodic orbits of* $\mathcal{F}_\mathfrak{w}$. *The product converges for* $\operatorname{Re} s > D$.

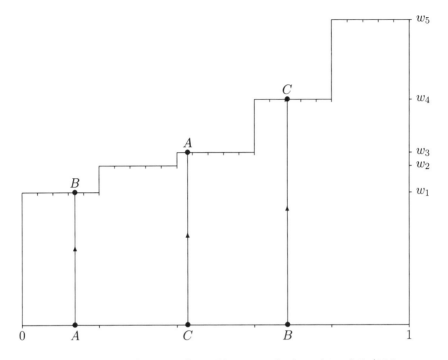

FIGURE 1. A self-similar flow, $N = 5$, with the orbit of $17/124$.

PROOF. In (2.8), we sum over k to obtain

$$(2.12) \qquad \frac{\zeta'_\mathfrak{w}}{\zeta_\mathfrak{w}}(s) = - \sum_{\mathfrak{p} \in \sigma \backslash \Omega} \frac{\mathfrak{w}_t(\mathfrak{p}) e^{-s\mathfrak{w}_t(\mathfrak{p})}}{1 - e^{-s\mathfrak{w}_t(\mathfrak{p})}} = - \sum_{\mathfrak{p} \in \sigma \backslash \Omega} \frac{d}{ds} \log\left(1 - e^{-s\mathfrak{w}_t(\mathfrak{p})}\right).$$

The theorem then follows upon integrating and taking exponentials. $\qquad \square$

In Section 4, we combine the above Euler product representation of $-\zeta'_\mathfrak{w}/\zeta_\mathfrak{w}$ with our explicit formulas of Section 3 to derive a Prime Orbit Theorem for primitive periodic orbits.

REMARK 2.8. We use $\psi_\mathfrak{w}$ instead of the more direct counting function

$$\pi_\mathfrak{w}(x) = \sum_{\mathfrak{w}_t(\mathfrak{p}) \leq \log x} 1.$$

However, setting $\theta_\mathfrak{w}(x) = \sum_{\mathfrak{w}_t(\mathfrak{p}) \leq \log x} \mathfrak{w}_t(\mathfrak{p})$, so that $\psi_\mathfrak{w}(x) = \theta_\mathfrak{w}(x) + \theta_\mathfrak{w}(x^{1/2}) + \theta_\mathfrak{w}(x^{1/3}) + \ldots$ and $\theta_\mathfrak{w}(x) = \psi_\mathfrak{w}(x) + O\left(\sqrt{\psi_\mathfrak{w}(x)}\right)$, as $x \to \infty$, we find

$$\pi_\mathfrak{w}(x) = \int_0^x \frac{1}{\log t} \, d\theta_\mathfrak{w}(t) = \frac{\theta_\mathfrak{w}(x)}{\log x} + \int_0^x \frac{\theta_\mathfrak{w}(t)}{\log^2 t} \frac{dt}{t},$$

from which it is easy to derive the corresponding theorems for $\pi_\mathfrak{w}$ from those for $\psi_\mathfrak{w}$.

2.2. Self-Similar Flows. A self-similar flow is best viewed as the following dynamics on the region of Figure 1. A point $x = x_1 N^{-1} + x_2 N^{-2} + \cdots = .x_1 x_2 \ldots$ on the unit interval moves vertically upward with unit speed until it reaches the graph, at which moment it jumps to $\{Nx\} = Nx - [Nx] = .x_2 x_3 \ldots$, the fractional part of Nx, and continues from there. In Figure 1, $N = 5$, and the expansions of A, B and C in base 5 are $A = 17/124 = .\overline{032}$, $B = 85/124 = .\overline{320}$, $C = 53/124 = .\overline{203}$.

DEFINITION 2.9. A flow $\mathcal{F}_{\mathfrak{w}}$ is *self-similar* if $N \geq 2$ and the weight function $\mathfrak{w}$ depends only on the first letter of the sequence on which it is evaluated. We then put

$$(2.13) \qquad\qquad w_j = \mathfrak{w}(j, j, j, \dots),$$

and

$$(2.14) \qquad\qquad r(\mathfrak{x}) = e^{-\mathfrak{w}_t(\mathfrak{x})}, \qquad r_j = e^{-w_j} = r(j, j, j, \dots),$$

for $j = 1, \dots, N$. The numbers r_j are called the *scaling ratios* of $\mathcal{F}_{\mathfrak{w}}$.

Note that $0 < r_j < 1$. We will assume that the *weights* $w_j = \log r_j^{-1}$ are ordered in increasing order, $0 < w_1 \leq w_2 \leq \cdots \leq w_N$, so that $1 > r_1 \geq r_2 \geq \cdots \geq r_N > 0$. When $N = 2$, the flow is called a *Bernoulli flow*. Such flows play an important role in ergodic theory (see [BKS, Chapters 2, 6 and 8]).

THEOREM 2.10. *The dynamical zeta function associated with a self-similar flow has a meromorphic continuation to the whole complex plane, given by*

$$(2.15) \qquad\qquad \zeta_{\mathfrak{w}}(s) = \frac{1}{1 - \sum_{j=1}^{N} r_j^s}.$$

Its logarithmic derivative is given by

$$(2.16) \qquad\qquad -\frac{\zeta'_{\mathfrak{w}}}{\zeta_{\mathfrak{w}}}(s) = \frac{\sum_{j=1}^{N} w_j r_j^s}{1 - \sum_{j=1}^{N} r_j^s}.$$

The dimension $D > 0$ *of the flow is the unique real solution of the equation* $1 = \sum_{j=1}^{N} r_j^s$.

PROOF. The sum over periodic sequences of fixed length l can be computed as follows:

$$\sum_{\mathfrak{x}: l(\mathfrak{x})=l} r(\mathfrak{x})^s = \sum_{a_1=1}^{N} \sum_{a_2=1}^{N} \cdots \sum_{a_l=1}^{N} r_{a_1}^s \cdots r_{a_l}^s$$
$$= (r_1^s + \cdots + r_N^s)^l.$$

Hence, for $\operatorname{Re} s > D$, the sum over all periodic sequences is equal to

$$\sum_{l=1}^{\infty} \frac{1}{l} \sum_{\mathfrak{x}: l(\mathfrak{x})=l} r(\mathfrak{x})^s = \sum_{l=1}^{\infty} \frac{1}{l} (r_1^s + \cdots + r_N^s)^l = -\log\left(1 - \sum_{j=1}^{N} r_j^s\right).$$

The theorem follows upon exponentiation and analytic continuation. Since the function $1 - \sum_{j=1}^{N} r_j^s$ is holomorphic, $\zeta_{\mathfrak{w}}$ is meromorphic. $\qquad\square$

REMARK 2.11. Because of Theorem 2.10, for a self-similar flow we can take the full complex plane for the window, $W = \mathbb{C}$; in that case, there is no screen. However, in applying our explicit formulas, we sometimes choose a screen to obtain information about the error of an approximation.

COROLLARY 2.12. *The set of complex dimensions $\mathcal{D}_{\mathfrak{w}} = \mathcal{D}_{\mathfrak{w}}(\mathbb{C})$ of the self-similar flow $\mathcal{F}_{\mathfrak{w}}$ is the set of solutions of the equation*

$$(2.17) \qquad \sum_{j=1}^{N} r_j^{\omega} = 1, \qquad \omega \in \mathbb{C}.$$

Moreover, the complex dimensions are simple (that is, the pole of $-\zeta_{\mathfrak{w}}'/\zeta_{\mathfrak{w}}$ at ω is simple). The residue at ω equals $-\operatorname{ord}(\zeta_{\mathfrak{w}}; \omega)$.

2.2.1. *Connection with Self-Similar Fractal Sets.* Given an open interval I of length L, we construct a self-similar one-dimensional fractal set $\mathcal{L}$ with scaling ratios $r_1, r_2, \ldots, r_N$. Such a set is called a *fractal string* (see [LP, LM, LvF1, 2]). The following construction is reminiscent of the construction of the Cantor set. Let N scaling factors $r_1, r_2, \ldots, r_N$ be given $(N \geq 2)$, with

$$1 > r_1 \geq r_2 \geq \ldots \geq r_N > 0.$$

Assume that

$$(2.18) \qquad R := \sum_{j=1}^{N} r_j < 1.$$

Subdivide I into intervals of length $r_1 L, \ldots, r_N L$. The remaining piece of length $(1 - R)L$ is the first member of the string, denoted by l_1, also called the first length in Remark 2.14 below. Repeat this process with the remaining intervals, to obtain N new lengths $l_2, \ldots, l_{N+1}$ in the next step, and N^{k-1} new lengths in the k-th step. As a result, we obtain a self-similar string $\mathcal{L}$ consisting of intervals of length $L(1-R)r_1^{k_1} \ldots r_N^{k_N}$ $(k_1, \ldots, k_N \in \mathbb{N})$, and a sequence $l_1 \geq l_2 \geq l_3 \geq \ldots$ of positive numbers, called the *lengths* of the string. We let $\zeta_{\mathcal{L}}(s) = \sum_{j=1}^{\infty} l_j^s$, the *geometric zeta function* of $\mathcal{L}$ (see [LvF1, 2]).

THEOREM 2.13. *Let $\mathcal{L}$ be a self-similar string, constructed as above with scaling ratios $r_1 = e^{-w_1}, \ldots, r_N = e^{-w_N}$. Then the geometric zeta function of this string has a meromorphic continuation to the whole complex plane, given by*

$$(2.19) \qquad \zeta_{\mathcal{L}}(s) = (L(1-R))^s \zeta_{\mathfrak{w}}(s), \quad \text{for } s \in \mathbb{C}.$$

Here, L is the total length of $\mathcal{L}$, and R is given by (2.18).

This follows from Theorem 2.10 combined with [LvF2, Theorem 2.3, p. 25].

REMARK 2.14. For a self-similar string, the total length of $\mathcal{L}$ is also the length of the initial interval I in the above construction. We can always normalize a self-similar string in such a way that $\zeta_{\mathcal{L}} = \zeta_{\mathfrak{w}}$ (equivalently, that the first length of $\mathcal{L}$ is 1), by choosing $L(1-R) = 1$. This does not affect the complex dimensions of the string.

Note that we need to assume that $R = \sum_{j=1}^{N} r_j < 1$, which corresponds to a lower bound on the weights $w_j = -\log r_j$. There is no analogue of this condition for general suspended flows.

REMARK 2.15. The Euler product does not seem to have a clear geometric interpretation in the language of fractal strings. There is, however, a natural self-similar flow on $\mathcal{L}$: the flow

$$(2.20) \qquad \mathcal{F}_{\mathcal{L}}(t, j, x) = \begin{cases} (0, j, xe^t) & \text{if } xe^t < l_j, \\ \mathcal{F}_{\mathcal{L}}(t - \log l_j, j, 1) & \text{otherwise.} \end{cases}$$

The lengths l_j correspond to the periodic sequences $\mathfrak{x}$ of the flow $\mathcal{F}_\mathfrak{w}$ via the formula

$$(2.21) \qquad\qquad l_j = \prod_{k=0}^{l(\mathfrak{x})-1} r(\sigma^k \mathfrak{x}).$$

REMARK 2.16 (Geometric and dynamical complex dimensions). In [LvF2], the geometric complex dimensions of a fractal string are defined as the poles of its geometric zeta function. Thus the complex dimensions are counted with a multiplicity, and the zeros of the geometric zeta function are unimportant. On the other hand, the dynamical complex dimensions are defined as the poles of the logarithmic derivative of the dynamical zeta function. Thus the complex dimensions are simple, and both the zeros and the poles of the dynamical zeta function are counted. For self-similar flows, the dynamical zeta function and the geometric zeta function of the corresponding string coincide (up to normalization), and this zeta function has no zeros. Hence, as sets (without multiplicity), the geometric and dynamical complex dimensions coincide for self-similar flows and strings.

REMARK 2.17 (Higher-dimensional case). We have discussed above the case of fractal strings (i.e., the one-dimensional case) because it is the one studied in most detail in [LvF2]. However, it is clear that our results can be applied to higher-dimensional self-similar fractals [F,Mn] as well. This allows us to obtain information about the symbolic dynamics of self-similar fractals. On the other hand, as in the previous remark, it does not give information about the actual geometry of such fractals.

2.3. The Lattice and Nonlattice Case. Let $\mathcal{F}_\mathfrak{w}$ be a self-similar flow. Recall that $\mathfrak{w}$ depends only on the first symbol and $w_j = \mathfrak{w}(j, j, \dots)$ for $j = 1, \dots, N$. Consider the subgroup G of $\mathbb{R}$ generated by these weights, $G = \sum_{j=1}^{N} \mathbb{Z} w_j$.

DEFINITION 2.18. The case when G is dense in $\mathbb{R}$ is called the *nonlattice case*. We then say that $\mathcal{F}_\mathfrak{w}$ is a *nonlattice flow*.

The case when G is not dense (and hence discrete) in $\mathbb{R}$ is called the *lattice case*. We then say that $\mathcal{F}_\mathfrak{w}$ is a *lattice flow*. In this situation there exists a unique positive real real number w, called the *generator* of the flow, and positive integers $k_1, \dots, k_N$ without common divisor, such that $1 \le k_1 \le \dots \le k_N$ and

$$(2.22) \qquad\qquad w_j = k_j w,$$

for $j = 1, \dots, N$.

The generator of $\mathcal{F}_\mathfrak{w}$ generates the flow in the sense that the weight of every periodic sequence of $\mathcal{F}_\mathfrak{w}$ is an integer multiple of w.

We introduce a real number D_0 as follows: Let m be the number of integers j in $1, \dots, N$ such that $r_j = r_N$, and let $D_0 \in \mathbb{R}$ be defined by

$$(2.23) \qquad\qquad 1 + \sum_{j=1}^{N-m} r_j^{D_0} = m r_N^{D_0}.$$

The dynamical complex dimensions of a self-similar flow are described in the following theorem. For brevity, we will usually refer to them as the complex dimensions of $\mathcal{F}_\mathfrak{w}$.

THEOREM 2.19. *Let $\mathcal{F}_{\mathfrak{w}}$ be a self-similar flow of dimension D and with scaling ratios $1 > r_1 \geq \cdots \geq r_N > 0$. Then the value $s = D$ is the only complex dimension of $\mathcal{F}_{\mathfrak{w}}$ on the real line, all complex dimensions are simple, and the residue at a complex dimension (i.e., $\mathrm{res}(-\zeta'_{\mathfrak{w}}/\zeta_{\mathfrak{w}}; \omega)$) is a positive integer. The set of complex dimensions in $\mathbb{C}$ (see Remark 2.11) of $\mathcal{F}_{\mathfrak{w}}$ is contained in the bounded strip $D_0 \leq \mathrm{Re}\, s \leq D$:*

$$(2.24) \qquad \mathcal{D}_{\mathfrak{w}} = \mathcal{D}_{\mathfrak{w}}(\mathbb{C}) \subset \{s \in \mathbb{C} \colon D_0 \leq \mathrm{Re}\, s \leq D\}.$$

It is symmetric with respect to the real axis and infinite, with density bounded by

$$(2.25) \qquad \#(\mathcal{D}_{\mathfrak{w}} \cap \{\omega \in \mathbb{C} \colon |\mathrm{Im}\, \omega| \leq T\}) \leq \frac{w_N}{\pi} T + O(1),$$

as $T \to \infty$.

In the *lattice case, $\zeta_{\mathfrak{w}}(s)$ is a rational function of e^{-ws}, where w is the generator of $\mathcal{F}_{\mathfrak{w}}$. So, as a function of s, it is periodic with period $2\pi i/w$. The complex dimensions ω are obtained by finding the complex solutions z of the polynomial equation (of degree k_N)*

$$(2.26) \qquad \sum_{j=1}^{N} z^{k_j} = 1, \quad \text{with } e^{-w\omega} = z.$$

Hence there exist finitely many poles $\omega_1(= D), \omega_2, \ldots, \omega_q$, such that

$$(2.27) \qquad \mathcal{D}_{\mathfrak{w}} = \{\omega_u + 2\pi i n/w \colon n \in \mathbb{Z}, u = 1, \ldots, q\}.$$

In other words, the poles lie periodically on finitely many vertical lines, and on each line they are separated by $2\pi/w$. The residue of the complex dimensions corresponding to one value of $z = e^{-w\omega}$ is the multiplicity of z as a solution of (2.26).

In the *nonlattice case, D is simple and is the unique pole of $\zeta_{\mathfrak{w}}$ on the line $\mathrm{Re}\, s = D$. Further, there is an infinite sequence of complex dimensions of $\mathcal{F}_{\mathfrak{w}}$ coming arbitrarily close (from the left) to the line $\mathrm{Re}\, s = D$. There exists a screen S to the left of the line $\mathrm{Re}\, s = D$, such that $-\zeta'_{\mathfrak{w}}/\zeta_{\mathfrak{w}}$ satisfies $(\mathbf{H}_1)$ and $(\mathbf{H}_2)$ with $\kappa = 0$ (see Equations (3.2) and (3.3) below), and the residue of $-\zeta'_{\mathfrak{w}}/\zeta_{\mathfrak{w}}$ at the pole ω in W is equal to 1. Finally, the complex dimensions of $\mathcal{F}_{\mathfrak{w}}$ can be approximated (via an explicit algorithm, as described in [LvF2, §2.6]) by the complex dimensions of a sequence of lattice strings, with smaller and smaller generator. Hence the complex dimensions of a nonlattice string have an almost periodic structure.*

COROLLARY 2.20. *Every self-similar flow has infinitely many complex dimensions with positive real part.*

PROOF OF THEOREM 2.19. For a proof of these facts, see [LvF2, Theorem 2.13, pp. 37–40]. The density estimate (2.25) follows from the fact that the right-hand side of (2.25) gives the asymptotic density of the number of poles of $\zeta_{\mathfrak{w}}$, counted *with* multiplicity. The $O(1)$ estimate improves [LvF2, Theorem 2.22, p. 47]. It is proved in [LvF3]. □

2.4. Examples of Complex Dimensions of Self-Similar Flows.

EXAMPLE 2.21 (The Cantor flow). This is the self-similar flow on the alphabet $\{0, 1\}$, with two equal weights $w_1 = w_2 = \log 3$. It has 2^n periodic sequences of weight $n \log 3$, for $n = 1, 2, \ldots$. The dynamical zeta function of this flow is

$$(2.28) \qquad \zeta_{\mathrm{CF}}(s) = \frac{1}{1 - 2 \cdot 3^{-s}}.$$

After taking the logarithmic derivative, one finds that the dynamical complex dimensions are the solutions of the equation

$$(2.29) \qquad 2 \cdot 3^{-\omega} = 1 \qquad (\omega \in \mathbb{C}).$$

We find

$$(2.30) \qquad \mathcal{D}_{\mathrm{CF}} = \left\{ D + \frac{2\pi i}{w} n \colon n \in \mathbb{Z} \right\},$$

with $D = \log_3 2$ and $w = \log 3$.

EXAMPLE 2.22 (The Fibonacci flow). Next we consider a self-similar flow with two lines of complex dimensions. The *Fibonacci flow* is the flow Fib on the alphabet $\{0, 1\}$ with weights $w_1 = \log 2$, $w_2 = 2 \log 2$. Its periodic sequences have weight $\log 2, 2 \log 2, \ldots, n \log 2, \ldots$, with multiplicity respectively $1, 2, \ldots, F_{n+1}, \ldots$, the Fibonacci numbers. Recall that these numbers are defined by the following recursive equation:

$$(2.31) \qquad F_{n+1} = F_n + F_{n-1}, \text{ with } F_0 = 0, \ F_1 = 1.$$

The dynamical zeta function of the Fibonacci flow is

$$(2.32) \qquad \zeta_{\mathrm{Fib}}(s) = \frac{1}{1 - 2^{-s} - 4^{-s}}.$$

The complex dimensions are found by solving the quadratic equation

$$(2.33) \qquad (2^{-\omega})^2 + 2^{-\omega} = 1 \qquad (\omega \in \mathbb{C}).$$

We find $2^{-\omega} = \left(-1 + \sqrt{5} \right)/2 = \phi^{-1}$ and $2^{-\omega} = -\phi$, where $\phi = (1 + \sqrt{5})/2$ is the golden ratio. Hence

$$(2.34) \qquad \mathcal{D}_{\mathrm{Fib}} = \left\{ D + \frac{2\pi i}{w} n \colon n \in \mathbb{Z} \right\} \cup \left\{ -D + \frac{2\pi i}{w} (n + 1/2) \colon n \in \mathbb{Z} \right\},$$

with $D = \log_2 \phi$ and $w = \log 2$.

EXAMPLE 2.23 (The Golden flow). We consider the nonlattice flow GF with weights $w_1 = \log 2$ and $w_2 = \phi \log 2$, where $\phi = (1 + \sqrt{5})/2$ is the golden ratio. We call this flow the *golden flow*. Its dynamical zeta function is

$$(2.35) \qquad \zeta_{\mathrm{GF}}(s) = \frac{1}{1 - 2^{-s} - 2^{-\phi s}},$$

and its complex dimensions are the solutions of the transcendental equation

$$(2.36) \qquad 2^{-\omega} + 2^{-\phi\omega} = 1 \qquad (\omega \in \mathbb{C}).$$

A diagram of the complex dimensions of the golden flow is given in Figure 2. To obtain it, we chose the approximation $\phi \approx 987/610$ to approximate the flow by the lattice flow with weights $w_1 = 610w$, $w_2 = 987w$, where $w = (1/610) \log 2$. We then used Maple to solve the corresponding polynomial equation. In particular, the dimension D of the golden flow is approximately equal to $D = .77921 \ldots$. See also Example 6.8.

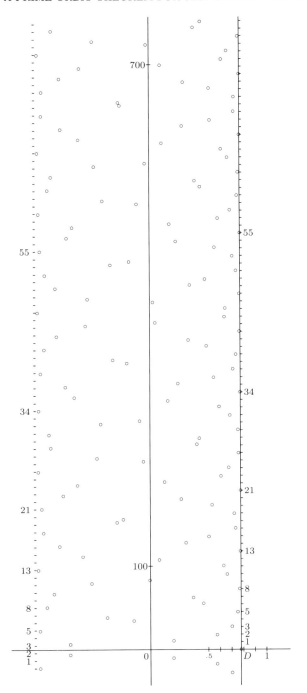

FIGURE 2. The almost periodic behavior of the complex dimensions of the golden flow.

3. Explicit Formulas

We will formulate our explicit formulas in the more general framework of [LvF2, Chapter 4]. We refer to this book for the proofs and much additional information.

Let η be a positive measure on $(0, \infty)$, supported away from 0. Its Mellin transform is

$$(3.1) \qquad \zeta_\eta(s) = \int_0^\infty x^{-s} d\eta,$$

the *geometric zeta function* of η. We assume that ζ_η is convergent for some s, and we write D for the abcissa of convergence. We assume that there exists a screen to the left of $\operatorname{Re} s = D$ such that ζ_η has a meromorphic continuation to the corresponding window. To simplify the exposition, we also assume that the poles of ζ_η are simple. This is the case, for example, if $\zeta_{\mathfrak{w}}$ has a meromorphic continuation to a neighborhood of W. The general case, when the poles of ζ_η may have arbitrary multiplicities, is treated in [LvF2, Chapter 4].

The screen S is given as the graph of a bounded function S, with the horizontal and vertical axes interchanged:

$$S = \{S(t) + it \colon t \in \mathbb{R}\}.$$

We will write $\inf S = \inf_{t \in \mathbb{R}} S(t)$ and $\sup S = \sup_{t \in \mathbb{R}} S(t)$. We assume in addition that S is a Lipschitz continuous function; i.e., there exists a nonnegative real number, denoted by $\|S\|_{\mathrm{Lip}}$, such that

$$|S(x) - S(y)| \le \|S\|_{\mathrm{Lip}} |x - y| \qquad \text{for all } x, y \in \mathbb{R}.$$

Further, recall from Section 2 that the window W is the part of the complex plane to the right of S; see formula (2.5).

Assume that ζ_η satisfies the following growth conditions:

There exist real constants $\kappa \ge 0$ and $C > 0$ and a sequence $\{T_n\}_{n \in \mathbb{Z}}$ of real numbers tending to $\pm\infty$ as $n \to \pm\infty$, with $T_{-n} < 0 < T_n$ for $n \ge 1$ and $\lim_{n \to +\infty} T_n / |T_{-n}| = 1$, such that

($\mathbf{H_1}$): For all $n \in \mathbb{Z}$ and all $\sigma \ge S(T_n)$,

$$(3.2) \qquad |\zeta_\eta(\sigma + iT_n)| \le C \cdot |T_n|^\kappa,$$

($\mathbf{H_2}$): For all $t \in \mathbb{R}$, $|t| \ge 1$,

$$(3.3) \qquad |\zeta_\eta(S(t) + it)| \le C \cdot |t|^\kappa.$$

Hypothesis ($\mathbf{H_1}$) is a polynomial growth condition along horizontal lines (necessarily avoiding the poles of ζ_η), while hypothesis ($\mathbf{H_2}$) is a polynomial growth condition along the vertical direction of the screen.

In the following, we denote by $\operatorname{res}(g(s); \omega)$ the residue of a meromorphic function $g = g(s)$ at ω. It vanishes unless ω is a pole of g. Also,

$$(3.4) \qquad (s)_k = \frac{\Gamma(s + k)}{\Gamma(s)},$$

for $k \in \mathbb{Z}$. Thus, $(s)_0 = 1$ and, for $k \ge 1$, $(s)_k = s(s + 1) \dots (s + k - 1)$.

Let

$$(3.5) \qquad N_\eta(x) = N_\eta^{[1]}(x) = \eta(0, x) + \frac{1}{2}\eta(\{x\}),$$

and more generally, let $N_\eta^{[k]}(x)$ be the $(k-1)$-st antiderivative of this function, for $k = 1, 2, \ldots$. Our explicit formula expresses this function as a sum over the poles of ζ_η.

THEOREM 3.1 (The pointwise explicit formula). *Let η be a generalized fractal string, satisfying hypotheses* $(\mathbf{H}_1)$ *and* $(\mathbf{H}_2)$. *Let k be a positive integer such that $k > \kappa + 1$, where $\kappa \geq 0$ is the exponent occurring in the statement of* $(\mathbf{H}_1)$ *and* $(\mathbf{H}_2)$. *Then, for all $x > 0$, the pointwise explicit formula is given by the following equality:*

$$
\begin{aligned}
N_\eta^{[k]}(x) = &\sum_{\substack{\omega \in \mathcal{D}_\eta(W) \\ \omega \notin \{0, -1, \ldots, -(k-1)\}}} \operatorname{res}\left(\zeta_\eta(s); \omega\right) \frac{x^{\omega+k-1}}{(\omega)_k} \\
&+ \sum_{j=0}^{k-1} \operatorname{res}\left(\frac{x^{s+k-1}\zeta_\eta(s)}{(s)_k}; -j\right) + R_\eta^{[k]}(x).
\end{aligned}
$$
(3.6)

Here, for $x > 0$, $R(x) = R_\eta^{[k]}(x)$ is the error term, given by the absolutely convergent integral

$$
R(x) = R_\eta^{[k]}(x) = \frac{1}{2\pi i} \int_S x^{s+k-1}\zeta_\eta(s) \frac{ds}{(s)_k}.
$$
(3.7)

Further, for all $x > 0$, we have

$$
R(x) = R_\eta^{[k]}(x) \leq C(1 + \|r\|_{\mathrm{Lip}}) \frac{x^{k-1}}{k - \kappa - 1} \max\{x^{\sup S}, x^{\inf S}\} + C',
$$
(3.8)

where C is the positive constant occurring in $(\mathbf{H}_1)$ *and* $(\mathbf{H}_2)$ *and C' is some suitable positive constant. The constants $C(1 + \|r\|_{\mathrm{Lip}})$ and C' depend only on η and the screen, but not on k.*

In particular, we have the following pointwise error estimate:

$$
R(x) = R_\eta^{[k]}(x) = O\left(x^{\sup S+k-1}\right),
$$
(3.9)

as $x \to \infty$. Moreover, if $S(t) < \sup S$ for all $t \in \mathbb{R}$ (i.e., if the screen lies strictly to the left of the line $\operatorname{Re} s = \sup S$), then $R(x)$ is of order less than $x^{\sup S+k-1}$ as $x \to \infty$:

$$
R(x) = R_\eta^{[k]}(x) = o\left(x^{\sup S+k-1}\right),
$$
(3.10)

as $x \to \infty$.

To formulate our second explicit formula, a distributional formula, we view η as a distribution, acting on a test function φ defined on $(0, \infty)$ by

$$
\left\langle N_\eta^{[0]}, \varphi \right\rangle = \int_0^\infty \varphi \, d\eta.
$$
(3.11)

We then define $N_\eta^{[k]}$ as the distribution obtained by integrating this one k times, so that

$$
\left\langle N_\eta^{[k]}, \varphi \right\rangle = \int_0^\infty \int_y^\infty \frac{(x-y)^{k-1}}{(k-1)!} \varphi(x) \, dx \, \eta(dy).
$$
(3.12)

It is easily verified that this definition coincides with formula (3.5) and the next line above when $k \geq 1$. For $k \leq 0$, we extend this definition by differentiating $|k|$ times the distribution $N_\eta^{[0]}$.

We shall denote by $\widetilde{\varphi}$ the *Mellin transform* of a (suitable) function φ on $(0, \infty)$; it is defined by

$$(3.13) \qquad \widetilde{\varphi}(s) = \int_0^\infty \varphi(x) x^{s-1} \, dx.$$

THEOREM 3.2 (The distributional explicit formula). *Let η be a generalized fractal string satisfying hypotheses $(\mathbf{H}_1)$ and $(\mathbf{H}_2)$. Then, for every $k \in \mathbb{Z}$, the distribution $N_\eta^{[k]}$ is given by formula (3.6), interpreted as a distribution. That is, the action of $N_\eta^{[k]}$ on a test function φ is given by*

$$
\left\langle N_\eta^{[k]}, \varphi \right\rangle = \sum_{\substack{\omega \in \mathcal{D}_\eta(W) \\ \omega \notin \{0, -1, \ldots, -(k-1)\}}} \mathrm{res}\left(\zeta_\eta(s); \omega\right) \frac{\widetilde{\varphi}(\omega + k)}{(\omega)_k}
$$

$$(3.14)$$

$$
+ \sum_{j=0}^{k-1} \mathrm{res}\left(\frac{\zeta_\eta(s)\widetilde{\varphi}(s+k)}{(s)_k}; -j\right) + \left\langle R_\eta^{[k]}, \varphi \right\rangle.
$$

Here, the distribution $R = R_\eta^{[k]}$ is the error term, given by

$$(3.15) \qquad \langle R, \varphi \rangle = \left\langle R_\eta^{[k]}, \varphi \right\rangle = \frac{1}{2\pi i} \int_S \zeta_\eta(s)\widetilde{\varphi}(s+k) \frac{ds}{(s)_k}.$$

DEFINITION 3.3. We will say that a distribution R on $(0, \infty)$ is of *asymptotic order* at most x^α (respectively, *less than* x^α)—and we will write $R(x) = O(x^\alpha)$ (respectively, $R(x) = o(x^\alpha)$), as $x \to \infty$—if applied to a test function φ, we have that

$$(3.16) \qquad \langle R, \varphi_a \rangle = O\left(a^\alpha\right) \quad \text{(respectively, } \langle R, \varphi_a \rangle = o\left(a^\alpha\right)\text{)}, \quad \text{as } a \to \infty,$$

where $\varphi_a(x) = a^{-1}\varphi(x/a)$.

THEOREM 3.4. *Fix $k \in \mathbb{Z}$. Assume that the hypotheses of Theorem 3.2 are satisfied, and let the distribution $R = R_\eta^{[k]}$ be given by (3.15). Then R is of asymptotic order at most $x^{\sup S + k - 1}$ as $x \to \infty$:*

$$(3.17) \qquad R_\eta^{[k]}(x) = O\left(x^{\sup S + k - 1}\right), \quad \text{as } x \to \infty,$$

in the sense of Definition 3.3.

Moreover, if $S(t) < \sup S$ for all $t \in \mathbb{R}$ (i.e., if the screen lies strictly to the left of the line $\mathrm{Re}\, s = \sup S$), then R is of asymptotic order less than $x^{\sup S + k - 1}$ as $x \to \infty$:

$$(3.18) \qquad R_\eta^{[k]}(x) = o\left(x^{\sup S + k - 1}\right), \quad \text{as } x \to \infty.$$

We refer to [LvF2, Chapter 4] for a proof of Theorems 3.1, 3.2 and 3.4.

REMARK 3.5 (Oscillatory terms in the explicit formula). Our explicit formulas give expansions of various functions associated with a measure η as a sum over the poles of ζ_η. The term corresponding to the pole ω of multiplicity one is of the form Cx^ω, where C is a constant depending on ω. If ω is real, the function x^ω simply has a certain asymptotic behavior as $x \to \infty$. If, on the other hand, $\omega = \beta + i\gamma$ has a nonzero imaginary part γ, then $x^\omega = x^\beta \cdot x^{i\gamma}$ is of order $O(x^\beta)$ as $x \to \infty$, with a multiplicatively periodic behavior: The function $x^{i\gamma} = \exp(i\gamma \log x)$ takes the same value at the points $e^{2\pi n/\gamma} x$ $(n \in \mathbb{Z})$. Thus, the term corresponding to ω will be called an oscillatory term. If there are poles with higher multiplicity, there will also be terms of the form $Cx^\omega(\log x)^m$, $m \in \mathbb{N}^*$, which have a similar oscillatory behavior.

EXAMPLE 3.6 (The classical Prime Number Theorem). Let $\zeta(s) = 1 + 2^{-s} + 3^{-s} + \ldots$ (for $\operatorname{Re} s > 1$) be the Riemann zeta function. This function has an Euler product $\zeta(s) = \prod_p (1 - p^{-s})^{-1}$ (for $\operatorname{Re} s > 1$), where p runs over the prime numbers. Analogously to Corollary 2.6, we obtain $-\zeta'/\zeta(s) = \int_0^\infty x^{-s}\, d\psi(x) = \zeta_{\mathfrak{P}}(s)$, where $\psi(x) = \sum_{p^k \le x} \log p = N_{\mathfrak{P}}^{[1]}(x)$, and

$$(3.19) \qquad \mathfrak{P} = \sum_{m \ge 1,\, p} (\log p)\delta_{\{p^m\}}$$

is the *prime string* (see [LvF2]). We apply Theorem 3.2 to $\eta = \mathfrak{P}$ to obtain the explicit formula for ψ:

$$(3.20) \qquad \psi(x) = x - \sum_{\rho \in W} \operatorname{res}(\zeta'/\zeta(s);\rho)\frac{x^\rho}{\rho} - \frac{1}{2\pi i}\int_S \frac{\zeta'}{\zeta}(s)x^s\frac{ds}{s},$$

where ρ runs through the sequence of critical zeros of ζ: $\zeta(\rho) = 0$, $0 < \operatorname{Re} \rho < 1$.

By means of classical arguments [I, Theorem 19], it is known that ζ has a zero free region of the form

$$\{\sigma + it \in \mathbb{C}\colon \sigma > 1 - C/\log t\},$$

for some positive constant C. Also, $-\zeta'/\zeta$ is not too large on the boundary of this region. In our language, this means that we can choose a screen to the left of $\operatorname{Re} s = 1$ such that there are no zeros of ζ in W and $-\zeta'/\zeta$ satisfies $(\mathbf{H_1})$ and $(\mathbf{H_2})$. Then (3.20) becomes

$$(3.21) \qquad \psi(x) = x + o(x),$$

as $x \to \infty$. This is equivalent to the classical Prime Number Theorem.

Using the existence of the zero free region, one can derive the stronger estimate

$$(3.22) \qquad \psi(x) = x + O\left(xe^{-c\sqrt{\log x}}\right),$$

as $x \to \infty$, for some positive constant c (see [E; I, Theorem 23]). This is the classical Prime Number Theorem, with Error Term.

4. The Prime Orbit Theorem for Flows

Let $\mathcal{F}_\mathfrak{w}$ be a suspended flow as in Section 2. In Corollary 2.6, we have written the logarithmic derivative of $\zeta_\mathfrak{w}(s)$ as the Mellin transform of the counting function $\psi_\mathfrak{w}$ of the weighted periodic orbits of σ, as defined in (2.9). Put $\eta = d\psi_\mathfrak{w}$, so that $\zeta_\eta = -\zeta'_\mathfrak{w}/\zeta_\mathfrak{w}$. The poles of $-\zeta'_\mathfrak{w}/\zeta_\mathfrak{w}$ are the complex dimensions of $\mathcal{F}_\mathfrak{w}$ and the residue at ω is $-\operatorname{ord}(\zeta_\mathfrak{w};\omega)$. By Theorems 3.2 and 3.4, we obtain the following explicit formula for the counting function of weighted periodic orbits of σ.

THEOREM 4.1 (The Prime Orbit Theorem with Error Term). *Let $\mathcal{F}_\mathfrak{w}$ be a suspended flow that satisfies conditions $(\mathbf{H_1})$ and $(\mathbf{H_2})$. Then we have the following equality between distributions:*

$$(4.1) \quad \psi_\mathfrak{w}(x) = \frac{x^D}{D} + \sum_{\omega \in \mathcal{D}_\mathfrak{w}\setminus\{D,0\}} -\operatorname{ord}(\zeta_\mathfrak{w};\omega)\frac{x^\omega}{\omega} + \operatorname{res}\left(-\frac{x^s\zeta'_\mathfrak{w}(s)}{s\zeta_\mathfrak{w}(s)};0\right) + R(x),$$

where $\mathrm{ord}\,(\zeta_{\mathfrak{w}};\omega) < 0$ *denotes the order of* $\zeta_{\mathfrak{w}}$ *at* ω, *and*

$$(4.2) \qquad R(x) = -\int_S \frac{\zeta_{\mathfrak{w}}'}{\zeta_{\mathfrak{w}}}(s)x^s\,\frac{ds}{s} = O\left(x^{\sup S}\right),$$

as $x \to \infty$.

    If 0 *is not a complex dimension of the flow, then the third term on the right-hand side of* (4.1) *simplifies to* $-\zeta_{\mathfrak{w}}'/\zeta_{\mathfrak{w}}(0)$. *In general, this term is of the form* $p + q\log x$, *for some constants* p *and* q.
    If D *is the only complex dimension on the line* $\mathrm{Re}\,s = D$, *then the error term,*

$$(4.3) \qquad \sum_{\omega \in \mathcal{D}_{\mathfrak{w}}\setminus\{D,0\}} -\mathrm{ord}\,(\zeta_{\mathfrak{w}};\omega)\frac{x^\omega}{\omega} + \mathrm{res}\left(-\frac{x^s\zeta_{\mathfrak{w}}'(s)}{s\zeta_{\mathfrak{w}}(s)};0\right) + R(x),$$

is estimated by $o(x^D)$, *as* $x \to \infty$. *If this is the case, then we obtain a Prime Orbit Theorem for* $\mathcal{F}_{\mathfrak{w}}$ *as follows:*

$$(4.4) \qquad \psi_{\mathfrak{w}}(x) = \frac{x^D}{D} + o\left(x^D\right),$$

as $x \to \infty$.

PROOF. The first part of the theorem follows from the distributional explicit formula (Theorem 3.2) and from the first part of Theorem 3.4, while the second part follows from the second part of Theorem 3.4.   □

    The explicit formula holds for every flow satisfying our conditions on $-\zeta_{\mathfrak{w}}'/\zeta_{\mathfrak{w}}$. In particular, we can apply it to the 'axiom A flows' considered in [PP2, Chapter 6], in view of [PP2, pp. 100–101]. We hope to do so more explicitly in a later work. For simplicity and for the sake of concision, however, we will focus in the rest of this paper on the important example of 'self-similar flows' (in the sense of Section 2.2 above).

5. The Prime Orbit Theorem for Self-Similar Flows

    For self-similar flows, $\zeta_{\mathfrak{w}}$ does not have any zeros (see (2.3)). Hence every contribution to (4.1) comes from a pole of $\zeta_{\mathfrak{w}}$, and each coefficient $-\mathrm{ord}(\zeta_{\mathfrak{w}};\omega)$ is positive. Furthermore, 0 is never a complex dimension, so the third term on the right-hand side of (4.1) in the explicit formula is

$$(5.1) \qquad -\frac{\zeta_{\mathfrak{w}}'(0)}{\zeta_{\mathfrak{w}}(0)} = -\frac{1}{N-1}\sum_{j=1}^N w_j.$$

We can obtain information about $\psi_{\mathfrak{w}}$ by choosing a suitable screen.

    5.1. Lattice Flows. In the lattice case, we obtain the Prime Orbit Theorem for lattice self-similar flows:

$$(5.2) \qquad \psi_{\mathfrak{w}}(x) = g_1(\log x)x^D - \frac{1}{N-1}\sum_{j=1}^N w_j + O\left(x^{D-\alpha}\right),$$

as $x \to \infty$. Here, $D - \alpha$ is the abcissa of the first vertical line of complex dimensions next to D, and the periodic function g_1, of period w, is given by[2]

$$
(5.3) \qquad g_1(y) = \sum_{n=-\infty}^{\infty} \frac{e^{2\pi i n y/w}}{D + 2\pi i n/w} = \frac{b_1 w}{b_1 - 1} b_1^{-\{y/w\}},
$$

where $b_1 = e^{wD}$. By choosing a screen located to the left of all the complex dimensions of $\mathcal{F}_{\mathfrak{w}}$, we can even obtain more precise information about $\psi_{\mathfrak{w}}$. In the notation of Theorem 2.19, we obtain

$$
(5.4) \qquad
\begin{aligned}
\psi_{\mathfrak{w}}(x) &= \sum_{u=1}^{q} -\operatorname{ord}\left(\zeta_{\mathfrak{w}}; \omega_u\right) \sum_{n \in \mathbb{Z}} \frac{x^{\omega_u + 2\pi i n/w}}{\omega_u + 2\pi i n/w} - \frac{1}{N-1} \sum_{j=1}^{N} w_j \\
&= \sum_{u=1}^{q} -\operatorname{ord}\left(\zeta_{\mathfrak{w}}; \omega_u\right) g_u(\log x) x^{\omega_u} - \frac{1}{N-1} \sum_{j=1}^{N} w_j,
\end{aligned}
$$

where for each $u = 1, \ldots, q$, the function g_u is periodic of period w, given by

$$
(5.5) \qquad g_u(y) = \sum_{n \in \mathbb{Z}} \frac{e^{2\pi i n y/w}}{\omega_u + 2\pi i n/w} = \frac{b_u w}{b_u - 1} b_u^{-\{y/w\}},
$$

where $b_u = e^{w\omega_u}$. Here, $\omega_1(= D), \omega_2, \ldots, \omega_q$ are given as in the lattice case of Theorem 2.19 and $\operatorname{ord}(\zeta_{\mathfrak{w}}; \omega_1) = -1$.

For instance, the Cantor flow (with $D = \log_3 2$ and $w_1 = w_2 = w = \log 3$, see Example 2.21) has

$$
(5.6) \qquad \psi_{\mathrm{CF}}(x) = g_1(\log x) x^D - 2 \log 3,
$$

with $g_1(y) = w 2^{1 - \{y/w\}}$, while the Fibonacci flow[3] of Example 2.22 (with $D = \log_2 \phi$ and $w_1 = w = \log 2$, $w_2 = 2w$) has:

$$
(5.7) \qquad \psi_{\mathrm{Fib}}(x) = g_1(\log x) x^D + g_2(\log x) x^{\pi i/w} x^{-D} - 3 \log 2,
$$

where $g_1(y) = w \phi^{2 - \{y/w\}}$ and

$$
g_2(y) = \sum_{n \in \mathbb{Z}} \frac{e^{2\pi i n y/w}}{-D + 2\pi i (n + 1/2)/w} = w \phi^{\{y/w\} - 2} e^{-\pi i \{y/w\}}.
$$

In the second asymptotic term, the product $e^{-\pi i \{(\log x)/w\}} x^{\pi i/w}$ combines to give the sign $(-1)^{[(\log x)/w]}$.

5.2. Nonlattice Flows. In the nonlattice case, we use Theorem 2.19 according to which there exists $\delta > 0$ and a screen S lying to the left of the vertical line $\operatorname{Re} s = D - \delta$ such that $-\zeta'_{\mathfrak{w}}/\zeta_{\mathfrak{w}}$ is bounded on S and all the complex dimensions ω to the right of S have residue $\operatorname{res}(-\zeta'_{\mathfrak{w}}/\zeta_{\mathfrak{w}}; \omega) = 1$. Then $R(x) = O(x^{D-\delta})$, as $x \to \infty$. There are no complex dimensions with $\operatorname{Re} \omega = D$ except for D itself. Hence, the assumptions of Theorem 4.1 are satisfied. Therefore, in view of Theorem 4.1, we deduce by a classical argument (see the proof of Theorems 6.7 and 6.14

---

[2]We use the notation $\{u\}$ for the fractional part, and $[u]$ for the integer part of the real number u, so that $\{u\} = u - [u] \in [0, 1)$.

[3]Also called the golden mean flow in the literature (see, e.g., [BKS, p. 59]), but not to be confused with the golden flow in our present paper, which is a nonlattice self-similar flow.

on page 24) the Prime Orbit Theorem for nonlattice suspended flows:

$$(5.8) \qquad \psi_{\mathfrak{w}}(x) = \frac{x^D}{D} + \sum_{\omega \in \mathcal{D}_{\mathfrak{w}} \setminus \{D\}} \frac{x^\omega}{\omega} + O(x^{D-\delta}) = \frac{x^D}{D} + o\left(x^D\right),$$

as $x \to \infty$. (See also Theorem 6.14, and when $N = 2$, Theorem 6.7 below for a better estimate of the error.) We note that this estimate is always best possible, since by the nonlattice case of Theorem 2.19, there always exist complex dimensions of $\mathfrak{w}$ arbitrarily close to the vertical line $\operatorname{Re} s = D$.

REMARK 5.1. It would be interesting to apply Theorem 4.1 to suspended flows that are more general than self-similar flows: for example, those considered by Lalley in [Lal1, 2], such as the 'approximately self-similar flows' naturally associated with limit sets of suitable Kleinian groups. This would require a more detailed study of the dynamical zeta function of each of these flows. It is known that the lattice-nonlattice dichotomy applies in these more general cases; see [Lal1, 2]. We hope to investigate this question in a later work.

6. The Error Term in the Nonlattice Case

A nonlattice flow has weights $w_1 \leq \cdots \leq w_N$, where at least one ratio w_j/w_k is irrational. Let

$$(6.1) \qquad f(s) = 1 - \sum_{j=1}^{N} e^{-w_j s}.$$

Then $D > 0$ is the unique real solution of the equation $f(s) = 0$. Moreover, the derivative

$$(6.2) \qquad f'(s) = \sum_{j=1}^{N} w_j e^{-w_j s}$$

does not vanish at D.

When $N = 2$, the flow is called a Bernoulli flow (see Section 2.2). We write $\alpha = w_2/w_1$, so $\alpha > 1$ is irrational. In this case, we obtain very detailed information about the growth of $-\zeta'_{\mathfrak{w}}/\zeta_{\mathfrak{w}}$ on the line $\operatorname{Re} s = D$, and we can compute a pole free region for this function, by considering the continued fraction expansion of α. We briefly collect here the facts that we will use. See [HaW, O], and [Ba, vF1, vF2] for a connection with the Riemann Hypothesis.

6.1. Continued Fractions. Let α be an irrational real number with a continued fraction expansion $\alpha = [[a_0, a_1, a_2, \dots]] = a_0 + 1/(a_1 + 1/(a_2 + \dots))$. We recall that the two sequences $a_0, a_1, \dots$ and $\alpha_0, \alpha_1, \dots$ are defined by $\alpha_0 = \alpha$ and, for $n \geq 0$, $a_n = [\alpha_n]$, the integer part of α_n, and $\alpha_{n+1} = 1/(\alpha_n - a_n)$. The *convergents* of α,

$$(6.3) \qquad \frac{p_n}{q_n} = [[a_0, a_1, a_2, \dots, a_n]],$$

are successively computed by $p_{-2} = q_{-1} = 0$, $p_{-1} = q_{-2} = 1$, and $p_{n+1} = a_{n+1}p_n + p_{n-1}$, $q_{n+1} = a_{n+1}q_n + q_{n-1}$. We also define $q'_n = \alpha_1 \cdot \alpha_2 \cdots \cdots \alpha_n$, and note the formula $q'_{n+1} = \alpha_{n+1}q_n + q_{n-1}$. Then

$$(6.4) \qquad q_n \alpha - p_n = \frac{(-1)^n}{q'_{n+1}}.$$

We have $q_n \geq \phi^{n-1}$, where $\phi = (1 + \sqrt{5})/2$ is the golden ratio.

Let $n \in \mathbb{N}$ and choose l such that $q_{l+1} > n$. We can successively compute (see [O])

$$n = d_l q_l + n_l, \ n_l = d_{l-1} q_{l-1} + n_{l-1}, \ldots, n_1 = d_0 q_0,$$

where d_ν is the quotient and $n_\nu < q_\nu$ is the remainder of the division of $n_{\nu+1}$ by q_ν. We set $d_{l+1} = d_{l+2} = \ldots = 0$. Then

$$(6.5) \qquad n = \sum_{\nu=0}^{\infty} d_\nu q_\nu.$$

We call this the α-*adic expansion* of n. Note that $0 \leq d_\nu \leq a_{\nu+1}$ and that if $d_\nu = a_{\nu+1}$, then $d_{\nu-1} = 0$. Also $d_0 < a_1$. It is not difficult to show that these properties uniquely determine the sequence $d_0, d_1, \ldots$ of α-*adic digits* of α.

LEMMA 6.1. *Let n be given by* (6.5). *Let $k \geq 0$ be such that $d_k \neq 0$ and $d_{k-1} = \cdots = d_0 = 0$. Put $m = \sum_{\nu=k}^{\infty} d_\nu p_\nu$. Then $n\alpha - m$ lies strictly between $(-1)^k/q'_{k+2}$ and $(-1)^k/q'_k$.*

PROOF. We have $n\alpha - m = \sum_{\nu=k}^{\infty} d_\nu(\alpha q_\nu - p_\nu)$. Since $\alpha q_\nu - p_\nu = (-1)^\nu/q'_{\nu+1}$, the terms in this sum are alternately positive and negative, and it follows that $n\alpha - m$ lies between the sum of the odd terms and the sum of the even terms. To bound these terms, we use $d_\nu \leq a_{\nu+1}$. Moreover, $d_k \geq 1$, hence $d_{k+1} \leq a_{k+2} - 1$. It follows that $n\alpha - m$ lies strictly between

$$a_{k+1}(\alpha q_k - p_k) + a_{k+3}(\alpha q_{k+2} - p_{k+2}) + a_{k+5}(\alpha q_{k+4} - p_{k+4}) + \ldots$$

and

$$(\alpha q_k - p_k) + (a_{k+2} - 1)(\alpha q_{k+1} - p_{k+1}) + a_{k+4}(\alpha q_{k+3} - p_{k+3}) + \ldots.$$

Now $a_{\nu+1}(\alpha q_\nu - p_\nu) = \alpha(q_{\nu+1} - q_{\nu-1}) - (p_{\nu+1} - p_{\nu-1})$. So both sums are telescopic. The first sum immediately evaluates to $-\alpha q_{k-1} + p_{k-1} = (-1)^k/q'_k$. The second sum equals $(\alpha q_k - p_k) - (\alpha q_{k+1} - p_{k+1}) - (\alpha q_k - p_k) = (-1)^k/q'_{k+2}$. $\square$

6.2. Two Generators: the Bernoulli Flow. Assume that $N = 2$, and let f be defined as in (6.1) with weights w_1 and $w_2 = \alpha w_1$, for some irrational number $\alpha > 1$. We want to study the complex solutions to the equation $f(\omega) = 0$ that lie close to the line $\mathrm{Re}\, s = D$. First of all, such solutions must have $e^{-w_1 \omega}$ close to $e^{-w_1 D}$, so we take ω to be close to $D + 2\pi i q/w_1$, for an integer q. Then we write $\alpha q = p + x/(2\pi i)$, for an integer p, which we will specify below, and $\omega = D + 2\pi i q/w_1 + \Delta$. With these substitutions, the equation $f(\omega) = 0$ transforms to $1 - e^{-w_1 D} e^{-w_1 \Delta} - e^{-w_2 D} e^{-x} e^{-w_2 \Delta} = 0$. This equation defines Δ as a function of x.

LEMMA 6.2. *Let $w_1, w_2 > 0$ and $\alpha = w_2/w_1 > 1$; let D be such that $e^{-w_1 D} + e^{-w_2 D} = 1$, and let $\Delta = \Delta(x)$ be the function of x, defined implicitly by*

$$(6.6) \qquad e^{-w_1 D} e^{-w_1 \Delta} + e^{-w_2 D} e^{-x} e^{-w_2 \Delta} = 1,$$

and $\Delta(0) = 0$. Then Δ is analytic in x, in a disc of radius at least π around $x = 0$, with power series

$$\Delta(x) = -\frac{e^{-w_2 D}}{f'(D)} x + \frac{w_1^2 e^{-w_1 D} e^{-w_2 D}}{2 f'(D)^3} x^2 + O(x^3), \qquad as \ x \to 0.$$

The coefficients of this power series are real. The coefficient of x is negative and that of x^2 is positive.

PROOF. Define $y = y(x)$ by $e^{-w_1 D} y + e^{-w_2 D} e^{-x} y^\alpha = 1$. Then $y(0) = 1$ and $w_1 \Delta = -\log y$. Since y does not vanish, it follows that if $y(x)$ is analytic in a disc centered at $x = 0$, then Δ will be analytic in that same disc. Moreover, y is real-valued and positive when x is real, so Δ is real-valued as well when x is real. Further, $y(x)$ is locally analytic in x, with derivative

$$y'(x) = \frac{e^{-w_1 D} y^\alpha e^{-x}}{e^{-w_1 D} + \alpha e^{-w_2 D} y^{\alpha-1} e^{-x}}.$$

Hence there is a singularity at those values of x at which the denominator vanishes, which is at $y = (\alpha/(\alpha-1))e^{w_1 D}$ and $e^{-x} = -\alpha^{-\alpha}(\alpha-1)^{\alpha-1}$. Since this value is negative, the disc of convergence of the power series for $y(x)$ is

$$|x| < |-\alpha \log \alpha + (\alpha - 1)\log(\alpha - 1) + \pi i|,$$

which is a disc of radius at least π. The first two terms of the power series for $\Delta(x)$ are now readily computed. □

Applying this, we find

(6.7) $$\omega = D + 2\pi i \frac{q}{w_1} - \frac{e^{-w_2 D}}{f'(D)} x + \frac{w_1^2 e^{-w_1 D} e^{-w_2 D}}{2 f'(D)^3} x^2 + O(x^3),$$

as $x = 2\pi i(q\alpha - p) \to 0$. We view this formula as expressing ω as an initial approximation $D + 2\pi i q/w_1$, which is corrected by each term in the power series. The first corrective term is in the imaginary direction, as are all the odd ones, and the second corrective term, along with all the even ones, are in the real direction. The second term decreases the real part of ω.

THEOREM 6.3. *Let α be irrational and let p_ν and q_ν be defined by (6.3). Let q be a positive integer, and let $q = \sum_{\nu=k}^\infty d_\nu q_\nu$ be the α-adic expansion of q, as in Lemma 6.1. Assume $k \geq 2$ or $k = 1$ and $a_1 \geq 2$, and put $p = \sum_{\nu=k}^\infty d_\nu p_\nu$. Then there exists a complex dimension of $\mathcal{F}_\mathfrak{w}$ at*

(6.8)
$$\omega = D + 2\pi i \frac{q}{w_1} - 2\pi i \frac{e^{-w_2 D}}{f'(D)}(q\alpha - p)$$
$$- 2\pi^2 \frac{w_1^2 e^{-w_1 D} e^{-w_2 D}}{f'(D)^3}(q\alpha - p)^2 + O\left((q\alpha - p)^3\right).$$

The imaginary part of this complex dimension is approximately $2\pi i q/w_1$, and its distance to the line $\mathrm{Re}\, s = D$ is at least $C/q_{k+2}'^2$, where $C = 2\pi^2 w_1^2 e^{-(w_1+w_2)D}/f'(D)^3$ depends only on w_1 and w_2.

Moreover, $|\zeta_\mathfrak{w}(s)| \ll q_{k+2}'^2$ around $s = D + 2\pi i q/w_1$ on the line $\mathrm{Re}\, s = D$, and $|\zeta_\mathfrak{w}(s)|$ reaches a maximum of size $C'(q\alpha - p)^{-2}$, where C' depends only on the weights w_1 and w_2.

PROOF. By Lemma 6.1, the quantity $q\alpha - p$ lies between $(-1)^k/q_{k+2}'$ and $(-1)^k/q_k'$. Under the given conditions on k, $q_k' > q_k \geq 2$, hence $x = 2\pi i(q\alpha - p)$ is less than π in absolute value. Then (6.7) gives the value of ω.

Since the derivative of f is bounded on the line $\mathrm{Re}\, s = D$, this also implies that $f(s)$ reaches a minimum of order $(q\alpha - p)^2$ on an interval around $s = D + 2\pi i q/w_1$ on the line $\mathrm{Re}\, s = D$. It follows that $|\zeta_\mathfrak{w}(s)| \ll q_{k+2}'^2$ on the line $\mathrm{Re}\, s = D$, with a maximum of order $(q\alpha - p)^{-2}$. □

We obtain more precise information when $q = q_k$.

THEOREM 6.4. *For every $k \geq 0$ (or $k \geq 1$ if $a_1 = 1$), there exists a complex dimension ω of $\mathcal{F}_{\mathfrak{w}}$ of the form*

$$(6.9) \quad \omega = D + 2\pi i \frac{q_k}{w_1} - 2\pi i (-1)^k \frac{e^{-w_2 D}}{f'(D) q'_{k+1}} - 2\pi^2 w_1^2 \frac{e^{-(w_1+w_2)D}}{f'(D)^3 q'^2_{k+1}} + O\left(q'^{-3}_{k+1}\right),$$

as $k \to \infty$.

Moreover, $|\zeta_{\mathfrak{w}}(s)| \ll q'^2_{k+1}$ around $s = D + 2\pi i q_k/w_1$ on the line $\operatorname{Re} s = D$, and $|\zeta_{\mathfrak{w}}(s)|$ reaches a maximum of size $C' q'^2_{k+1}$, where C' is as in Theorem 6.3.

PROOF. Put $p = p_k$. Then $x = 2\pi i (-1)^k / q'_{k+1}$, which is less than π in absolute value. The rest of the proof is the same as in the proof of Theorem 6.3. $\qquad \square$

DEFINITION 6.5. *A domain in the complex plane containing the line $\operatorname{Re} s = D$ is a* dimension free region *for the flow $\mathcal{F}_{\mathfrak{w}}$ if the only pole of $-\zeta'_{\mathfrak{w}}/\zeta_{\mathfrak{w}}$ in that region is $s = D$.*

COROLLARY 6.6. *Assume that the coefficients $a_0, a_1, \ldots$ of α are bounded by M. Put $B = \pi^4 e^{-(w_1+w_2)D}/(2f'(D)^3)$. Then $\mathcal{F}_{\mathfrak{w}}$ has a dimension free region of the form*

$$(6.10) \qquad \left\{\sigma + it \in \mathbb{C} \colon \sigma > D - \frac{B}{M^2 t^2}\right\}.$$

The function $-\zeta'_{\mathfrak{w}}/\zeta_{\mathfrak{w}}$ satisfies hypotheses $(\mathbf{H}_1)$ and $(\mathbf{H}_2)$ with $\kappa = 2$.

More generally, let $a \colon \mathbb{R}^+ \to [1, \infty)$ be a function such that the coefficients $\{a_k\}_{k=0}^{\infty}$ of the continued fraction of α satisfy $a_{k+1} \leq a(q_k)$ for every $k \geq 0$. Then $\mathcal{F}_{\mathfrak{w}}$ has a dimension free region of the form

$$(6.11) \qquad \left\{\sigma + it \in \mathbb{C} \colon \sigma > D - \frac{B}{t^2 a^2(tw_1/(2\pi))}\right\}.$$

If a grows at most polynomially, then $-\zeta'_{\mathfrak{w}}/\zeta_{\mathfrak{w}}$ satisfies hypotheses $(\mathbf{H}_1)$ and $(\mathbf{H}_2)$ with κ such that $t^\kappa \geq t^2 a^2(tw_1/(2\pi))$.

PROOF. This follows from Theorem 6.4, if we note that for $t = 2\pi q_k/w_1$, we have $q'_{k+1} = \alpha_{k+1} q'_k \leq 2a(q_k) q'_k \leq 4a(q_k) q_k$. So the complex dimension close to $D + it$ is located at $D + i(t + O(q'^{-1}_{k+1})) - (w_1^2/\pi^2) B q'^{-2}_{k+1} + O(q'^{-4}_{k+1})$, where the orders denote real-valued functions. The real part of this complex dimension is less than $D - Bt^{-2} a^{-2}(tw_1/(2\pi))$. $\qquad \square$

This has the following consequence for the Prime Orbit Theorem:

THEOREM 6.7 (Prime Orbit Theorem with Error Term, for Bernoulli flows). *Let $\alpha = w_2/w_1$ have bounded coefficients in its continued fraction. Then*

$$(6.12) \qquad \psi_{\mathfrak{w}}(x) = \frac{x^D}{D} + O\left(x^D \left(\frac{\log \log x}{\log x}\right)^{1/4}\right),$$

as $x \to \infty$.

If α is 'polynomially approximable', with coefficients in its continued fraction satisfying $a_{k+1} \leq a(q_k)$, for some increasing function a with $a(x) = O(x^l)$, as $x \to \infty$, then

$$(6.13) \qquad \psi_{\mathfrak{w}}(x) = \frac{x^D}{D} + O\left(x^D \left(\frac{\log \log x}{\log x}\right)^{\frac{1}{4l+4}}\right),$$

as $x \to \infty$.

The proof will be given in the next section, see Theorem 6.14.

EXAMPLE 6.8. For the Golden flow, we have $\alpha = \phi$ and $w_1 = \log 2$ (see Example 2.23). The continued fraction of ϕ is $[1, 1, 1, \ldots]$, hence $q_k = F_{k+1}$, the $(k+1)$-st Fibonacci number, and $q'_k = \phi^k$. Numerically, we find $D \approx .7792119034$ and the following approximation of $\Delta(x)$:

$$-.47862\,x + .08812\,x^2 + .00450\,x^3 - .00205\,x^4 - .00039\,x^5 + .00004\,x^6 + \ldots.$$

For every k, we find a complex dimension close to $D + 2\pi i q_k / \log 2$. For example, $q_9 = 55$, and we find a complex dimension at $D - .00023 + 498.58i$. More generally, for numbers like $q = 55 + 5$ or $q = 55 - 5 = 34 + 13 + 3$, we find a complex dimension close to $D + 2\pi i q / \log 2$, in this case respectively at $D - .023561 + 543.63i$ and at $D - .033919 + 453.53i$. In both these cases, the distance to the line $\operatorname{Re} s = D$ is comparable to the distance of the complex dimension for $q = 5$ to this line, which is located at $D - .028499 + 45.05i$. See Figure 2, where the markers are at the Fibonacci numbers. The pattern persists for other complex dimensions as well. Indeed, every complex dimension repeats itself according to the Fibonacci numbers.

6.3. More than Two Generators. The following lemma replaces the continued fraction construction.

LEMMA 6.9. *Let $w_1, w_2, \ldots, w_N$ be weights such that at least one ratio w_j / w_k is irrational. Then for every $Q > 1$, there exist integers $q < Q^{N-1}$ and p_j such that $|qw_j - p_j w_1| \leq w_1 / Q$ for $j = 1, \ldots, N$. In particular, $|qw_j - p_j w_1| < w_1 q^{-1/(N-1)}$ for $j = 1, \ldots, N$.*

REMARK 6.10. Note that the condition implies that at least one ratio w_j / w_1 is irrational. Also, $|qw_j - p_j w_1| \neq 0$ when w_j / w_1 is irrational, so $q \to \infty$ when $Q \to \infty$.

The construction of such integers q and p_j is much less explicit than for $N = 2$, since there does not exist a continued fraction algorithm for simultaneous approximation.[4] The number Q plays the role of q'_{k+1} in Theorem 6.4 above. In particular, if q is often much smaller than Q, then $w_1, \ldots, w_N$ is well approximable by rationals, and we find a small dimension free region.

Again, we are looking for a solution of $f(\omega) = 0$ close to $s = D + 2\pi i q / w_1$, where f is defined by (6.1). We write $\omega = D + 2\pi i q / w_1 + \Delta$ and $w_j q = w_1 p_j + w_1 x_j / (2\pi i)$. For $j = 1$, we take $p_1 = q$ and consequently $x_1 = 0$. In general, $x_j = 2\pi i (qw_j / w_1 - p_j)$. Then $f(\omega) = 0$ is equivalent to $1 - \sum_{j=1}^{N} e^{-w_j D} e^{-x_j - w_j \Delta} = 0$.

The following lemma is the several variable analogue of Lemma 6.2. However, in this case we do not know the radius of convergence with respect to each of the variables involved.

---

[4]However, the L^3-algorithm can be used as a substitute for the continued fraction algorithm. We thank H. W. Lenstra, Jr. for guiding us to the following information: The L^3-algorithm [LLL] can be used to find fractions p_j / q that approximate w_j / w_1 for $j = 1, \ldots, N$. This algorithm works in polynomial time, like the continued fraction algorithm, but it does not give the best possible value for q (given a certain error of approximation). The problem of finding the best value for q is NP-complete [Lag]. (See also [GLS].)

LEMMA 6.11. *Let $w_1 \leq w_2 \leq \cdots \leq w_N$, let D be such that $\sum_{j=1}^{N} e^{-w_j D} = 1$, and let $\Delta = \Delta(x_2, \ldots, x_N)$ be implicitly defined by*

$$(6.14) \qquad \sum_{j=1}^{N} e^{-w_j D} e^{-x_j - w_j \Delta} = 1,$$

and $x_1 = 0$. Then Δ is analytic in $x_2, \ldots, x_N$, with power series

$$
\begin{aligned}
(6.15) \qquad \Delta = {} & -\sum_{j=2}^{N} \frac{e^{-w_j D}}{f'(D)} x_j + \frac{1}{2} \sum_{j=2}^{N} \frac{e^{-w_j D}}{f'(D)} x_j^2 \\
& - \frac{1}{2} \sum_{j,k=2}^{N} \left(\frac{f''(D)}{f'(D)^3} + \frac{w_j + w_k}{f'(D)^2} \right) e^{-(w_j + w_k)D} x_j x_k + O\left(\sum_{j=2}^{N} |x_j|^3 \right).
\end{aligned}
$$

This power series has real coefficients. The terms of degree two form a positive definite quadratic form.

PROOF. The positive definiteness follows from the fact that the complex dimensions lie to the left of $\operatorname{Re} s = D$, see Theorem 2.19. It can also be verified directly. □

Applying this, we find

$$
\begin{aligned}
(6.16) \qquad \omega = {} & D + 2\pi i \frac{q}{w_1} - \sum_{j=2}^{N} \frac{e^{-w_j D}}{f'(D)} x_j + \frac{1}{2} \sum_{j=2}^{N} \frac{e^{-w_j D}}{f'(D)} x_j^2 \\
& - \frac{1}{2} \sum_{j,k=2}^{N} \left(\frac{f''(D)}{f'(D)^3} + \frac{w_j + w_k}{f'(D)^2} \right) e^{-(w_j + w_k)D} x_j x_k + O\left(\sum_{j=2}^{N} |x_j|^3 \right),
\end{aligned}
$$

where $x_j = 2\pi i(qw_j/w_1 - p_j)$. Again, this formula expresses ω as an initial approximation $D + 2\pi i q/w_1$, which is corrected by each term in the power series. The corrective terms of degree one are again in the imaginary direction, as are all the odd degree ones, and the corrective terms of degree two, along with all the even ones, are in the real direction. The degree two terms decrease the real part of ω.

THEOREM 6.12. *Let $N \geq 2$ and let $w_1, \ldots, w_N$ be weights. Let Q and q be as in Lemma 6.9. Then $\mathcal{F}_\mathfrak{w}$ has a complex dimension close to $D + 2\pi i q/w_1$ at a distance of at most $O(Q^{-2})$ from the line $\operatorname{Re} s = D$, as $Q \to \infty$. The function $|\zeta'_\mathfrak{w}/\zeta_\mathfrak{w}|$ reaches a maximum of order Q^2.*

PROOF. Again, the numbers x_j are purely imaginary, so the corrective terms of degree 1 (and of every odd degree) give a correction in the imaginary direction, and only the corrective terms of even degree will give a correction in the real direction. Since $|x_j| < 2\pi/Q$, the theorem follows. □

COROLLARY 6.13. *The best dimension free region that $\mathcal{F}_\mathfrak{w}$ can have is of size*

$$(6.17) \qquad \left\{ \sigma + it : \sigma \geq D - O\left(t^{-2/(N-1)} \right) \right\}.$$

The implied constant depends only on $w_1, \ldots, w_N$.

If $w_1, \ldots, w_N$ is 'a-approximable', then the dimension free region has the form

$$(6.18) \qquad \left\{ \sigma + it : \sigma \geq D - O\left(a^{-2}(w_1 t/(2\pi)) t^{-2/(N-1)} \right) \right\},$$

where $a: [1, \infty) \to \mathbb{R}^+$ is an increasing function such that for every integer $q \geq 1$, $|qw_j - p_jw_1| \geq (w_1/a(q))q^{-1/(N-1)}$ for $j = 1, \ldots, N$.

This has the following consequence for the Prime Orbit Theorem:

THEOREM 6.14 (Prime Orbit Theorem with Error Term). *Suppose $w_1, \ldots, w_N$ are badly approximable, in the sense that $|qw_j - p_jw_1| \gg q^{-1/(N-1)}$ for $j = 1, \ldots, N$ and every $q \geq 1$. Then*

$$(6.19) \qquad \psi_{\mathfrak{w}}(x) = \frac{x^D}{D} + O\left(x^D \left(\frac{\log\log x}{\log x}\right)^{\frac{N-1}{4}}\right),$$

as $x \to \infty$.

If $w_1, \ldots, w_N$ is 'polynomially approximable', in the sense that $|qw_j - p_jw_1| \geq (w_1/a(q))q^{-1/(N-1)}$ for $j = 1, \ldots, N$ and every $q \geq 1$, for some increasing function a on $[1, \infty)$ such that $a(x) = O(x^l)$ as $x \to \infty$, then

$$(6.20) \qquad \psi_{\mathfrak{w}}(x) = \frac{x^D}{D} + O\left(x^D \left(\frac{\log\log x}{\log x}\right)^{\frac{N-1}{4l(N-1)+4}}\right),$$

as $x \to \infty$.

PROOF OF THEOREMS 6.7 AND 6.14. We apply the pointwise explicit formula at level $k = 2$ (see Theorem 3.1) to obtain

$$\psi_{\mathfrak{w}}^{[2]}(x) = \frac{x^{D+1}}{D(D+1)} + \sum_{\omega \in \mathcal{D}_{\mathfrak{w}} \setminus \{D\}} \frac{x^{\omega+1}}{\omega(\omega+1)} + R^{[2]}(x).$$

The error term is estimated by $R^{[2]}(x) = O(x^{D+1-c})$ for some positive c. We will estimate the sum by an argument which is classical in the theory of the Riemann zeta function and the Prime Number Theorem, under the assumptions that the ω have a linear density, and that every $\omega = \sigma + it$ satisfies $\sigma \leq D - Ct^{-\rho}$ for some positive number ρ. We then obtain Theorem 6.14 by taking $\rho = 2/(N-1) + 2l$, and Theorem 6.7 corresponds to the case when $N = 2$.

The sum $\sum_\omega \frac{x^{\omega+1}}{\omega(\omega+1)}$ is absolutely convergent. We split this sum into the parts with $|\operatorname{Im}\omega| > T$ and with $|\operatorname{Im}\omega| \leq T$. Put $A = \sum_\omega |\omega(\omega+1)|^{-1}$. From the fact that the complex dimensions have a linear density, it follows that there exists a constant B is such that $\sum_{|\operatorname{Im}\omega| \geq T} |\omega(\omega+1)|^{-1} \leq B/T$ for every T. Then $\left|\sum_\omega \frac{x^{\omega+1}}{\omega(\omega+1)}\right| \leq Ax^{D+1-CT^{-\rho}} + Bx^{D+1}/T$. For $T = (\rho C \log x / \log\log x)^{1/\rho}$, we find

$$\left|\sum_\omega \frac{x^{\omega+1}}{\omega(\omega+1)}\right| = O\left(x^{D+1} \left(\frac{\log\log x}{\log x}\right)^{1/\rho}\right).$$

We then apply a Tauberian argument to deduce a similar error estimate for $\psi_{\mathfrak{w}}(x)$; see [I, p. 64]. Let $h = x(\log\log x / \log x)^{1/(2\rho)}$. Thus

$$\psi_{\mathfrak{w}}(x) \leq \frac{1}{h} \int_x^{x+h} \psi_{\mathfrak{w}}(t)\, dt = \frac{\psi_{\mathfrak{w}}^{[2]}(x+h) - \psi_{\mathfrak{w}}^{[2]}(x)}{h}.$$

Now $\frac{(x+h)^{D+1} - x^{D+1}}{hD(D+1)} = x^D/D + O(x^{D-1}h) = x^D/D + x^D O((\log\log x/\log x)^{1/(2\rho)})$. Further, $O(x^{D+1}(\log\log x/\log x)^{1/\rho}/h) = x^D O((\log\log x/\log x)^{1/(2\rho)})$. $\square$

REMARK 6.15. Note that by using the Tauberian argument, we lose a factor two in the exponent. Indeed, the estimate

$$\sum_{\omega \in \mathcal{D}_\mathfrak{w} \setminus \{D\}} \frac{x^\omega}{\omega} + R(x) = O\left(x^D \left(\frac{\log \log x}{\log x}\right)^{\frac{N-1}{2l(N-1)+2}}\right)$$

holds distributionally.

REMARK 6.16. If $a(q)$ grows more than polynomially, we obtain a bound of the form $x^D/a^{\mathrm{inv}}(\log x)$ for the error in the Prime Orbit Theorem, where a^{inv} is the inverse function of a.

6.4. Conclusion. If $l > 0$ in the exponent $(N-1)/(4l(N-1)+4)$ of $\log x$ in the error term of Theorems 6.7 and 6.14, then the error term is independent of N, essentially of order $x^D(\log x)^{-1/4l}$ (ignoring the factor of $\log \log x$). Thus, if the weights are well approximable, the error term is never better than x^D divided by a fixed power of the logarithm of x. On the other hand, when $l = 0$, that is, roughly speaking, when the weights are never close to rational numbers, then the error term is essentially of order $x^D(\log x)^{-(N-1)/4}$. Hence, the larger N, the smaller the error term in that case.

We may compare this, somewhat superficially in view of the Riemann Hypothesis, with the situation of the Riemann zeta function. In view of Example 3.6, the weights are $w_p = \mathfrak{w}_t(p) = \log p$, for each prime number p, and there are infinitely many of them. Since it is expected that $\{\log p\}_{p:\mathrm{prime}}$ is badly approximable, one expects an error term of order "$x^D(\log x)^{-\infty}$". Indeed, in (3.22), we have $e^{-c\sqrt{\log x}} = O\left((\log x)^{-N}\right)$ for every $N > 0$. The corresponding pole free region has width $A/\log t$ at height t (see [I, Theorem 19]), which is "$t^{-1/\infty}$". This lends credibility to the conjecture that $\{\log p\}_{p:\mathrm{prime}}$ is badly approximable by rational numbers.

Acknowledgement. The authors wish to thank Gabor Elek for helpful conversations about dynamical zeta functions.

References

[Ba]   L. Báez-Duarte, A class of invariant unitary operators, *Adv. Math.* **144** (1999), 1–12.

[BKS]   T. Bedford, M. Keane and C. Series (eds.), *Ergodic Theory, Symbolic Dynamics and Hyperbolic Spaces*, Oxford Univ. Press, Oxford, 1991.

[Bo]   R. Bowen, Symbolic dynamics for hyperbolic flows, *Amer. J. Math.* **95** (1973), 429–460.

[E]   H. M. Edwards, *Riemann's Zeta Function*, Academic Press, New York, 1974.

[F]   K. J. Falconer, *Fractal Geometry: Mathematical Foundations and Applications*, Wiley, Chichester, 1990.

[GLS]   M. Grötschel, L. Lovász, A. Schrijver, *Geometric Algorithms and Combinatorial Optimization*, Springer-Verlag, Berlin, 1993.

[HaW]   G. H. Hardy, E. M. Wright, *An Introduction to the Theory of Numbers*, Oxford University Press, Oxford, 1960.

[Hu]   H. Huber, Zur analytischen Theorie hyperbolischer Raumformen und Bewegungsgruppen, *Math. Ann.* **138** (1959), 1–26.

[I]   A. E. Ingham, *The Distribution of Prime Numbers*, Cambridge University Press, 1992.

[Lag]   J. C. Lagarias, The computational complexity of simultaneous diophantine approximation problems, *SIAM J. Comput.* **14** (1985) 196–209.

[Lal1] S. P. Lalley, Renewal theorems in symbolic dynamics, with applications to geodesic flows, noneuclidean tessellations and their fractal limits, *Acta Math.* **163** (1989), 1–55.

[Lal2] ___, *Probabilistic counting methods in certain counting problems of ergodic theory*, in: [BKS, pp. 223–258].

[LM]   M. L. Lapidus and H. Maier, The Riemann hypothesis and inverse spectral problems for fractal strings, *J. London Math. Soc.* (2) **52** (1995), 15–34.

[LP]   M. L. Lapidus and C. Pomerance, The Riemann zeta-function and the one-dimensional Weyl–Berry conjecture for fractal drums, *Proc. London Math. Soc.* (3) **66** (1993), 41–69.

[LvF1] M. L. Lapidus and M. van Frankenhuysen, *Complex dimensions of fractal strings and oscillatory phenomena in fractal geometry and arithmetic*, Contemporary Mathematics **237** (1999), 87–105.

[LvF2] ___ , *Fractal Geometry and Number Theory (Complex dimensions of fractal strings and zeros of zeta functions)*, Research Monograph, Birkhäuser, Boston, 2000.

[LvF3] M. L. Lapidus and M. van Frankenhuysen, *Complex dimensions of self-similar fractal strings and Diophantine approximation*, preprint, 2001.

[LLL]   A. K. Lenstra, H. W. Lenstra, Jr., L. Lovász, Factoring polynomials with rational coefficients, *Math. Ann.* **261** (1982), 515–534.

[Mn]   B. B. Mandelbrot, *The Fractal Geometry of Nature*, Freeman, New York, 1983.

[Mr]   G. Margulis, Certain applications of ergodic theory to the investigation of manifolds of negative curvature, *Functional Anal. Appl.* **3** (1969), 89–90.

[O]   A. Ostrowski, *Bemerkungen zur Theorie der Diophantischen Approximationen*, Abh. Math. Sem. Hamburg Univ. **1** (1922), 77–98.

[PP1]   W. Parry and M. Pollicott, An analogue of the prime number theorem and closed orbits of Axiom A flows, *Annals of Math.* **118** (1983), 573–591.

[PP2]   ___ , *Zeta Functions and the Periodic Orbit Structure of Hyperbolic Dynamics*, Astérisque, vols. 187–188, Soc. Math. France, Paris, 1990.

[R]   D. Ruelle, Zeta functions for expanding maps and Anosov flows, *Invent. Math.* **34** (1978), 231–242.

[S]   Y. G. Sinai, The asymptotic behaviour of the number of closed geodesics on a compact manifold of negative curvature, *Transl. AMS* **73** (1968), 227–250.

[vF1]   M. van Frankenhuysen, *Over het vermoeden van Riemann*, afstudeerscriptie (graduate thesis), Katholieke Universiteit Nijmegen, The Netherlands, 1990.

[vF2]   ___ , *Zero-free regions for the Riemann zeta-function, density of invariant subspaces of functions, and the theory of equal distribution*, preprint, IHES, 1997.

DEPARTMENT OF MATHEMATICS, UNIVERSITY OF CALIFORNIA, SPROUL HALL, RIVERSIDE, CALIFORNIA 92521-0135, USA

E-mail address: `lapidus@math.ucr.edu`

DEPARTMENT OF MATHEMATICS, UNIVERSITY OF CALIFORNIA, SPROUL HALL, RIVERSIDE, CALIFORNIA 92521-0135, USA

Current address: Rutgers University, Department of Mathematics, 110 Frelinghuysen Road, Piscataway, NJ 08854-8019, USA

E-mail address: `machiel@math.rutgers.edu`

Contemporary Mathematics
Volume **290**, 2001

The 10^{22}-nd zero of the Riemann zeta function

A. M. Odlyzko

ABSTRACT. Recent and ongoing computations of zeros of the Riemann zeta function are described. They include the computation of 10 billion zeros near zero number 10^{22}. These computations verify the Riemann Hypothesis for those zeros, and provide evidence for additional conjectures that relate these zeros to eigenvalues of random matrices.

1. Introduction

This is a brief report on computations of large numbers of high zeros of the Riemann zeta functions. It provides pointers to sources of more detailed information.

There have been many calculations that verified the Riemann Hypothesis (RH) for initial sets of zeros of the zeta function. The first were undertaken by Riemann himself almost a century and a half ago. Those calculations did not become known to the scientific community until Siegel deciphered Riemann's unpublished notes [**Sie**]. The first published computation, by Gram in 1903, verified that the first 10 zeros of the zeta function are on the critical line. (Gram calculated values for the first 10 zeros accurate to 6 decimal places, and showed that these were the only zeros below height 50. He also produced much less accurate values for the next 5 zeros. See [**Edw**] for more details.) Gram's work was extended by a sequence of other investigators, who were aided by improvements in both hardware and algorithms, with the two contributing about equally to the improvements that have been achieved. The latest published result is that of van de Lune, te Riele, and Winter [**LRW**]. They checked that the first 1.5×10^9 nontrivial zeros all lie on the critical line. Their computations used about 1500 hours on one of the most powerful computers in existence at that time. Since then, better algorithms have been developed, and much more computing power has become available. With some effort at software and at obtaining access to the idle time on a large collection of computers, one could hope to verify the RH for the first 10^{12} zeros in the next year or so, Jan van de Lune has been extending his earlier work with te Riele and Winter, using the algorithms of [**LRW**] and very modest computational resources. By the end of the year 2000, he had checked that the first 5.3×10^9 zeros of the

---

1991 *Mathematics Subject Classification.* Primary 11M26; Secondary 81Q50, 15A52, 65D20.

Key words and phrases. Riemann zeta function, Riemann Hypothesis, pair-correlation conjecture, zero computations.

zeta function lie on the critical line and are simple, even though he was relying on just three not very modern PCs (unpublished).

Starting in the late 1970s, I carried out a series of computations that not only verified that nontrivial zeros lie on the critical line (which was the sole aim of most of the computations, including those of van de Lune, te Riele, and Winter [**LRW**]), but in addition obtained accurate values of those zeros. These calculations were designed to check the Montgomery pair-correlation conjecture [**Mon**], as well as further conjectures that predict that zeros of the zeta function behave like eigenvalues of certain types of random matrices. Instead of starting from the lowest zeros, these computations obtained values of blocks of consecutive zeros high up in the critical strip. The motivation for studying high ranges was to come closer to observing the true asymptotic behavior of the zeta function, which is often approached slowly.

The initial computations, described in [**Od1**], were done on a Cray supercomputer using the standard Riemann-Siegel formula. This formula was invented and implemented by Riemann, but remained unknown to the world until the publication of Siegel's paper [**Sie**]. The highest zeros covered by [**Od1**] were around zero # 10^{12}. Those calculations stimulated the invention, jointly with Arnold Schönhage [**Od2, OS**], of an improved algorithm for computing large sets of zeros. This algorithm, with some technical improvements, was implemented in the late 1980s and used to compute several hundred million zeros at large heights, many near zero # 10^{20}, and some near zero # 2×10^{20}. Implementation details and results are described in [**Od3, Od4**]. These papers have never been published, but have circulated widely.

During the last few years, the algorithms of [**Od3, Od4**] have been ported from Cray supercomputers to Silicon Graphics workstations. They have been used to compute several billion high zeros of the zeta function, and computations are continuing, using spare cycles on machines at AT&T Labs. Some of those zeros are near zero # 10^{22}, and it has been established (not entirely rigorously, though, as is explained in [**Od3, Od4**]) that the imaginary parts of zeros number $10^{22} - 1$, 10^{22}, and $10^{22} + 1$ are

$$1,370,919,909,931,995,308,226.490240...$$
$$1,370,919,909,931,995,308,226.627511...$$
$$1,370,919,909,931,995,308,226.680160...$$

These values and many others can be found at

⟨http://www.research.att.com/∼amo/zeta_tables/index.html⟩.

Further computations are under way and planned for the future. Very soon 10^{10} zeros near zero # 10^{22} will be available. It is likely that some billions of zeros near zero # 10^{23} will also be computed. A revision of [**Od3, Od4**] that describes them is planned for the future [**Od7**]. Results will be available through my home page,

⟨http://www.research.att.com/∼amo⟩.

Finally, let me mention that many other computations of zeros of various zeta and L-functions have been done. Many are referenced in [**Od5**]. There are also interesting new results for other classes of zeta functions in the recent Ph.D. thesis of Michael Rubinstein [**Ru**].

The next section describes briefly the highlights of the recent computations.

2. High zeros and their significance

No counterexamples to the RH have been found so far. Heuristics suggest that if there are counterexamples, then they lie far beyond the range we can reach with currently known algorithms (cf. [Od7]). However, there is still an interest in undertaking additional computations in the ranges we can reach. The main motivation is to obtain further insights into the Hilbert-Pólya conjecture, which predicts that the RH is true because zeros of the zeta function correspond to eigenvalues of a positive operator. When this conjecture was formulated about 80 years ago, it was apparently no more than an inspired guess. Neither Hilbert nor Pólya specified what operator or even what space would be involved in this correspondence. Today, however, that guess is increasingly regarded as wonderfully inspired, and many researchers feel that the most promising approach to proving the RH is through proving some form of the Hilbert-Pólya conjecture. Their confidence is bolstered by several developments subsequent to Hilbert's and Pólya's formulation of their conjecture. There are very suggestive analogies with Selberg zeta functions. There is also the extensive research stimulated by Hugh Montgomery's work on the pair-correlation conjecture for zeros of the zeta function [Mon]. Montgomery's results led to the conjecture that zeta zeros behave asymptotically like eigenvalues of large random matrices from the GUE ensemble that has been studied extensively by mathematical physicists. This was the conjecture that motivated the computations of [Od1, Od3, Od4] as well as those described in this note. Although this conjecture is very speculative, the empirical evidence is overwhelmingly in its favor.

To describe some of the numerical results, we recall standard notation. We consider the nontrivial zeros of the zeta function (i.e., those zeros that lie in the critical strip $0 < \mathrm{Re}\,(s) < 1$), and let the ones in the upper half of the critical strip be denoted by $\frac{1}{2} + i\gamma_n$, where the γ_n are positive real numbers arranged in increasing order. (We are implicitly assuming the RH here for simplicity. We do not have to consider the zeros in the lower half plane since they are the mirror images of the ones in the upper half plane.) Since spacings between consecutive zeros decrease as one goes up in the critical strip, we consider the normalized spacings

$$(2.1) \qquad \delta_n = (\gamma_{n+1} - \gamma_n)\frac{\log(\gamma_n/(2\pi))}{2\pi} \ .$$

It is known that the average value of the δ_n is 1. The conjecture is that the distribution of the δ_n is asymptotically the same as the Gaudin distribution for GUE matrices.

Figure 1 compares the empirical distribution of δ_n for $1,006,374,896$ zeros of the zeta function starting with zero # $13,048,994,265,258,476$ (at height approximately $2.51327412288 \cdot 10^{15}$). The smooth curve is the probability density function for the normalized gaps between consecutive eigenvalues in the GUE ensemble. The scatter plot is the histogram of the δ_n. The point plotted at $(0.525, w)$ means that the probability that δ_n is between 0.5 and 0.55 is w, for example. As we can see, the empirical distribution matches the predicted one closely.

The paper [Od1] presented similar graphs based on the first million zeros, where the agreement was much poorer, as well as on $100,000$ zeros starting at zero # 10^{12}, where the empirical and GUE distributions matched pretty closely. The graphs in [Od3, Od4], based on large sets of zeros as high as zero # 10^{20} showed far better agreement, even better than that of Figure 1.

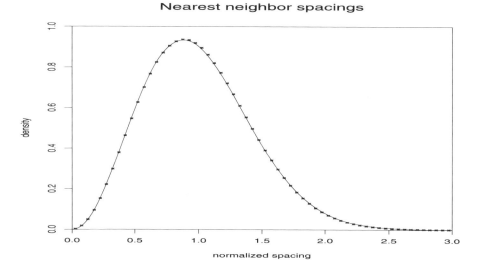

FIGURE 1. Probability density of the normalized spacings δ_n. Solid line: Gue prediction. Scatterplot: empirical data based on a billion zeros near zero # $1.3 \cdot 10^{16}$.

One motivation for continuing the computations is to obtain more detailed pictures of the evolution of the spacing distribution. Graphs such as that of Figure 1 are convincing, but are often inadequate. These graphs do not convey a good quantitative idea of the speed with which the empirical distribution of the δ_n converges to the GUE. They can be misleading, since in the steeply rising parts of the curve, substantial differences can be concealed from the human eye. It is often more valuable to consider graphs such as Figure 2, which shows the difference between the empirical and GUE distributions. This time the bins are of size 0.01, and not the larger 0.05 bins used in Figure 1. It is the large sample size of a billion zeros that allows the use of such small bins, and leads to a picture of a continuous curve. (With small data sets, say of 100,000 data points, which is all that was available in [**Od1**], sampling errors would have obscured what was going on.) Clearly there is structure in this difference graph, and the challenge is to understand where it comes from.

There are many other numerical comparisons between the zeta function and various conjectures that can be performed with large sets of zeros. For example, one can compute moments of the zeta function on the critical line, and compare them with the predictions of the fascinating conjectures of Keating and Snaith [**KeaS**] that relate the behavior of the zeta function at a fixed height to that of eigenvalues of random GUE matrices of a fixed dimension. (The basic Montgomery conjecture only suggested that the asymptotic limits would be the same.) There is also the

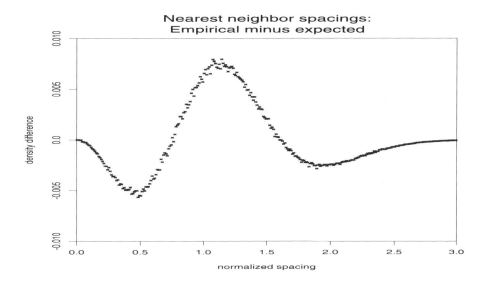

FIGURE 2. Probability density of the normalized spacings δ_n. Diffrence between empirical distribution for a billion zeros near zero # $1.3 \cdot 10^{16}$ and the GUE prediction.

general fact that convergence of some properties of the zeta function to asymptotic limits is fast (for example, for the distribution of δ_n), while for others it is slow.

Large scale computations of zeros can also be used in other contexts. In particular, they can be used to improve known bounds on the de Bruijn-Newman constant, as is done in [**Od6**].

Ideally, of course, one would like to use numerical evidence to help in the search for the Hilbert-Pólya operator, and thereby prove the RH. Unfortunately, so far theoretical progress has been limited. Some outstanding results have been obtained, such as the Katz-Sarnak proof that the GUE distribution does apply to zero spacings of zeta functions of function fields [**KatzS1, KatzS2**]. However, these results so far have not been extended to the regular Riemann zeta function.

Acknowledgements. I thank Jeff Lagarias and Jan van de Lune for their comments on earlier versions of this note.

References

[BerK]   M. V. Berry and J. P. Keating, The Riemann zeros and eigenvalue asymptotics, *SIAM Rev.* 41 (1999), 236–266.

[Edw]   H. M. Edwards, *Riemann's Zeta Function*, Academic Press, 1974.

[KatzS1] N. M. Katz and P. Sarnak, *Random Matrices, Frobenius Eigenvalues, and Monodromy*, Colloquium Publications # 45, Amer. Math. Soc., 1999.

[KatzS2] N. M. Katz and P. Sarnak, Zeroes of zeta functions and symmetry, *Bull. Amer. Math. Soc. (N.S.)* 36 (1999), 1–26.

[KeaS]   J. Keating and N. Snaith, Random matrix theory and $\zeta(1/2 + it)$, *Comm. Math. Phys.*
         214 (2000), 57–89.

[LRW]   J. van de Lune, H. J. J. te Riele, and D. T. Winter, On the zeros of the Riemann zeta
         function in the critical strip. IV., *Math. Comp. 46* (1986), 667–681.

[Mon]   H. L. Montgomery, The pair correlation of zeros of the zeta function, pp. 181–193 in
         Analytic Number Theory, H. G. Diamond, ed., Proc. Symp. Pure Math. *24*, Amer. Math.
         Soc., Providence 1973.

[Od1]   A. M. Odlyzko, On the distribution of spacings between zeros of
         the zeta function, *Math. Comp. 48* (1987), 273–308. Available at
         ⟨http://www.research.att.com/~amo/doc/zeta.html⟩.

[Od2]   A. M. Odlyzko, New analytic algorithms in number theory, pp. 466–475 in
         Proc. Intern. Congress Math. 1986, Amer. Math. Soc. 1987. Available at
         ⟨http://www.research.att.com/~amo/doc/zeta.html⟩.

[Od3]   A. M. Odlyzko, The 10^{20}-th zero of the Riemann zeta function and
         70 million of its neighbors, unpublished manuscript, 1989. Available at
         ⟨http://www.research.att.com/~amo/unpublished/index.html⟩.

[Od4]   A. M. Odlyzko, The 10^{20}-th zero of the Riemann zeta function and
         175 million of its neighbors, unpublished manuscript, 1992. Available at
         ⟨http://www.research.att.com/~amo/unpublished/index.html⟩.

[Od5]   A. M. Odlyzko, Analytic computations in number theory, pp. 451–463 in *Mathe-*
         matics of Computation 1943-1993: A Half-Century of Computational Mathe matics,
         Gautschi, W., ed. Amer. Math. Soc., Proc. Symp. Appl. Math. # 48 (1994). Available
         at ⟨http://www.research.att.com/~amo/doc/zeta.html⟩.

[Od6]   A. M. Odlyzko, An improved bound for the de Bruijn-Newman constant, *Numerical Al-*
         gorithms, to appear. Available at ⟨http://www.research.att.com/~amo/doc/zeta.html⟩.

[Od7]   A. M. Odlyzko, *The Riemann Zeta Function: The 10^{22}-nd Zero and Ten Billion of its*
         Neighbors, (tentative title), manuscript in preparation.

[OS]   A. M. Odlyzko and A. Schönhage, Fast algorithms for multiple evaluations of the
         Riemann zeta function, *Trans. Amer. Math. Soc. 309* (1988), 797–809. Available at
         ⟨http://www.research.att.com/~amo/doc/zeta.html⟩.

[Ru]   M. Rubinstein, Evidence for a spectral interpretation of the zeros of L-functions, Ph.D.
         thesis, Math. Dept., Princeton Univ., 1998.

[Rum]   R. Rumely, Numerical computations concerning the ERH, *Math. Comp.* 61 (1993) 415–
         440, S17–S23.

[Sie]   C. L. Siegel, Über Riemanns Nachlass zur analytischen Zahlentheorie, *Quellen und Stu-*
         dien zur Geschichte der Math. Astr. Phys. 2 (1932), 45–80. Reprinted in *C. L. Siegel,*
         Gesammelte Abhandlungen, Springer, 1966, Vol. 1, pp. 275–310.

AT&T LABS - RESEARCH, FLORHAM PARK, NEW JERSEY 07932
E-mail address: amo@research.att.com

Contemporary Mathematics
Volume **290**, 2001

Spectral Theory, Dynamics, and Selberg's Zeta Function for Kleinian Groups

Peter Perry

ABSTRACT. We review recent developments in the theory of Selberg's zeta function for discrete groups of infinite co-volume which characterize its divisor in terms of spectral and topological data. Applications include asymptotics of the length spectrum and bounds on the distribution scattering resonances for the associated hyperbolic manifolds.

1. Introduction

The celebrated Elstrodt-Patterson-Sullivan Theorem [**13, 46, 73**] connects the lowest eigenvalue and eigenfunction of a hyperbolic manifold of infinite volume to dynamical invariants associated with the covering group Γ of hyperbolic isometries. To state it, first recall that a discrete group Γ of hyperbolic isometries acting on real hyperbolic space $\mathbf{H}^{n+1}$ is called *geometrically finite* if Γ admits a finite-sided fundamental domain in $\mathbf{H}^{n+1}$, and has infinite co-volume if the orbit space $X = \Gamma\backslash\mathbf{H}^{n+1}$ has infinite metric volume. The boundary at infinity of $\mathbf{H}^{n+1}$ may be visualized as S^n as in the ball model of hyperbolic space; Γ-orbits of points in $\mathbf{H}^{n+1}$ accumulate (as visualized in the ball model) on S^n in the limit set $\Lambda(\Gamma)$, which for infinite co-volume discrete groups is a proper subset of S^n.

The Patterson-Sullivan theorem connects the lowest eigenvalue of the Laplacian on X, denoted $\lambda_0(X)$, with Hausdorff measure on the limit set $\Lambda(\Gamma)$. Let $\delta(\Gamma)$ be the exponent of convergence of the Poincare series

$$\sum_{\gamma\in\Gamma} \exp\left(-s\operatorname{dist}\left(\gamma(0),0\right)\right),$$

where 0 is a chosen 'origin' in $\mathbf{H}^{n+1}$ and $\operatorname{dist}(\cdot,\cdot)$ is hyperbolic distance. The exponent $\delta(\Gamma)$ is also the Hausdorff dimension of $\Lambda(\Gamma)$. The Elstrodt-Patterson-Sullivan Theorem asserts that for $\delta(\Gamma) > n/2$, $\lambda_0(\Gamma) = \delta(\Gamma)(n-\delta(\Gamma))$ (if $\delta(\Gamma) \leq n/2$, the

2000 *Mathematics Subject Classification.* Primary (11M36,58J50); Secondary (35P25).

Key words and phrases. Selberg zeta function, hyperbolic manifolds, length spectrum, scattering resonances, scattering theory.

Supported in part by NSF grant DMS-9707051.

This paper is in final form and no version of it will be submitted for publication elsewhere.

Laplacian on X has no L^2 eigenvalues). Moreover, it asserts that the associated positive eigenfunction can be recovered by integrating the non-Euclidean Poisson kernel against an invariant measure, the so-called Patterson-Sullivan measure, which is supported in $\Lambda(\Gamma)$.

Here we would like to report on recent work [10, 51, 58, 59] which deepens the relationship between spectral theory and dynamics of geometrically finite discrete groups Γ acting on $\mathbf{H}^{n+1}$. The work to be described concerns convex co-compact discrete groups: these are groups Γ which possess a finite-sided fundamental domain that does not touch the limit set of Γ. Equivalently, these groups do not contain parabolic elements, so that the quotient X does not have cusp singularities. For this class, the spectral and scattering theory of the Laplacian is well-understood, thanks to the work of a number of mathematicians [1, 11, 17, 21, 35, 36, 37, 38, 42, 47, 52, 53] (see also the survey [32]). Although the Laplacian on X has finitely many eigenvalues, it has an infinite number of scattering resonances which play a very similar role to that played by eigenvalues of the Laplacian in Selberg's theory for compact surfaces. This is of particular interest in scattering theory since resonances are in general far less well-understood than eigenvalues: in the present case, the zeta function and trace formula provide powerful tools for analyzing their geometric content. Among the consequences of the trace formula are leading asymptotics of the length spectrum of closed geodesics [59] and various bounds on the distribution of scattering resonances [58] which build on previous techniques and results of Zworski and Guillopé [29, 30].

What emerges is a striking similarity between the cases when X is a compact hyperbolic manifold and when X is convex co-compact: although the spectral theory of the Laplacian is very different, the theory of the zeta function is very similar. Patterson [49] conjectured that the singularities of the zeta function could be understood in all cases in terms of distributions supported in the limit set of the discrete group Γ–just as, in the Elstrodt-Patterson-Sullivan theorem, the first eigenvalue of the Laplacian is understood in terms of a measure supported in the limit set of Γ. Patterson's conjecture, formulated in terms of group cohomology of Γ with coefficients in distributions on the limit set of Γ, was proved in somewhat modified form by Bunke and Olbrich [10].

We begin by reviewing the theory of the zeta function on a compact hyperbolic manifold in section 2. Next, in section 3, we discuss the geometry of the non-compact manifolds of infinite volume to be studied. In section 4 we review the spectral and scattering theory of the Laplacian on these manifolds, and in section 5 we characterize the divisor of the dynamical zeta function. Finally, in section 6, we give several applications.

1.1. Acknowledgements. Some of the work described here was carried out jointly with S. J. Patterson, whose papers introduced me to this entire subject. It is a pleasure to acknowledge our fruitful collaboration.

2. Selberg's Zeta Function on a Compact Manifold

We begin by recalling the theory of the zeta function when X is a compact surface of constant negative curvature; we draw on ideas of [7, 13, 18] (see especially [7] where explicit trace formulas of the type discussed below are derived for zeta functions associated to locally homogeneous vector bundles over any compact locally symmetric spaces of rank one). By the uniformization theorem we have

$X = \Gamma \backslash \mathbf{H}^2$ where Γ is a discrete group of isometries of real hyperbolic space $\mathbf{H}^2$, consisting only of hyperbolic elements. An element $\gamma \in \Gamma$ is called *primitive* if it is not conjugate to σ^n for some $\sigma \in \Gamma$ and $n > 1$. Let $[\sigma]$ denote the conjugacy class of $\sigma \in \Gamma$ in the group Γ. The conjugacy classes of Γ (other than the identity class) take the form $[\gamma^n]$ where γ is a primitive element and $n \in \mathbf{N}$. If we choose a set of inconjugate primitive elements (but don't "double count" γ and γ^{-1} which correspond to opposite orientations for the same closed geodesic) there is a one-to-one correspondence between this set and the closed geodesics of X. We denote by $\ell(\gamma)$ the length of the closed geodesic corresponding to $\gamma \in \Gamma$; in fact, γ is conjugate to a dilation (in the upper half-space model) by $\exp \ell(\gamma)$. By a crude lattice-point estimate, the counting function for closed geodesics,

$$\pi_X(t) = \#\{\gamma : \ell(\gamma) \le t\}$$

grows at most like $\exp(t)$ as $t \to \infty$.

We can now define Selberg's zeta function for $\Re(s) > 1$ as the Euler product

$$(2.1) \qquad Z_X(s) = \prod_{\{\gamma\}} \prod_{k=0}^{\infty} [1 - \exp(-(s+k)\ell(\gamma))]$$

where $\{\gamma\}$ denotes a listing of inconjugate primitive elements of Γ. Equivalently

$$(2.2) \qquad Z_X(s) = \exp\left(-\sum_{\{\gamma\}} \sum_{m=1}^{\infty} \frac{1}{m} \det\left(I - P_{\gamma^m}^-\right)^{-1/2} \exp(-s\ell(\gamma^m))\right)$$

which exhibits $Z_X(s)$ as a dynamical zeta function for the closed orbits of geodesic flow on X. Here P_γ^- is the contracting part of the Poincaré once-return map for the closed geodesic γ.

The singularities of Selberg's zeta function are completely determined by geometric and spectral data of X. To describe them, we first recall that the positive Laplacian Δ_X on X has purely discrete spectrum consisting of eigenvalues λ_k with at most finitely many exceptional eigenvalues $\lambda_k < 1/4$.

It will be useful to write the eigenvalue problem

$$\Delta_X u = \zeta(1 - \zeta)u$$

for a complex parameter ζ; we denote by Z_p the set of all $\zeta \in \mathbf{C}$ with $\zeta(1 - \zeta) \in Z_p$.

By Weyl's law,

$$(2.3) \qquad \sum_{\zeta \in Z_p} \langle \zeta \rangle^{-2-\varepsilon} < \infty$$

for any $\varepsilon > 0$ where, here and in what follows, $\langle z \rangle = \left(1 + |z|^2\right)^{1/2}$ for $z \in \mathbf{C}$. Secondly, let $\chi(X)$ denote the Euler characteristic of X, and recall that, by the Gauß-Bonnet theorem,

$$(2.4) \qquad \mathrm{vol}(X) = -2\pi \chi(X).$$

In this section, we will sketch proofs of the following two well-known theorems. The point of the sketch is to parallel developments for convex co-compact hyperbolic manifolds that occur later on.

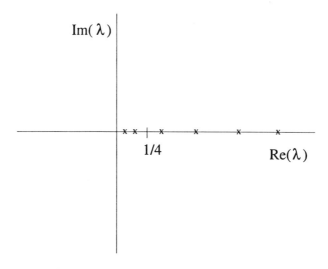

FIGURE 1. The spectrum of Δ_X in the complex λ-plane if X is
compact and $\dim(X) = 2$

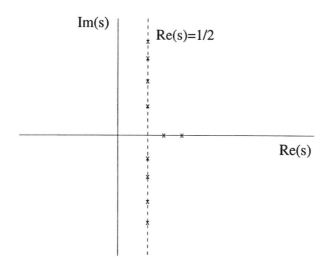

FIGURE 2. The spectrum of Δ_X in the complex s-plane if X is
compact and $\dim(X) = 2$

THEOREM 2.1. *Selberg's zeta function $Z_X(s)$ has an analytic continuation to
the complex s-plane with zeros at $s \in Z_p$ and zeros at $s = -k$, $k = 0, 1, 2, \cdots$ of
multiplicity $-(2k+1)\chi(X)$.*

The second theorem gives a Hadamard product development for $Z_X(s)$. Let

$$(2.5) \qquad Z_0(s) = s \prod_{k=1}^{\infty} (1 + s/k)^{(2k+1)} \exp\left[(2k+1)\left(-s/k + \frac{1}{2}s^2/k^2 \right) \right],$$

and set

$$(2.6) \qquad Z_p(s) = \prod_{\zeta \in Z_p} (1 - s/\zeta) \exp \left(s/\zeta + \frac{1}{2}(s/\zeta)^2 \right).$$

THEOREM 2.2. *The function $Z_X(s)$ is an entire function of order two with*

$$(2.7) \qquad Z_X(s) = \exp(Q(s)) Z_0(s)^{-\chi(X)} Z_p(s)$$

where $Q(s)$ is a polynomial of degree at most two.

The equality of the Euler product (2.1) and Hadamard product (2.7) is essentially equivalent to Selberg's celebrated trace formula. To see this, let

$$\Phi(t) = (Z_X'/Z_X)(1/2 + it) + (Z_X'/Z_X)(1/2 - it)$$

and let $\varphi \in \mathcal{S}(\mathbf{R})$ with Fourier transform $\hat{\varphi}$ belonging to $\mathcal{C}_0^\infty(\mathbf{R}^+)$. Finally, let $\{\zeta_j\}$ be a listing of the elements of Z_p and write $\zeta_j = 1/2 + it_j$. Computing the integral $\int_{-\infty}^\infty \varphi(t)\Phi(t)\,dt$ using the respective formulas (with a careful analytic continuation in the first case since the Euler product only converges for $\Re(s) > 1$) leads to the identity

$$\tfrac{1}{2\pi} \sum_\gamma \sum_{m=1}^\infty \tfrac{\ell(\gamma^m)}{m} \det \left(I - P_{\gamma^m}^- \right)^{-1/2} \exp\left(-\ell(\gamma^m)/2\right) \hat{\varphi}(\ell(\gamma^m)) =$$

$$\sum_j \varphi(t_j) + \tfrac{1}{2\pi} \int_{-\infty}^\infty \Phi_{top}(t)\varphi(t)\,dt,$$

where

$$\phi_{top}(t) = (Z_0'(/Z_0)(1/2 + it) + (Z_0'/Z_0)(1/2 - it).$$

The proof of Theorems 2.1 and 2.2 given here is based on a remarkable identity [13, 49] that relates the logarithmic derivative of $Z_X(s)$ to a renormalized trace of the resolvent

$$R_X(s) = (\Delta_X - s(1-s))^{-1}$$

(the renormalization is necessary since the resolvent is not a trace-class operator). For $\Re(s) > 1$, the integral kernel of the operator $R_X(s)$ is given by the series

$$(2.8) \qquad G_X(\pi(w), \pi(w'), s) = \sum_{\gamma \in \Gamma} G_0(\gamma(w), w', s)$$

where $\pi : \mathbf{H}^2 \to X$ is the natural projection, and $G_0(w, w', s)$ is the integral kernel for the operator

$$R_0(s) = (\Delta_{\mathbb{H}^2} - s(1-s))^{-1}$$

on the covering space $\mathbf{H}^2$ and is smooth for $w \neq w'$. The function $G_0(w, w's)$ is given by the explicit formula (for $w = (x, y)$ and $w' = (x', y')$ in the upper half-plane model)

$$(2.9) \qquad G_0(w, w'; s) = \pi^{-1/2} 2^{-2s-1} \frac{\Gamma(s)}{\Gamma(s+1/2)} \sigma^{-s} F\left(s; s; 2s; \sigma^{-1}\right)$$

where

$$\sigma(w, w') = \frac{|x - x'|^2 + (y + y')^2}{4yy'}$$

and

$$F(a; b; c; z) = 1 + \frac{ab}{c}z + \frac{a(a-1)b(b-1)}{c(c-1)}z^2 + \cdots$$

is a hypergeometric function. The kernel $G_0(w, w'; s)$ is a point-pair invariant, i.e.,

$$(2.10) \qquad G_0(\gamma(w), \gamma(w'), s) = G_0(w, w', s);$$

this property guarantees that the right-hand side of (2.8) descends to a function on $X \times X$. By direct computation, one can prove the following relationship between the resolvent and zeta function: if $\mathcal{F}$ is a fundamental domain for the action of Γ on $\mathbf{H}^2$ and $\Re(s) > 1$,

(2.11)

$$Z_X'(s)/Z_X(s) = (2s - 1) \int_{\mathcal{F}} [G_X(\pi(w), \pi(w'), s) - G_0(w, w', s)]|_{w=w'} \, d\mathrm{vol}(w).$$

The restriction to the diagonal makes sense because $G_0(w, w'; s)$ has the same singularity as $G_X(\pi(w), \pi(w'); s)$, so the difference is actually a smooth function, as the series (2.8) shows. By (2.8) and (2.10), the integrand is actually a Γ-invariant function as well.

We give some hints for the proof of (2.11). From equations (2.1) and (2.8), both the logarithmic derivative of the zeta function and the integrand of (2.11) can be expressed as a sum over conjugacy classes of primitive hyperbolic elements in Γ and their iterates, excluding the identity class: the proof of (2.11) reduces to a computation for the zeta function and resolvent of the abelian group generated by a single hyperbolic motion γ (see, for example, [49] for details). While the left-hand side is only defined for $\Re(s) > 1$, the right-hand side defines a meromorphic function on the complex plane.

To prove Theorem 2.1, we first extend $Z_X(s)$ to $\Re(s) > 0$. From the explicit formula (2.9) for $G_0(w, w', s)$, it is known that $G_0(w, w', s)$ is holomorphic for $\Re(s) > 0$, so that the only singularities of $Z_X'(s)/Z_X(s)$ come from poles of $R_X(s)$ at $\zeta \in Z_p$. The residues are finite-rank projections with rank ν_ζ equal to the dimension of the corresponding eigenspace. It follows that $Z_X'(s)/Z_X(s)$ has simple poles at $\zeta \in Z_p$ with integral residues ν_ζ, so that one can integrate $Z_X'(s)/Z_X(s)$ and exponentiate to obtain a continuation of $Z_X(s)$ to $\Re(s) > 0$ with zeros as claimed.

To continue the zeta function to $\mathbf{C}$, we obtain a functional equation using the self-adjointness of Δ_X and some explicit computations involving $G_0(w, w'; s)$. Using (2.11), the formula

$$(2s - 1) [G_0(w, w'; s) - G_0(w, w'; 1 - s)]|_{w=w'} = \frac{\Gamma(s)\Gamma(1 - s)}{\Gamma(s - 1/2)\Gamma(1/2 - s)}$$

(see, for example, [49]), the identity

$$G_X(\pi(w), \pi(w'), s) - G_X(\pi(w), \pi(w'), 1 - s) = 0$$

for $s \notin Z_p$ (which follows from self-adjointness of Δ_X), and the Gauß-Bonnet formula (2.4), we conclude that

(2.12)

$$Z_X'(s)/Z_X(s) + Z_X'(1 - s)/Z_X(1 - s) = -2\pi \frac{\Gamma(s)\Gamma(1 - s)}{\Gamma(s - 1/2)\Gamma(1/2 - s)} \chi(X)$$

where $\chi(X)$ is the Euler characteristic of X. Note that

$$(2.13) \qquad \frac{\Gamma(s)\Gamma(1 - s)}{\Gamma(s - 1/2)\Gamma(1/2 - s)} = (s - 1/2)\pi \cot(\pi s).$$

From this it follows that $Z'_X(s)/Z_X(s)$ has a meromorphic continuation to the complex plane having poles at $s = -k$, $k = 0, 1, 2, \cdots$ with residue $-(2k+1)\chi(X)$. Thus $Z_X(s)$ has an analytic extension to the complex plane with 'spectral' zeros ζ where $\zeta(1-\zeta)$ is an eigenvalue of the Laplacian, and 'topological' zeros at $\zeta = -k$, $k = 0, 1, 2, \cdots$.

Next we prove Theorem 2.2. Let $H(s) = Z_0(s)Z_p(s)$, an entire function of order two whose zeros are equal with multiplicity to those of $Z(s)$ (see (2.5) and (2.6)). To prove Theorem 2.2, it suffices to show that the logarithmic derivative

$$\Xi(s) = Z'_X(s)/Z_X(s) - H'(s)/H(s),$$

an entire function of s, is in fact a linear polynomial. We will first show that $\Xi(s)$ is a polynomial of order at most five, and then show that $\Xi(s)$ is actually of degree one by studying its growth as $s \to \infty$ for s real.

We will actually carry out the estimates away from a set $\mathcal{D}$ which contains the singularities of $Z'_X(s)/Z_X(s)$. To describe it we need the following lemma.

LEMMA 2.3. *There is a family $\{D_k\}_{k=1}^{\infty}$ of disjoint discs in $\mathbb{C}$ such that (i) the radii of the D_k are uniformly bounded, and (ii) if $s \in \mathbb{C}\backslash\{\cup_k D_k\}$, dist$(s, \zeta) > \langle s \rangle^{-2-\varepsilon}$ for all $\zeta \in Z_p$.*

PROOF. As a preliminary step we consider the (not necessarily disjoint) family of discs $\bar{D}_j$ with center s_j and radius $\langle \zeta_j \rangle^{-2-\varepsilon}$, where $\{\zeta_j\}_{j=1}^{\infty}$ is a listing of elements in Z_p.

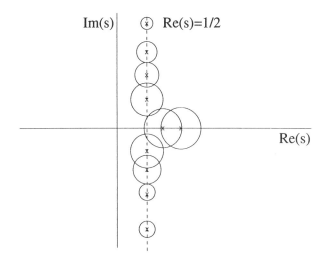

FIGURE 3. The discs $\overline{D}_j$

We claim that this set can be covered by a set of disks $\{D_k\}$ with the claimed properties, although one D_k may contain many $\bar{D}_j$'s. Clearly we can ignore the finitely many ζ which do not lie on the line $\Re(s) = 1/2$, corresponding to the finitely many exceptional eigenvalues.

To prove the claim, note that the intersection of the $\bar{D}_j$ with the line $\Re(s) = 1/2$ gives a collection of open intervals I_j with $\sum_j |I_j| < \infty$ by (2.3). These intervals

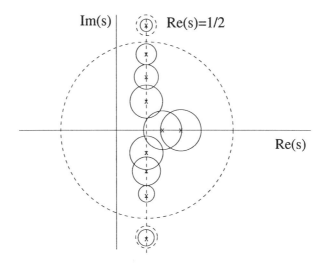

FIGURE 4. The discs D_k (dashed lines)

may overlap but any connected component of $\cup_j I_j$ is contained in a bounded interval of length no greater than $\sum_j |I_j|$. Labelling the connected components by $k = 1, 2, \cdots$, we cover them by a disc D_k whose radius is no greater than $\sum_j |I_j|$. The collection $\{D_k\}$ now has the required properties. $\qquad \square$

Since $\Xi(s)$ is an entire function, by the maximum modulus theorem, it suffices to prove an estimate of the form

$$(2.14) \qquad |\Xi(s)| \le C \langle s \rangle^{4+\varepsilon}$$

for any $\varepsilon > 0$ in the region $\mathbf{C} \backslash \mathcal{D}$ where $\mathcal{D}$ is the set

$$\tilde{\mathcal{D}} = (\cup_{k=0}^{\infty} \{s : |s + k| < 1/4\}) \cup (\cup_k D_k)$$

together with its image under the reflection $s \mapsto 1 - s$ (this set is a countable union of discs of bounded radii). Observing that $|H'(s)/H(s)|$ can be bounded by the right-hand side of (2.14) in $\mathbf{C} \backslash \mathcal{D}$, it suffices to prove the same estimate for $|Z_X'(s)/Z_X(s)|$.

To do so, we first note that the right-hand side of (2.12) is bounded by $C \langle s \rangle^{1+\varepsilon}$ away from $\mathcal{D}$ by the formula (2.13) and the explicit series

$$\cot(\pi s) = \frac{1}{s} + \sum_{n \neq 0} \left(\frac{1}{s - n} + \frac{1}{n} \right),$$

so that we need only estimate $|Z_X'(s)/Z_X(s)|$ in the intersection of $\mathbf{C} \backslash \mathcal{D}$ and the closed half-plane $\Re(s) \ge 1/2$.

Recall that

$$R_X(s) = (\Delta_X - s(1 - s))^{-1}$$

on $L^2(X)$, and that

$$R_0(s) = (\Delta_{\mathbf{H}^2} - s(1 - s))^{-1}$$

on $L^2(\mathbf{H}^2)$. Note that, in contrast to $R_X(s)$, the resolvent $R_0(s)$ is well-defined as an operator on $L^2(\mathbf{H}^2)$ only in the right half-plane $\Re(s) > 1/2$, since the line

$\Re(s) = 1/2$ corresponds to the continuous spectrum of $\Delta_{\mathbf{H}^2}$ in $[1/4, \infty)$. However, if

$$\rho(w) = y/\left(x^2 + (y+1)^2\right),$$

the operator $\rho R_0(s)\rho$ extends to a bounded operator on $L^2(\mathbf{H}^2)$ for s in the half-plane $\Re(s) > -1/2$ with bound

$$\|\rho R_0(s)\rho\| \leq C_\varepsilon$$

for $\Re(s) > -1/2 + \varepsilon$, as may be seen from the explicit formula (2.9) (see, for example, [58]). By using (2.11), the resolvent equation

$$R_X(s) = R_X(s_0) + (s - s_0) R_X(s) R_X(s_0)$$

and a similar equation for $R_0(s)$, we obtain for fixed $s_0 \in \mathbf{C}\backslash Z_p$, $s_0 \neq 1/2$, and any $s \in \mathbf{C}\backslash Z_p$ with $\Re(s) \geq 1/2$ that

$$(2.15) \qquad Z_X'(s)/Z_X(s) = (2s-1)(2s_0-1)^{-1} Z_X'(s_0)/Z_X(s_0)$$
$$+ (2s-1)(s-s_0)\,\mathrm{Tr}\,(R_X(s) R_X(s_0))$$
$$- (2s-1)(s-s_0)\,\mathrm{Tr}\,(\chi R_0(s) R_0(s_0)\chi).$$

Here χ is the characteristic function of $\mathcal{F}$, and the traces in the second and third terms are respectively over operators on $L^2(X)$ and $L^2(\mathbf{H}^2)$. The first term is a linear polynomial in s while the second can be estimated spectrally: choosing s_0 so that $\lambda_0 = s_0(n - s_0) < 0$ and using the fact that $R_X(s_0)^{1+\varepsilon}$ is trace-class for any $\varepsilon > 0$, we have

$$|\mathrm{Tr}\,(R_X(s) R_X(s_0))| = \left|\mathrm{Tr}\left((\Delta_X - \lambda_0)^\varepsilon R_X(s) R_X(s_0)^{1+\varepsilon}\right)\right|$$

$$\leq C_\varepsilon \|(\Delta_X - \lambda_0)^\varepsilon R_X(s)\|$$

$$\leq C_\varepsilon \langle s\rangle^{2+3\varepsilon}$$

where $\|\cdot\|$ denotes the operator norm (the "bad points" are those s close to points of Z_p). This gives a crude estimate $C_\varepsilon \langle s\rangle^{4+3\varepsilon}$ for the second term. Finally we can estimate the third term as follows. By choosing $s_0 > 0$ large we can suppose that $B = \chi R_0(s_0)\rho^{-1}$ is a Hilbert-Schmidt operator. We can then write

$$(2.16) \quad (s - s_0) R_0(s) R_0(s_0) = (s - s_0) R_0(s_0)^2 + (s - s_0)^2 R_0(s_0) R_0(s) R_0(s_0)$$

and note that the first term is locally trace-class while the second term is a polynomial in s times

$$BC(s)B$$

where $C(s)$ is bounded in the closed half-plane $\Re(s) \geq 1/2$. Since B is Hilbert-Schmidt this shows that the second term in (2.16) is a polynomial in s times an operator with trace-norm bounded uniformly in $\Re(s) \geq 1/2$, and hence is estimated by $\langle s\rangle^2$.

Assembling these remarks we see that $|Z_X'(s)/Z_X(s)|$ is bounded by $\langle s\rangle^{4+\varepsilon}$ on $\mathbf{C}\backslash\mathcal{D}$ so that the same is true of the difference. It now follows that the zeta function is entire of order at most five. One can improve this result by estimating $|Z_X'(s)/Z_X(s) - H'(s)/H(s)| = |Q'(s)|$ for real s with $s \to +\infty$; it is not difficult to see in this way that $|Q'(s)| \leq \langle s \rangle^{1+\varepsilon}$ (one estimates $|\text{Tr}(R_X(s)R_X(s_0))| \leq$ $\langle s \rangle^{\varepsilon-1}$ away from the eigenvalues of Δ_X). Theorem 2.2 now follows.

3. Convex Co-Compact Hyperbolic Manifolds

In the preceding section, a discrete group Γ was chosen so that the quotient $\Gamma \backslash \mathbf{H}^2$ was compact. Now we consider cases where Γ is a discrete group of isometries acting on the $(n+1)$-dimensional real hyperbolic space $\mathbf{H}^{n+1}$ (visualized, for instance, as the upper half-space $\mathbf{R}_+^{n+1}$ with the Poincaré metric

$$ds^2 = y^{-2}\left(d\mathbf{x}^2 + dy^2\right)$$

where $\mathbf{x} \in \mathbf{R}^n$ and $y > 0$) so chosen that the orbit space $\Gamma \backslash \mathbf{H}^{n+1}$ has infinite volume. Some simple examples of such groups include: (1) the abelian cyclic group of dilations $(x,y) \mapsto \left(\mu^k x, \mu^k y\right)$ for $k \in \mathbf{Z}$ acting on $\mathbf{H}^2$, (2) cocompact groups Γ_2 acting on $\mathbf{H}^2$, lifted to act on $\mathbf{H}^3$ (see, for example, Appendix B of [**51**]), and (3) Schottky groups acting on $\mathbf{H}^3$ (see, for example, [**6**]). The quotients $X = \Gamma \backslash \mathbf{H}^{n+1}$ for these groups are respectively an infinite hyperbolic cylinder, diffeomorphic to $(0,1) \times S^1$, a warped product $(0,1) \times X_2$ where $X_2 = \Gamma_2 \backslash \mathbf{H}^2$ is a compact hyperbolic surface, and the interior of a g-holed torus if Γ is a Schottky group with g generators. In all of these examples, the quotients have a natural compactification to manifolds $\bar{X}$ with boundary, namely $[0,1] \times S^1$, $[0,1] \times X_2$, and the closed solid g-holed torus, respectively. We will restrict attention to so-called convex co-compact discrete groups Γ for which the orbit space $\Gamma \backslash \mathbf{H}^{n+1}$ is diffeomorphic to the interior of a manifold with boundary. The quotient $X = \Gamma \backslash \mathbf{H}^{n+1}$ will be called a convex co-compact hyperbolic manifold.

The geometry at infinity of such manifolds is easy to describe. If H denotes the hemispherical neighborhood

$$H = \left\{(\mathbf{x},y) \in \mathbf{R}_+^{n+1} : |\mathbf{x}|^2 + y^2 < 1\right\}$$

then there is a finite covering of X in a neighborhood of infinity by charts isometric to H, whose transition maps are hyperbolic isometries. That is, infinity 'looks like' the infinity of hyperbolic space. On the other hand, the space is not simply connected and unlike hyperbolic space its boundary at infinity is not the sphere but may have several connected components. In the examples above, the infinite boundary $\partial_\infty X$ is respectively two copies of S^1, two copies of X_2, and the surface of the g-holed torus. Note that the 'boundary at infinity' of the hyperbolic manifold is diffeomorphic to the boundary of the compactification $\bar{X}$: that is, $\partial_\infty X = \partial \bar{X}$.

If $\partial_\infty X$ consists of connected components C_i, the manifold X then consists of a compact manifold with boundary, N, together with 'ends' E_i diffeomorphic to $C_i \times (0,1)$. The manifold with boundary N can be chosen to be a compact, convex manifold with smooth, compact boundary. Letting $\Sigma = \partial N$, we can construct a map ψ from Σ onto $\partial_\infty X$ as follows. Let $p \in \Sigma$ and let $\gamma_p(t)$ be the unit speed geodesic with $\gamma_p'(0) = \nu_p$, the outward normal to Σ at p. We then set $\psi(p) = \lim_{t \to \infty} \gamma_p(t)$. It can be shown that this map is actually a diffeomorphism; see for example [**14**] or section 3 and Appendix A of [**51**].

A *defining function* for $\partial \bar{X}$ is a nonnegative C^∞ function ρ with $\rho^{-1}(0) = \partial \bar{X}$ and $d\rho(m) \neq 0$ for each $m \in \partial \bar{X}$: that is, ρ vanishes exactly to first order on $\partial \bar{X}$. The hyperbolic metric on a convex co-compact hyperbolic manifold takes the form $\rho^{-2}h$ where ρ is a defining function and h is the restriction to X of a nondegenerate smooth metric on $\bar{X}$. A manifold of this type whose sectional curvatures $K(m)$ converge to -1 as $m \to \partial_\infty X$ is called *conformally compact*; the spectral theory of conformally compact manifolds was studied by Mazzeo [39] and Mazzeo and Melrose [42], motivated by the example of convex co-compact hyperbolic manifolds.

4. Spectral and Scattering Theory

The spectral theory of the Laplacian on convex co-compact hyperbolic manifolds is very different from the compact case. The Laplacian Δ_X has at most finitely many discrete eigenvalues in $[0, n^2/4)$ [35] and continuous spectrum of infinite multiplicity in the interval $[n^2/4, \infty)$ (first proved in [36]). This means that the operator-valued function $(\Delta_X - \lambda)^{-1}$ is meromorphic in a *cut* plane $\mathbf{C} \backslash [n^2/4, \infty)$.

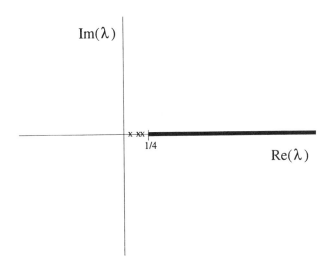

FIGURE 5. The spectrum of Δ_X in the complex λ-plane if X is con vex co-compact and $\dim(X) = 2$

Thus the resolvent operator

$$R_X(s) = (\Delta_X - s(n-s))^{-1}$$

is an operator-valued meromorphic function in the half-plane $\Re(s) > n/2$, with finitely many poles corresponding to the finitely many eigenvalues. We denote the set of such poles by Z_p. The cut on the half-line $[n^2/4, \infty)$ in the λ-plane corresponds to the line $\Re(s) = n/2$ in the complex s-plane: the operator norm of $R_X(s)$ as an operator from $L^2(X)$ to itself actually diverges as $\Re(s) \downarrow n/2$, owing to the continuous spectrum of Δ_X.

On the other hand, the resolvent can be continued through the line $\Re(s) = n/2$ if viewed as a map between different spaces. For a real number N, let $\rho^N L^2(X)$ denote the space of locally square-integrable functions u on X of the form $u = \rho^N v$ where $v \in L^2(X)$, and ρ is a defining function for X. Thus $\rho^N L^2(X)$ consists of functions with rapid decay at infinity if $N > 0$, and rapid growth at infinity if $N < 0$. It is a deep result of Mazzeo and Melrose [42] that for any $N > 0$, the operator $R_X(s)$ extends to a meromorphic family of bounded operators from $\rho^N L^2(X)$ to $\rho^{-N} L^2(X)$ in the half-plane $\Re(s) > n/2 - N$. This reflects the fact that the integral kernel of $R_X(s)$ has the asymptotic form $G_X(m, m'; s) \sim \rho(m)^s \rho(m')^s$ for m and m' approaching $\partial\bar{X}$, $m \neq m'$. The poles of the meromorphically continued resolvent are called *resolvent resonances* (see the surveys [**77**] for results on resonances in a variety of scattering problems and [**79**] for a general introduction). We shall see that the resolvent resonances play a role in the theory of the zeta function on convex co-compact X very similar to the role played by eigenvalues in the theory of the zeta function on compact X.

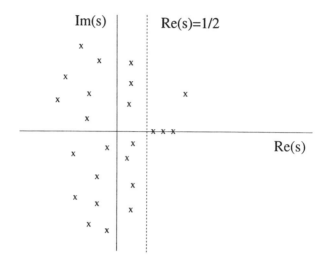

FIGURE 6. Eigenvalues $(\Re(s) > 1/2)$ and resonances $(\Re(s) < 1/2)$ of Δ_X in the complex-s plane if X is convex co-compact and $\dim(X) = 2$

To understand the physical meaning of resolvent resonances, one needs to understand the scattering theory associated to Δ_X. Scattering theory has its roots in acoustics and electromagnetics, where wave propagation is governed by a scalar equation or system in Euclidean space. The simplest Euclidean wave equation $\Delta u = u_{tt}$ (where Δ is the Euclidean Laplacian and so is negative as an operator!) has a natural analogue on any complete Riemannian manifold if one replaces the Euclidean Laplacian by the Laplace-Beltrami operator (with due care about signs!). Just as one can remove the time variable in the Euclidean case by assuming harmonic time dependence (i.e., look for solutions of the form $u(x, t) = \exp(i\omega t) v(x)$) so one can look for such solutions on a manifold. One is naturally lead to the reduced wave equation $(\Delta_X - \omega^2) u = 0$. Solutions of this equation describe spatial patterns for scattered waves and as such should reflect the geometry of the underlying manifold X.

If X is a convex co-compact hyperbolic manifold of dimension $n + 1$, it is more convenient to write the reduced wave equation as

(4.1) $$(\Delta_X - s(n - s)) u = 0$$

for a complex parameter s. On the covering space $\mathbf{H}^{n+1}$ the corresponding equation reads

$$\left[-(y\partial_y)^2 + n(y\partial_y) - y^2 \Delta_x - s(n - s) \right] u = 0$$

and has has solutions y^s and y^{n-s}; these solutions represent respectively incoming and outgoing waves in the direction of the boundary point $y = 0$. The wavefronts are level sets of these functions and are clearly hyperplanes $y =$constant.

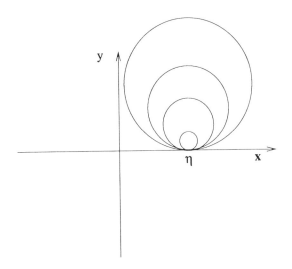

FIGURE 7. Wavefronts of $e_0(w, \eta, s)$ in the upper half-space model

More generally if η is a boundary point of $\mathbf{H}^{n+1}$ there are solutions of the reduced wave equation of the form

(4.2)
$$e_0(w, \eta, s) = \left(\frac{y}{|x - \eta|^2 + y^2} \right)^s ;$$

these solutions represent waves incoming from or outgoing towards the boundary point η of hyperbolic space. The wavefronts are spheres tangent to the boundary at the point η (horospheres). Note that these solutions are very singular near the boundary point η.

One would expect there to be corresponding solutions of the reduced wave equation on X with similar singularities near a prescribed boundary point $\eta \in \partial \bar{X}$, and this is indeed the case. One can obtain such solutions from the integral kernel of the resolvent operator $R_X(s)$. Heuristically one imagines solving the equation

$$(\Delta_X - s(n - s)) u = \delta_{m'}$$

where $\delta_{m'}$ is a δ-function at $m \in X$, and taking a limit as $m' \to \eta \in \partial \bar{X}$ in order to "put the pole at infinity"; this corresponds to taking the same limit for the integral kernel of $R_X(s)$, which we denote $G_X(m, m'; s)$. More precisely, we set

$$e_X(m, \eta; s) = \lim_{m' \to \eta} \rho(m')^{-s} G_X(m, m'; s)$$

where the prefactor guarantees the existence of the limit. The resulting solution has a leading singularity identical to that of (4.2). We denote these solutions by $e_X(m, \eta; s)$ for $m \in X$ and $\eta \in \partial \bar{X}$.

There are also smooth solutions of the reduced wave equation which play an important role emphasized by Melrose [43]. Suppose that $f_- \in C^\infty(\partial \bar{X})$; one should think of f_- as prescribing the amplitude, in each direction $\eta \in \partial \bar{X}$, of an incoming wave. The superposition

$$u(m) = \int e_X(m, \eta; s) f_-(\eta) \, d\eta$$

is also a solution to the reduced wave equation. If $\Re(s) = n/2$ and $s \neq n/2$, this solution has the asymptotic form

$$u = \rho^s u_+ + \rho^{n-s} u_-$$

where $u_\pm \in C^\infty(\bar{X})$, and $u_-|_{\partial \bar{X}} = f_-$. Moreover it turns out that this solution is the *unique* solution over $C^\infty(X)$ of this form obeying the asymptotic condition $u_-|_{\partial \bar{X}} = f_-$. It follows that $f_+ = u_+|_{\partial \bar{X}}$ is also uniquely determined. Exploiting the uniqueness we may define two important operators: the *Poisson operator*

$$P(s) : C^\infty(\partial \bar{X}) \ni f_- \mapsto u \in C^\infty(X)$$

and the *scattering operator*

$$S_X(s) : C^\infty(\partial \bar{X}) \ni f_- \mapsto f_+ \in C^\infty(\partial \bar{X}).$$

Heuristically, the scattering operator maps incoming (f_-) to outgoing (f_+) radiation patterns for a given 'energy' s. As stressed by Melrose [43], the data $f_\pm$ are analogous to Dirichlet and Neumann data for a boundary value problem and the scattering operator is analogous to the 'Dirichlet to Neumann operator' associated to second-order elliptic boundary value problems.

The scattering operator, initially defined for $\Re(s) = n/2$, has a meromorphic continuation to the complex plane. The meromorphic continuation defines a family of elliptic pseudodifferential operators of degree $2s - n$ and order $2\Re(s) - n$. With the normalization above, the scattering operator has infinite rank zeros and poles which are easily removed by setting

$$\mathcal{S}_X(s) = 2^{2s-n} \frac{\Gamma(s - n/2)}{\Gamma(n/2 - s)} S_X(s).$$

The principal symbol of this operator is exactly $|\xi|^{2s-n}$, and the operator has singularities with finite polar parts and finite-rank residues. Using the theory of Gohberg and Sigal [23] one can compute the order of the singularity at $\zeta \in \mathbf{C}$ by computing the integer

$$\nu_\zeta = \mathrm{Tr} \left(\frac{1}{2\pi i} \int_{\gamma_{\zeta, \varepsilon}} \mathcal{S}_X(s)^{-1} \mathcal{S}_X'(s) \, ds \right)$$

where $\gamma_{\zeta, \varepsilon}$ is a contour surrounding ζ and no other singularity of $\mathcal{S}_X(s)$; roughly speaking ν_ζ counts the 'number of zeros minus the number of poles' as in Rouché's theorem for scalar-valued functions. It can be shown that for all but finitely many ζ with $\Re(\zeta) < n/2$, ν_ζ is nonpositive, and it follows from the functional equation $\mathcal{S}_X(s) \mathcal{S}_X(n - s) = I$ that $\nu_{n-\zeta} = -\nu_\zeta$.

The scattering poles correspond to the resolvent resonances. More precisely, if the multiplicity m_ζ of a resolvent resonance at $\zeta \in \mathbf{C}$ is defined to be the rank of the operator

$$\frac{1}{2\pi i} \int_{\gamma_{\zeta, \varepsilon}} R_X(s) \, ds$$

where $\gamma_{\zeta,\varepsilon}$ is a contour surrounding ζ and no other resolvent resonance, then $\nu_\zeta = m_{n-\zeta} - m_\zeta$ with finitely many exceptions if $\dim X$ is even [4], and with the possible exception of the negative integers if $\dim X$ is odd (the restriction is probably unnecessary but there is not yet a proof!).

5. Divisor of the Zeta Function

We can now discuss the zeta function on a convex co-compact Riemannian manifold X of dimension $n+1$. The underlying discrete group Γ consists of elements γ conjugate to the composition of a dilation $e^{\ell(\gamma)}$ on $\mathbf{H}^{n+1}$ and an $O(n)$ rotation in the x-variables. The eigenvalues $\alpha_1(\gamma), \cdots, \alpha_n(\gamma)$ are conjugacy invariants. The correspondence between inconjugate primitive elements of Γ and closed geodesics of $X = \Gamma \backslash \mathbf{H}^{n+1}$ again holds, and a crude lattice-point counting argument shows that the counting function of closed geodesics is bounded above by $\mathcal{O}(e^{nt})$. Again letting $\{\gamma\}$ be a set of representatives of conjugacy classes of primitive elements, we define, following Patterson [49]

$$Z_X(s) = \prod_{\{\gamma\}} \prod_{k_1, \cdots k_n \geq 0} \left[1 - \alpha_1(\gamma)^{k_1} \cdots \alpha_n(\gamma)^{k_n} \exp\left(-(s + k_1 + \cdots + k_n)\ell(\gamma)\right) \right].$$

Equivalently,

$$(5.1) \qquad Z_X(s) = \exp\left(-\sum_{\{\gamma\}} \sum_{m=1}^{\infty} \frac{1}{m} \det\left(I - P_-^{\gamma^m}\right)^{-1/2} \exp\left(-s\ell(\gamma^m)\right) \right)$$

(cf. (2.2)). Both expressions converge for $\Re(s) > n$.

The basic formula (2.11) also remains valid for $\Re(s) > n$. If we set

$$\tilde\Psi(w) = (2s - n)\left[G_X(\pi(w), \pi(w'), s) - G_0(w, w', s)\right]\big|_{w=w'},$$

where $G_0(w, w'; s)$ is the resolvent kernel for the Laplacian on $\mathbf{H}^{n+1}$, then $\tilde\Psi$ is an automorphic function on $\mathbf{H}^{n+1}$. It therefore descends to a function Ψ on X. It turns out that for $\Re(s) > (n-1)/2$, Ψ takes the form $\rho^{2s}\psi$ where ρ is a defining function for ∂X, and ψ is the restriction to X of a smooth function on $\bar X$ which is meromorphic as a function of s. As before, we have the formula

$$(5.2) \qquad Z_X'(s)/Z_X(s) = \int_{\mathcal{F}} \tilde\Psi \, \mathrm{dvol}(w)$$

$$= \int_X \Psi \, \mathrm{dvol}.$$

Since the integration is now over an infinite volume, convergence depends on the asymptotic behavior of Ψ as $\rho \to 0$. We can choose local coordinates $(x_1, \cdots, x_n)$ for $\partial \bar X$ and obtain local coordinates $(x_1, \cdots, x_n, \rho)$ for $\bar X$ in a neighborhood of $\partial \bar X$; in such coordinates, the Riemannian volume measure on X takes the form $\rho^{-(n+1)} dx_1 \cdots dx_n \, d\rho$ up to a positive smooth factor. It follows that (5.2) makes sense for $\Re(s) > n/2$ but thereafter is not meaningful. Thus, although we can readily conclude from spectral theory and (5.2) that the zeta function continues to $\Re(s) > n/2$ with zeros at any eigenvalues of Δ_X, the continuation to $\Re(s) \leq n/2$ is subtler and depends on scattering theory.

From the remarks above it follows that for $\Re(s) \geq n/2$ the integral $\int_{\rho \geq \varepsilon} \Psi \, dvol$ has a small-ε asymptotic expansion of the form

$$a_0(s) + a_1(s) \varepsilon^{2s-n} + \mathcal{O}\left(\varepsilon^{2s-n+1}\right)$$

and that $Z_X'(s)/Z_X(s) = a_0(s)$ for $\Re(s) > n/2$. We can thus continue $Z_X'(s)/Z_X(s)$ to $\Re(s) > (n-1)/2$ and obtain a functional equation

$$Z_X'(s)/Z_X(s) + Z_X'(n-s)/Z_X(n-s) = a_0(s) + a_0(n-s).$$

To evaluate the right-hand side, one uses the identities

$$G_X(m, m'; s) - G_X(m, m'; n-s)|_{m=m'} =$$

$$(n - 2s) \int_{\partial \bar{X}} e_X(m, \eta; s) \, e_X(m, \eta, n-s) \, dh$$

and

$$G_0(w, w'; s) - G_0(w, w'; n-s)|_{w=w'} = \pi^{-n/2} \frac{\Gamma(s)\Gamma(n-s)}{\Gamma(s-n/2)\Gamma(n/2-s)} \frac{\Gamma(n/2)}{\Gamma(n)}$$

to obtain a functional equation taking the form

$$(5.3) \qquad (Z_X'/Z_X)(s) + (Z_X'/Z_X)(n-s) = -\text{``Tr''}\left(\mathcal{S}_X(s)^{-1}\mathcal{S}_{X'}(s)\right)$$

$$- \frac{\Gamma(s)\Gamma(n-s)}{\Gamma(s-n/2)\Gamma(n/2-s)} c$$

where the trace is in quotation marks since $\mathcal{S}_X(s)^{-1}\mathcal{S}_X(s)$ is far from trace-class! The trace in question is defined by a limiting procedure: the resulting meromorphic function of s has poles exactly at the poles of $\mathcal{S}_X(s)$ with residue ν_ζ. The proof of this formula uses the fact that the leading asymptotic behavior of the functions $e_X(m; \eta; s)$ is determined by the scattering operator.

If $\dim X$ is even, the second right-hand term in (5.2) has poles at the integers, and the constant c is a multiple of the Euler characteristic of the manifold with boundary $\bar{X}$. The situation if $\dim X$ is odd is somewhat more complicated and is described in [51]; results of [10] allow one to identify the poles in this case, also, as multiples of the Euler characteristic of X. The analysis discussed in [51] (using results of [10] if $\dim X$ is odd) leads to the following result on the divisor of $Z_X(s)$ which parallels the characterization given for compact X very closely. Let

$$h_n(k) = (2k+n)\frac{(n+k-1)!}{n!k!},$$

the dimension of the space of spherical harmonics of degree k on S^{n+1}.

THEOREM 5.1. *Let X be a convex co-compact hyperbolic manifold, $\dim X = n+1$, and let $Z_X(s)$ be the associated dynamical zeta function. Then $Z_X(s)$ has a meromorphic extension to $\mathbf{C}$ with the following singularities: (a) zeros at the eigenvalues of Δ_X whose multiplicity is the dimension of the corresponding eigenspace, (b) a zero of order $m_{n/2} = \dim(I + \mathcal{S}_X(n/2))$ at $\zeta = n/2$, (c) zeros at the scattering poles of $\mathcal{S}_X(s)$ with multiplicity ν_ζ, with at most finitely many exceptions, and (d) singularities at $s = -k$, $k = 0, 1, 2, \cdots$, of multiplicity $h_n(k)\chi(\bar{X})$, where a positive sign indicates a zero and a negative sign indicates a pole.*

The proof of this theorem is given in the case of $\dim X$ even in [51] and in the case of $\dim X$ odd in [10]; the result about $\zeta = n/2$ is due to [10] although a different proof is given in [51]. Unfortunately, it does not seem possible at present to deduce by these methods the fact that $Z_X(s)$ is a quotient of entire functions of finite order. On the other hand, the dynamical methods of Ruelle and Fried [19, 64] can be used to represent the zeta function as a quotient of determinants of transfer operators, from which it follows that $Z_X(s)$ is indeed such a quotient.

From this it follows that the zeta function has a Hadamard product representation. Let

$$E_N(z) = (1-z)\exp\left(\sum_{j=1}^{N}\frac{1}{j}z^j\right)$$

be the usual Hadamard factor,

(5.4) $$Z_0(s) = s\prod_{k=1}^{\infty}E_{n+1}\left(-s/k\right)^{h_n(k)},$$

let

(5.5) $$Z_p(s) = \prod_{\zeta\in Z_p}(1-s/\zeta)^{m_\zeta},$$

and let

(5.6) $$Z_{scatt}(s) = (s-n/2)^{m_{n/2}}\prod_{\zeta\in Z_s}E_{n+1}\left(s/\zeta\right)^{\nu_\zeta}$$

where Z_s is the set of scattering poles of $\mathcal{S}_X(s)$, and $Z_p = \{s : s(n-s) \in \sigma_p(\Delta_X)\}$.

THEOREM 5.2. *Let Γ be a convex co-compact, torsion-free, discrete group of hyperbolic isometries acting on $\mathbf{H}^{n+1}$, let $\chi(X)$ be the Euler characteristic of . Then Selberg's zeta function admits the Hadamard product representation*

$$Z_X(s) = \exp(P(s))Z_0(s)^{-\chi(\bar{X})}Z_p(s)Z_{scatt}(s)$$

where $P(s)$ is a polynomial of degree at most $n+1$.

6. Resonances and Length Spectrum

Theorem 5.2 together with the definition (2.7) of $Z_X(s)$ yield a version of Selberg's trace formula for convex co-compact hyperbolic manifolds [58]. Let $\{\zeta_j\}$ be a listing of the scattering resonances of Δ_X, and write $\zeta_j = n/2 + it_j$. Finally, let

$$\Phi_{top}(t) = (-\chi(X))\left[Z_0'(n/2+it)/Z_0(n/2+it) + Z_0'(n/2-it)/Z_0(n/2-it)\right].$$

where $Z_0(s)$ is defined in (5.4).

THEOREM 6.1. *Let $X = \Gamma\backslash\mathbf{H}^{n+1}$ where Γ is convex co-compact and torsion-free. Let $\varphi \in \mathcal{S}(\mathbf{R})$ where $\hat{\varphi}$ is a real-valued function belonging to $C_0^\infty(\mathbf{R})$. Then*

(6.1)
$$\frac{1}{2\pi}\sum_{\gamma}\sum_{m=1}^{\infty}\frac{\ell(\gamma^m)}{m}\det\left(I-P_{\gamma^m}^-\right)^{-1/2}\exp\left(-n\ell(\gamma^m)/2\right)\hat{\varphi}(\ell(\gamma^m)) =$$

$$\sum_j \varphi(t_j) + \frac{1}{2\pi}\int_{-\infty}^{\infty}\Phi_{top}(t)\varphi(t)\,dt$$

where P_γ^- is the restriction to the contracting direction of the Poincaré map for the closed geodesic γ.

A trace formula of this form was already proved by Guillopé and Zworski in [**30**]. By mimicking arguments from [**29, 30**], one can prove a lower bound on the distribution of resonances for these manifolds already obtained in [**30**] if dim $X = 2$.

THEOREM 6.2. [**58**] *Let* $N(r) = \#\{t_j : |t_j| \leq r\}$. *Then under the same hypotheses as above,*

$$N(r) \geq C_X \, |\chi(X)| \, r^{n+1}.$$

where $\chi(X)$ *is the Euler characteristic of* X.

Note that this bound is rather trivial if $\chi(X) = 0$! On the other hand the exponent is optimal as is shown by computable examples (see [**27**] and the second appendix to [**51**]).

For geodesic flow on compact manifolds, the first singularity of the zeta function determines the asymptotics of the counting function for geodesics. By Theorem 5.1, the first pole of the zeta function has a spectral characterization as either (1) the $\zeta \in Z_p$ corresponding to the first eigenvalue of Δ_X if $\delta(\Gamma) > n/2$, (2) the point $\zeta = n/2$ if there are no eigenvalues but $\delta(\Gamma) = n/2$, or (3) the point $\zeta = \delta(\Gamma) \in Z_{scatt}$ if $\delta(\Gamma) < n/2$. In all cases the first pole of the zeta function is exactly $\delta(\Gamma)$. By standard arguments, this leads to the following result for the counting function of closed geodesics on X. Let

$$\pi_X(t) = \#\{\{\gamma\} : \ell(\gamma) \leq t\}.$$

be the counting function for primitive closed geodesics γ.

THEOREM 6.3. [**59**] *Let* $X = \Gamma \backslash \mathbf{H}^{n+1}$ *be a convex co-compact hyperbolic manifold. Then*

$$\lim_{t \to \infty} (\pi_X(t) / [\exp(\delta t) / (\delta t)]) = 1$$

where $\delta = \delta(\Gamma)$.

For infinite-volume surfaces with finite geometry, including convex co-compact hyperbolic manifolds with $n = 1$, this result is due to Guillopé [**26**], and for Schottky groups similar results have been obtained by Lalley [**34**] and D'Albo and Peigné [**12**].

All of these results parallel very closely those for compact X. Patterson [**49**] proposed that the singularities of the zeta function (and hence the spectrum of the Laplacian as well as $\pi_X(t)$) could be understood in terms of group cohomology of Γ with coefficients in certain classes of distributions supported in the limit set of Γ. In a series of papers [**7, 9, 10**], Bunke and Olbrich developed and proved a refined form of Patterson's conjecture. The paper [**10**] characterizes the singularities of the scattering operator in terms of distributions supported in the limit set of Γ, much as the Patterson-Sullivan theorem characterizes the lowest eigenvalue of the Laplacian in terms of the Patterson-Sullivan measure; it then uses results of an earlier version of [**51**] for dim X even to connect singularities of the scattering operator to those of the zeta function. By a clever embedding trick (compare Mandouvalos [**38**]) they are able to extend the results on the scattering operator *and* the zeta function to dim X odd.

These results suggest that the convex co-compact hyperbolic manifolds provide an important class of examples for understanding the spectral geometry of scattering resonances. It would be very interesting to see what other aspects of the theory for compact surfaces can be developed in this setting, including determinants of Laplacians, behavior of resonances under deformations of the hyperbolic structure, and behavior under compact metric perturbtions. It would also be very helpful to have computations of resonances for models other than the cylindrical manifolds considered, for example, in [**27**] or Appendix B of [**51**]. We hope to address a number of these questions in the near future.

References

[1] S. Agmon. On the spectral theory of the Laplacian on non-compact hyperbolic manifolds. Journées "Equations aux dérivées partielles" (Saint Jean de Monts, 1987), Exposé No. XVII, Ecole Polytechnique, Palaiseau, 1987.

[2] M. Babillot, M. Peigné. Closed geodesics in homology classes on hyperbolic manifolds with cusps. *C. R. Acad. Sci. Paris* **324** (1997), 901–906.

[3] D. Borthwick. Scattering theory and deformation of asymptotically hyperbolic manifolds. Preprint, 1997.

[4] D. Borthwick, P. Perry. Scattering poles for hyperbolic manifolds. Submitted to *Trans. A. M. S.*

[5] R. Bowen. Symbolic dynamics for hyperbolic flows, *Amer. Math. J.* **95** (1973), 429–460.

[6] R. Brooks, R. Gornet, P. Perry. Isoscattering Schottky manifolds. *G. A. F. A.* **10** (2000), 307–326.

[7] U. Bunke, M. Olbrich. *Selberg Zeta and Theta Functions: A Differential Operator Approach.* Berlin: Akademie Verlag, 1995 (Mathematical Research **83**, 1995).

[8] U. Bunke, M. Olbrich. On the cohomology of Kleinian groups with coefficients in representations carried by the limit set. Preprint, Humboldt-Universität zu Berlin, Institut für Reine Mathematik, 1995.

[9] U. Bunke, M. Olbrich. Fuchsian groups of the second kind and representations carried by the limit set. *Inventiones Math.* **127** (1997), 137–154.

[10] U. Bunke, M. Olbrich. Group cohomology and the singularities of the Selberg zeta function associated to a Kleinian group. *Ann. Math.* **149** (1999), 627–689.

[11] U. Bunke, M. Olbrich. The spectrum of Kleinian manifolds. *J. Funct. Anal.* **172** (2000), no. 1, 76–164.

[12] F. D'Albo, M. Peigné. Groupes du ping-pong et géodésiques fermées en courbure −1. *Ann. Inst. Fourier (Grenoble)* **46** (1996), no. 3, 755–799.

[13] J. Elstrodt, F. Grunewald, J. Mennicke. *Groups Acting on Hyperbolic Space: Harmonic Analysis and Number Theory*, Berlin, Springer-Verlag, 1998.

[14] C. Epstein. Envelopes of horospheres and Weingarten surfaces in hyperbolic 3-space. Preprint, Princeton University, 1984.

[15] C. Epstein. Private communication.

[16] D. B. A. Epstein, A. Marden. Convex hulls in hyperbolic space, a Theorem of Sullivan, and measured pleated surfaces. *In Analytical and Geometrical Aspects of Hyperbolic Space*, edited by D. B. A. Epstein, *London. Math. Soc. Lect. Note Ser.* **111** (1984), 113–253.

[17] Fay, John D. Fourier coefficients of the resolvent for a Fuchsian group. *J. Reine Angew. Math.* **293/294** (1977), 143–203.

[18] J. Fischer. *An approach to the Selberg Trace Formula via the Selberg Zeta Function.* Berlin, Heidelberg, New York: Springer-Verlag, 1987 (Lect. Notes in Math. **1253**).

[19] D. Fried. The zeta functions of Ruelle and Selberg, I, *Ann. Sci. Éc. Norm. Sup.* **19** (1986), 491–517.

[20] R. Froese, P. Hislop. In preparation.

[21] R. Froese, P. Hislop, P. Perry. The Laplace operator on hyperbolic three-manifolds with cusps of non-maximal rank. *Inventiones Math.* **106** (1991), 295–333.

[22] R. Gangolli. Zeta functions of Selberg's type for compact space forms of symmetric spaces of rank one. *Ill. J. Math.* **21** (1977), 1–41.

[23] I. C. Gohberg, E. I. Sigal. An operator generalization of the logarithmic residue theorem and the theorem of Rouché. *Math. U. S. S. R. Sbornik* **13** (1971), 603–625.

[24] L. Guillopé. Théorie spectrale de quelques variétés à bouts. *Ann. Sci. Éc. Norm. Sup.* **22** (1989), 137–160.

[25] L. Guillopé. Fonctions zeta de Selberg et surfaces de géométrie finie. In *Zeta functions in geometry*, Adv. Stud. Pure Math. **21** (1990), 33–70.

[26] L. Guillopé. Entropies et spectres. *Osaka J. Math.* **31** (1994), 247–289.

[27] L. Guillopé, M. Zworski. Upper bounds on the number of resonances for non-compact Riemann surfaces. *J. Funct. Anal.* **129** (1995), 364–389.

[28] L. Guillopé, M. Zworski. Polynomial bounds on the number of resonances for some complete spaces of constant curvature. *Asymptotic Anal.* **11** (1995), 1–22.

[29] L. Guillopé, M. Zworski. Scattering asymptotics for Riemann surfaces. *Ann. Math.* **145** (1997), 597–660.

[30] L. Guillopé, M. Zworski. The wave trace for Riemann surfaces. *J. Geom. Anal.* 9 (1999), no. 6, 1156–1168.

[31] D. Hejhal. The Selberg trace formula for $PSL(2, \mathbf{R})$. *Lect. Notes in Math.* **548** (1976) and **1001** (1983).

[32] P. D. Hislop. The geometry and spectra of hyperbolic manifolds. Spectral and inverse spectral theory (Bangalore, 1993). *Proc. Indian Acad. Sci. Math. Sci.* **104** (1994), no. 4, 715–776.

[33] M. Joshi, A. Sá Barreto. Inverse scattering on asymptotically hyperbolic manifolds. *Acta Math.* **184** (2000), 41–86.

[34] S. P. Lalley. Renewal theorems in symbolic dynamics, with applications to geodesic flows, non-Euclidean tesselations and their fractal limits. *Acta. Math.* **163** (1989), 1–55.

[35] P. Lax, R. S. Phillips. The asymptotic distribution of lattice points in Euclidean and non-Euclidean spaces. *J. Funct. Anal.* **46**, 280-350 (1982).

[36] P. Lax, R. S. Phillips. Translation representation for automorphic solutions of the non-Euclidean wave equation I, II, III. *Comm. Pure. Appl. Math.* **37** (1984), 303–328, **37** (1984), 779–813, and **38** (1985), 179–208.

[37] N. Mandouvalos. Eisenstein series, inner product formula, and "Maass-Selberg" relations for Kleinian groups. *Mem. Amer. Math. Soc.* **400** (1989).

[38] N. Mandouvalos. Relativity of the spectrum and discrete groups on hyperbolic spaces. *Trans. A. M. S.* **350** (1998), 559-569.

[39] R. Mazzeo. The Hodge cohomology of a conformally compact metric. *J. Diff. Geom.* **28** (1988), 309–339.

[40] R. Mazzeo. Elliptic theory of edge operators. *Comm. P. D. E.* **16** (1991), 1615–1664.

[41] R. Mazzeo. Unique continuation at infinity and embedded eigenvalues for asymptotically hyperbolic manifolds. *American J. Math.* **113** (1991), 25–56.

[42] R. Mazzeo, R. Melrose. Meromorphic extension of the resolvent on complete spaces with asymptotically constant negative curvature. *J. Funct. Anal.* **75** (1987), 260–310.

[43] R. B. Melrose. *Geometric Scattering Theory.* New York, Melbourne: Cambridge University Press, 1995.

[44] R. B. Melrose, M. Zworski. Scattering metrics and geodesic flow at infinity. *Inventiones Math.* **124** (1996), 399–436.

[45] L. B. Parnovski. The Selberg zeta function for co-compact discrete subgroups of $SO^+(1, \nu)$. *Funktsional. Anal. i Prilozhen.* **26** (1992), 55-64 (Russian); Functional Anal. Appl. **26** (1992), 196–203.

[46] S. J. Patterson. The Laplacian operator on a Riemann surface I, II, III. Compositio Math. **31** (1975), 83–107, **32** (1976) 71–112, and **33** (1976), 227–259.

[47] S. J. Patterson. The limit set of a Fuchsian group. *Acta. Math.* **136** (1976), 241–273.

[48] S. J. Patterson. The exponent of convergence of Poincaré series. *Montash. Math.* **82** (1976), 297–315.

[49] S. J. Patterson. The Selberg zeta-function of a Kleinian group. In *Number Theory, Trace Formulas, and Discrete Groups: Symposium in Honor of Atle Selberg, Oslo, Norway, July 14–21, 1987*, New York, Academic Press, 1989.

[50] S. J. Patterson. On Ruelle's zeta-function. In *Festschrift in honor of I. I. Piatetski-Shapiro on the Occasion of his Sixtieth Birthday*, ed. S. Gelbart, R. Howe, P. Sanrak. Jeruslaem: Weisman Science Press, 1990.

[51] S. J. Patterson, P. A. Perry. The divisor of Selberg's zeta function for Kleinian groups. To appear in *Duke Math. J.*

[52] P. A. Perry. The Laplace operator on a hyperbolic manifold, II. Eisenstein series and the scattering matrix. *J. reine angew. Math.* **398** (1989), 67–91.

[53] P. A. Perry. The Selberg zeta function and a local trace formula for Kleinian groups. *J. reine angew. Math.* **410** (1990), 116–152.

[54] P. A. Perry. The Selberg zeta function and scattering poles for Kleinian groups. *Bull. Amer. Math. Soc.* **24** (1991), 327–333.

[55] P. A. Perry. The Selberg zeta function and scattering poles for Kleinian groups. In *Mathematical Quantum Theory, I. Schrödinger Operators (Vancouver, B.C. 1993)*, 243-251.

[56] P. A. Perry. A trace-class rigidity theorem for Kleinian groups. *Ann. Acad. Sci. Fenn.* **20** (1995), 251–257.

[57] P. A. Perry. Meromorphic continuation of the resolvent for Kleinian groups. In *Spectral Problems in Geometry and Arithmetic: NSF-CBMS Conference, August 18–22, 1997*, ed. Thomas Branson. Providence, Rhode Island: American Mathematical Society, 1999.

[58] P. A. Perry. A Poisson summation formula and lower bounds for resonances in hyperbolic manifolds. To appear in *J. Funct. Anal.*

[59] P. A. Perry. Asymptotics of the length spectrum for hyperbolic manifolds of infinite volume. To appear in *G. A. F. A.*

[60] I. R. Porteous. The normal singularities of a submanifold. *J. Diff. Geom.* **5** (1971), 543–564.

[61] M. Reed, B. Simon. *Methods of Modern Mathematical Physics, I: Functional Analysis*. New York: Academic Press, 1972.

[62] M. Reed, B. Simon. *Methods of Modern Mathematical Physics, IV: Analysis of Operators*. New York: Academic Press, 1978.

[63] A. Rocha. Meromorphic extension of the Selberg zeta function for Kleinian groups via thermodynamic formalism. *Math. Proc. Cambridge Philos. Soc.* **119** (1996), 179–190.

[64] D. Ruelle. Zeta functions for expanding maps and Anosov flows, *Inventiones Math.* **34** (1976), 231–242.

[65] P. Sarnak. Determinants of Laplacians. *Commun. Math. Phys.* **110** (1987), 113–120.

[66] H. Schlichtkrull. *Hyperfunctions and harmonic analysis on symmetric spaces. (Progress in Mathematics, 49)*. Boston: Birkhäuser, 1984.

[67] D. R. Scott. Selberg type zeta functions for the group of complex two by two matrices of determinant one. *Math. Ann.* **253** (1980), 177-194.

[68] R. T. Seeley. Complex powers of an elliptic operator. *Proc. Symp. Pure Math.* **10**, 288–307 (1967).

[69] A. Selberg. Harmonic analysis and discontinuous groups in weakly symmetric Riemannian spaces with applications to Dirichlet series. *J. Indian Math. Soc.* **20** (1956), 47–87.

[70] M. A. Shubin. *Pseudodifferential Operators and Spectral Theory*. New York: Springer-Verlag, 1985.

[71] B. Simon. Trace ideals and their applications. *London Math. Soc. Lecture Note Ser.* **35** (1979).

[72] J. Sjöstrand, M. Zworski. Lower bounds on the number of scattering poles. *Comm. P. D. E.* **18** (1993), 847–857.

[73] D. Sullivan. Entropy, Hausdorff measures old and new, and limit sets of geometrically finite Kleinian groups. *I. H. E. S. Publ. Math.* **50** (1979), 259–277.

[74] M. F. Vignéras. L'equation fonctionelle de la fonction zêta de Selberg du groupe modulaire *PSL(2, ℤ)*. In *Journées Arithmétiques de Luminy, 1978 (Astérisque 61 (1979))*, 235–249.

[75] M. Wakayama. Zeta functions of Selberg's type associated with homogeneous vector bundles. *Hiroshima J. Math.* **15** (1985), 235–29.

[76] M. Wakayama. A note on the Selberg zeta function for compact quotients of hyperbolic spaces. *Hiroshima Math. J.* **21** (1991), 539–555.

[77] M. Zworski. Counting scattering poles. In Proceedings of the Taniguchi International Workshop, *Spectral and Scattering Theory*, M. Ikawa, ed., Marcel Dekker, New York, Basel, Hong Kong, 1994.

[78] M. Zworski. Dimension of the limit set and density of resonances for convex co-compact hyperbolic quotients. *Inventiones Math.* **136** (1999), 353–409.

[79] M. Zworski. Resonances in physics and geometry. *Notices Amer. Math. Soc.* **46** (1999), no. 3, 319–328.

(Perry) DEPARTMENT OF MATHEMATICS, UNIVERSITY OF KENTUCKY, LEXINGTON, KENTUCKY 40506–0027, U. S. A.

E-mail address, Perry: perry@ms.uky.edu

Contemporary Mathematics
Volume **290**, 2001

On zeroes of automorphic L-functions

C. Soulé

We shall extend part of Connes' work on the Riemann hypothesis [C] to zeta functions of simple algebras [GJ]. To a cuspidal automorphic representation of the units in such an algebra we attach, in a natural way, an operator the spectrum of which consists of the critical zeroes of its L-function.

More precisely, let F be a number field, M a finite dimensional central simple algebra over F, and G the group of invertible elements in M. Let π be an irreducible cuspidal automorphic representation of its group of adelic points $G(\mathbb{A})$. Godement and Jacquet associated to π an L-function $L(s, \pi)$, and they showed that $L(s, \pi)$ is holomorphic in the whole complex plane and satisfies a functional equation [GJ]. Their proof is a very natural extension of Hecke's method for proving these facts for the L-function $L(s, \chi)$ of a Grössen-character χ. The main tool is the Poisson formula for theta series over $M(F)$.

On the other hand, Connes associated to any Grössen-character χ an operator D_χ, the spectrum of which is the set of imaginary parts of the zeroes of $L(s, \chi)$ on the line $\mathrm{Re}\,(s) = 1/2$.

Here we do the same for Jacquet-Godement's L-functions $L(s, \pi)$. Namely we define a Hilbert space $\mathcal{H}_\pi$ and a natural operator D_π on $\mathcal{H}_\pi$, the spectrum of which is the set of imaginary parts of the critical zeroes of $L(s, \pi)$ (Theorem 2).

The method of proof is quite similar to [C], Theorem III.1, and consists in "revisiting" [GJ] the same way Connes revisits the method of Hecke, Iwasawa, Tate and Weil for Grössen-characters. In [GJ], $L(s, \pi)$ is shown to be the greatest common divisor of some holomorphic functions $Z(\phi, f, s)$. We note that $Z(\phi, f, s)$ is the Mellin transform of a function $E(\Phi, f)$ on the positive real line (see (2)). Both $\mathcal{H}_\pi$ and D_π are defined using these functions $E(\Phi, f)$ (see (16), (17) and (18)).

As in [C], Theorem III.1, $\mathcal{H}_\pi$ also depends on the choice of a real number $\delta > 1$, and the multiplicities of the eigenvalues of D_π need not coincide with those of the corresponding zeroes of $L(s, \pi)$, since they are bounded by $(1 + \delta)/2$. However, one can hope to improve the situation, and to include non critical zeroes, by extending to matrices the method of regularization developed by Connes in [C], VIII, Lemma 3.

What looks more mysterious to the author is how to get trace formulae similar to [C], Th. 3, 4 and 5. Probably one cannot do so for a given π independently of a

---

2000 *Mathematics Subject Classification.* 11M26 11F55 11F70.

study of the full spectrum of square integrable functions on $G(F)\backslash G(\mathbb{A})$. In other words, the Selberg trace formula should come in, to develop some noncommutative geometry on $G(F)\backslash M(\mathbb{A})$.

Another difficulty in extending Connes' work to modular forms is that the explicit formulae seem better understood for Artin L-functions than they are for automorphic ones [M]. Langlands' correspondence is required to relate the two.

Nevertheless, it is worthwhile pursuing such a study for at least two reasons. First, Langlands' philosophy says that L-functions of arbitrary motives should be products of Godement-Jacquet L-functions (for matrices over $\mathbb{Q}$). Second, as several people suggested, the study of zeroes of Hecke L-functions might require an understanding of L-functions of groups of higher rank (note that Rankin's trick inspired Deligne's proof of the Weil conjectures).

The results presented here are also valid – mutatis mutandis – for global fields of positive characteristic. They are mentionned in [C], p.73, and were announced in [So].

I am very grateful to A. Connes for his help, and for telling me about his discoveries soon after he made them. I had also helpful discussions with L. Clozel and H. Jacquet.

1.1. Let F be a global field, i.e. a number field or an extension of transcendence degree one of a finite field. Denote by $\mathbb{A}$ the ring of adeles of F, by I its group of ideles, by

$$|\cdot| : I \to \mathbb{R}_+^*$$

the norm map, and by $N \subset \mathbb{R}_+^*$ its image. When F is a number field we have $N = \mathbb{R}_+^*$, and $N = q^{\mathbb{Z}}$ when F has positive characteristic p, where q is the order of the field of constants in F (a power of p). Let $I_1 \subset I$ be the group of ideles of norm one and $F^*\backslash I$ the idele class group. We choose a (non-canonical) group isomorphism

$$(1) \qquad\qquad F^*\backslash I \simeq (F^*\backslash I_1) \times N \,,$$

i.e. a section of the norm map $F^*\backslash I \to N$.

We fix integers $d \geq 1$ and $m \geq 1$, and we let $n = md$. We denote by H a central division algebra over F of rank d^2. Let $M = M(m, H)$ be the simple algebra of $m \times m$ matrices over H and let G be the multiplicative group of M, an algebraic group over F. The reduced norm induces a group morphism

$$\nu : G(\mathbb{A}) \to I \,.$$

We let $Z = \mathbb{G}_m$ be the center of G and we denote by

$$R = |\nu(Z(\mathbb{A}))| \subset \mathbb{R}_+^*$$

the image of its adelic points by ν composed with the norm map:

$$|\nu| : G(\mathbb{A}) \to \mathbb{R}_+^* \,.$$

When F is a number field $R = \mathbb{R}_+^*$, and $R = Q^{\mathbb{Z}}$, where $Q = q^n$, when F has positive characteristic. For any $t \in R$ we let

$$G_t = \{g \in G(\mathbb{A})/|\nu(g)| = t\}$$

(note that G_1 is denoted G_0 in [GJ]).

Let

$$\omega : F^* \backslash I_1 \to S^1$$

be a unitary character. We view ω as a central character on $Z(\mathbb{A})$ by composing it with the projection coming from (1):

$$Z(F) \backslash Z(\mathbb{A}) \simeq F^* \backslash I \to F^* \backslash I_1 \,.$$

1.2. On R we choose the Haar measure $d^\times t$ which is equal to dt/t when $R = R^*_+$ and to $\log(Q)$ times the counting measure when $R = Q^{\mathbb{Z}}$. We fix a Haar measure dg on $G(\mathbb{A})$, and we choose on G_1 the Haar measure dh such that $d^\times t$ is the quotient of dg by dh. For any $t \in R$, we equip the coset G_t of G_1 in $G(\mathbb{A})$ by the translate of dh (since G_1 is unimodular, we do not have to distinguish left from right). Finally, we endow $F^* \backslash I_1$ with the Haar measure of total volume one.

We denote by $L^2_0(G(F) \backslash G(\mathbb{A}), \omega)$ the space of cusp-forms on $G(\mathbb{A})$ with central character ω, i.e. the Hilbert space of measurable complex valued functions φ on $G(F) \backslash G(\mathbb{A})$ such that:

- $\varphi(a\,g) = \omega(a)\,\varphi(g)$ for any $a \in Z(\mathbb{A})$ and any $g \in G(\mathbb{A})$;

-
$$\int_{G(F)\,Z(\mathbb{A}) \backslash G(\mathbb{A})} |\varphi(g)|^2 \, dg < \infty;$$

- If U is the unipotent radical of a proper F-parabolic subgroup of G, one has

$$\int_{U(F) \backslash U(\mathbb{A})} \varphi(g\,u) \, du = 0.$$

When $m = 1$, the last condition is replaced (as in [GJ] p. 144) by the hypothesis that φ be orthogonal to $\omega' \circ \nu$ whenever $\omega'^n = \omega$.

1.3. Let f be an admissible coefficient of the representation of $G(\mathbb{A})$ on the Hilbert space $L^2_0(G(F) \backslash G(\mathbb{A}), \omega)$ ([GJ] (10.17), and 1.4 below), and $\Phi \in \mathcal{S}(M(\mathbb{A}))$ a Schwartz-Bruhat function on the adelic points of $M(\mathbb{A})$. We define a complex valued function $E(\Phi, f)$ on R by the formula

$$(2) \qquad\qquad E(\Phi, f)(t) = t^{n/2} \int_{G_t} \Phi(x) \, f(x) \, dx \,.$$

Theorem 1.
i) *The integral (2) converges absolutely.*
ii) *For any integer $k > 0$, there exists a positive constant C such that, for all $t \in R$,*

$$|E(\Phi, f)(t)| \leq C \, t^{-k}.$$

iii) *Let $\widehat{\Phi}$ be the Fourier transform of Φ, and let $\overset{\vee}{f}(g) = f(g^{-1})$. Then*

$$(3) \qquad\qquad E(\Phi, f)(t) = E(\widehat{\Phi}, \overset{\vee}{f})(t^{-1}) \,.$$

1.4. Theorem 1 is similar to [C], AI Lemma 2, in the rank one case and to the main result of [GJ], Theorem 13.8. We prove it by repeating the arguments of [GJ]. What we need is another way to write $E(\Phi, f)(t)$, and [GJ], pp. 176-177, tells us what to do.

First recall the definition of the subspace $\mathcal{A}_0(G, \omega) \subset L^2_0(G(F)\backslash G(\mathbb{A}), \omega)$ of cuspical automorphic forms with central character ω ([GJ] p. 146). Let   .

$$K = \prod_v K_v$$

be the standard maximal compact subgroup of $G(\mathbb{A})$ and let $\mathcal{Z}$ be the center of the enveloping algebra of the Lie group

$$G_\infty = \prod_{v \text{ archimedean}} G(F_v)$$

(where F_v is the completion of F at the place v). A function $\varphi \in L^2_0(G(F)\backslash G(\mathbb{A}), \omega)$ lies in $\mathcal{A}_0(G, \omega)$ when it is K-finite on the right and (in the number field case) $\mathcal{Z}$-finite ([GJ] (10.15) (10.16)). By definition, a function f on $G(\mathbb{A})$ is an admissible coefficient when there exist $\varphi \in \mathcal{A}_0(G, \omega)$ and $\widetilde{\varphi} \in \mathcal{A}_0(G, \omega^{-1})$ such that

$$(4) \qquad f(g) = \int_{Z(\mathbb{A}) G(F)\backslash G(\mathbb{A})} \varphi(hg)\, \widetilde{\varphi}(h)\, dh \,.$$

As in [GJ] p.165, let us choose some elements g_i in $G(\mathbb{A})$, $0 \leq i \leq u$, such that $G(\mathbb{A})$ is the disjoint union

$$G(\mathbb{A}) = \coprod_{1 \leq i \leq u} Z(\mathbb{A})\, G_1\, g_i \,,$$

i.e. the finite set $\{|\nu(g_i)|\}_{1 \leq i \leq u}$ is a set of representatives of $|\nu(G(\mathbb{A}))|$ modulo R. We let

$$G' = \coprod_{1 \leq i \leq u} G_1\, g_i \,.$$

If F is a number field, we simply take

$$G' = G_1 \,.$$

Proposition 1. *For any $t \in R$ the following identity holds*

$$(5) \qquad E(\Phi, f)(t) = t^{n/2} \int_{(G(F)\backslash G')^2} \widetilde{\varphi}(h)\, \varphi(g) \int_{F^* \backslash I_1} \omega(a)\, \theta(h, g, a, \tau)\, da\, dh\, dg \,,$$

where $\tau \in N \subset F^* \backslash I \simeq Z(F)\backslash Z(\mathbb{A})$ is the unique element such that

$$(6) \qquad |\nu(\tau)| = t\, |\nu(hg^{-1})| \,,$$

and

$$(7) \qquad \theta(g, h, a, \tau) = \sum_{\xi \in G(F)} \Phi(h^{-1}\, a\, \tau\, \xi\, g) \,.$$

1.5. We now prove Theorem 1, i), ii) and Proposition 1. First, from (2) and (4), we get

$$(8) \qquad E(\Phi, f)(t) = t^{n/2} \int_{G_t \times (Z(\mathbb{A}) \, G(F) \backslash G(\mathbb{A}))} \Phi(x) \, \varphi(hx) \, \widetilde{\varphi}(h) \, dh \, dx \, .$$

If we prove that this double integral converges absolutely, Theorem 1 i) will follow. By definition of G', given $h \in G(\mathbb{A})$ and $x \in G_t$ we can find $z \in Z(F) \backslash Z(\mathbb{A})$ and $g \in G'$ such that

$$hx = zg \, .$$

By (1) we can write $z = \tau a$, with $\tau \in N$ and $a \in Z(F) \backslash Z_1$. We compute

$$|\nu(\tau)| = |\nu(hxa^{-1}g^{-1})| = t|\nu(hg^{-1})| \, .$$

Since

$$\varphi(hx) = \omega(a) \, \varphi(g)$$

and $\varphi(g)$ depends only on the class of g in $G(F) \backslash G'$, we get

$$(9) \qquad E(\Phi, f)(t) = t^{n/2} \int_{(G(F) \backslash G')^2} \widetilde{\varphi}(h) \, \varphi(g) \int_{F^* \backslash I_1} \omega(a) \, \theta(g, h, a, \tau) \, da \, dh \, dg$$

where $\theta(g, h, a, \tau)$ is defined by (7) (compare [GJ] p.177).

We need to show that the right hand side of (9) converges absolutely. It is known that both φ and $\widetilde{\varphi}$ are rapidly decreasing in the number field case, and are compactly supported modulo $Z(\mathbb{A}) \, G(F)$ in the function field case ([GJ] Lemma 10.8 and Lemma 10.9). Therefore, it is enough to check that $\theta(g, h, a, \tau)$ is slowly increasing in the variables g, h and rapidly decreasing in t, uniformly in a.

To be more precise, let T be a maximal split torus of G over F and α_j, $1 \le j \le m$, a system of simple roots. For any $u \in T(\mathbb{A})$ we let

$$\eta(u) = \inf_j |\alpha_j(u)|$$

and

$$\epsilon(u) = \sup_j |\alpha_j(u)| \, .$$

All we need to show is that, for any real constant $c > 0$ and any integer $k \ge 1$, and for any compact subset C of $G(\mathbb{A})$, there exist $c' > 0$ and $k' > 0$ such that, if g and h lie in C, $u \in T(\mathbb{A}) \cap G_1$ and $u' \in T(\mathbb{A}) \cap G_1$ with $\epsilon(u) \le c$, $\epsilon(u') \le c$ and $t \ge c$, then

$$(10) \qquad |\theta(ug, u'h, a, \tau)| \le c' \, \eta(u)^{-k'} \, \eta(u')^{-k'} \, t^{-k} \, .$$

Indeed, (10) will imply that the integrand $\widetilde{\varphi}(h) \, \varphi(g) \, \theta(g, h, a, \tau)$ in (9) is rapidly decreasing in g and h, uniformly in a and, since $F^* \backslash I_1$ is compact, the integral will be absolutely convergent and it will define a function of t which is rapidly decreasing, as stated in Theorem 2, ii).

1.6. The inequality (10) follows from the proof of [GJ] Lemma 11.5. We let $V = M(F)$, a vector space of rank $r = n^2$. The group $G(F) \times G(F)$ acts on the right upon V by the formula

$$(g, h)(x) = h^{-1} x g.$$

Choose a basis $e_1, \ldots, e_r$ of V and characters χ_i, $1 \le i \le r$, of the maximal torus $S = T \times T$ of $G \times G$ such that the action of this torus on V is given by

$$s(e_i) = \chi_i(s) e_i$$

for any s in $S(F)$. We may assume that

$$\Phi\left(\sum_{i=1}^r x_i e_i\right) = \prod_{i=1}^r \Phi_i(x_i)$$

where, for each i, the function Φ_i lies in $\mathcal{S}(\mathbb{A})$ and is positive. It is shown in op.cit., p. 156, line 6, that, for any $k \ge 1$, there exists a real polynomial P in r variables with positive coefficients such that, under our assumptions,

$$|\theta(ug, u'h, a, \tau)| \le |\nu(\tau)|^{-k} \, P(|\chi_i(u, u')|^{-1}).$$

We know from (6) that
$$|\nu(\tau)| = t \, |\nu(hg^{-1})|$$

hence, since h and g lie in C, $|\nu(\tau)|$ is bounded by a constant multiple of t. On the other hand, since u and u' lie in $T(\mathbb{A}) \cap G_1$, as explained in op.cit., p.157, the quantity $P(|\chi_i(u, u')|^{-1})$ is bounded from above by a constant multiple of $\eta(u)^{-k'} \eta(u')^{-k'}$ for some constant $k' > 0$. The inequality (10) follows.

1.7. We shall now prove the functional equation

$$(3) \qquad\qquad E(\Phi, f)(t) = E(\widehat{\Phi}, \overset{\vee}{f})(t^{-1}).$$

First, recall the definition of the Fourier transform $\widehat{\Phi}$ ([GJ] pp. 158 and 160). Fix a non-trivial character ψ of $F \backslash \mathbb{A}$ and denote by

$$\tau_M : M(\mathbb{A}) \to \mathbb{A}$$

the reduced trace. Given $x, y \in M(\mathbb{A})$ we let

$$\langle x, y \rangle = \tau_M(xy)$$

so that $M(\mathbb{A})$ gets identified with its dual. Given a Schwartz-Bruhat function $\Phi \in \mathcal{S}(M(\mathbb{A}))$, its Fourier transform is defined by the formula

$$(11) \qquad\qquad \widehat{\Phi}(x) = \int_{M(\mathbb{A})} \Phi(y) \, \psi(\langle x, y \rangle) \, dy,$$

where the additive Haar measure dy is normalized by the condition that $M(F) \backslash M(\mathbb{A})$ be of measure one. To check (3) we use the expression (5) in Proposition 1. We first notice that, given an integer r such that $1 \le r < m$, since φ is cuspidal, the integral

$$\int_{G(F)\backslash G'} \varphi(g) \sum_{\substack{\xi \in M(F) \\ rk(\xi)=r}} \Phi(h^{-1} a \tau \xi g) \, dg$$

vanishes ([GJ] p. 171). Therefore, if we define

$$\theta_\Phi(g,h,a,\tau) = \sum_{\xi \in M(F)} \Phi(h^{-1} a \tau \xi g)$$

we get from (5) the formula

$$E(\Phi,f)(t) = t^{n/2} \int_{(G(F)\backslash G')^2 \times F^* \backslash I_1} \widetilde{\varphi}(h) \, \varphi(g) \, \omega(a) \, [\theta_\Phi(g,h,a,\tau) - \Phi(0)] \, dh \, dg \, da \, .$$

Note that

$$(12) \qquad \int \widetilde{\varphi}(h) \, \varphi(g) \, \omega(a) \, dh \, dg \, da = 0$$

since, by [GJ] Lemma 12.11, p. 146 (with $s = 0$) we have

$$\int_{G(F)\backslash G'} \varphi(g) \, dg = 0 \, .$$

Therefore

$$(13) \quad E(\Phi,f)(t) = t^{n/2} \int_{(G(F)\backslash G')^2 \times (F^* \backslash I_1)} \widetilde{\varphi}(h) \, \varphi(g) \, \omega(a) \, \theta_\Phi(g,h,a,\tau) \, dh \, dg \, da \, .$$

We may now apply the Poisson formula

$$(14) \qquad \sum_{\xi \in M(F)} \Psi(\xi) = \sum_{\xi \in M(F)} \widehat{\Psi}(\xi)$$

to the Schwartz-Bruhat function

$$\Psi(x) = \Phi(h^{-1} a \tau x g) \, .$$

Since h, a and g lie in the kernel of $|\nu|$, we find from (11) that

$$(15) \qquad \widehat{\Psi}(x) = t^{-n} \, \widehat{\Phi}(g^{-1} a^{-1} \tau^{-1} x h) \, .$$

From (13), (14) and (15) we deduce that

$$E(\Phi,f)(t) = t^{-n/2} \int_{(G(F)\backslash G')^2 \times (F^* \backslash I_1)} \widetilde{\varphi}(h) \, \varphi(g) \, \omega(a) \, \theta_{\widehat{\Phi}}(h,g,a^{-1},\tau^{-1}) \, dh \, dg \, da \, .$$

When we exchange φ with $\widetilde{\varphi}$ (hence ω with ω^{-1}), the function f becomes $\check{f}$. So, using (13), the identity above can be rewritten

$$E(\Phi,f)(t) = E(\widehat{\Phi},\check{f})(1/t) \, .$$

This ends the proof of Theorem 1.

1.8. Remark. As in [GJ], the reason why we took φ and $\widetilde{\varphi}$ to be cuspidal was to allow us to replace $G(F)$ by $M(F)$ when using the Poisson formula in 1.7.

2. A Hilbert space

2.1. When v is an archimedean place of F, we say that a Schwartz function $\Phi_v \in \mathcal{S}(M(F_v))$ is *gaussian* when there exists a polynomial function P on $M(F_v)$ and a positive real number $\alpha > 0$ such that

$$\Phi_v(m) = P(m) \exp(-\alpha |m|^2)$$

for all $m \in M(F_v)$. Here $|m|^2$ means the following (cf. [GJ] p. 115). Let ${}^t m$ be the transpose of the matrix m. If $F_v = H_v = \mathbb{R}$, we let

$$|m|^2 = \operatorname{Tr}\left(m \, {}^t m\right).$$

If $F_v = \mathbb{R}$, if H_v is the field of quaternions and if $z \mapsto z^\iota$ is the principal involution on H_v, we let

$$|m|^2 = \tau_M(m \, {}^t m^\iota).$$

Finally, if $F_v = H_v = \mathbb{C}$ and $\overline{m}$ is the complex conjugate of m, we let

$$|m|^2 = \operatorname{Tr}\left(m \, {}^t \overline{m}\right).$$

A global Schwartz-Bruhat function $\Phi \in \mathcal{S}(M(\mathbb{A}))$ is said to be *gaussian* when it can be written as a product

$$\Phi(x) = \prod_v \Phi_v(x_v),$$

over all places of F, where Φ_v is a Schwartz-Bruhat function on $M(F_v)$ which is equal to the characteristic function of the standard maximal compact subgroup for almost all finite places v and which is gaussian in the sense above when v is archimedean.

2.2. Fix a real number $\delta > 1$, and let $L^2_\delta(R)$ be the Hilbert space of measurable functions $\chi(t)$ on R such that

$$\int_R |\chi(t)|^2 \left(1 + (\log t)^2\right)^{\delta/2} d^\times t < \infty.$$

Theorem 1 implies that $E(\Phi, f)$ lies in $L^2_\delta(R)$, since it has rapid decay both when $t \to +\infty$ and when $t \to 0$.

2.3. Let ω be a unitary character of $F^*\backslash I_1$ and let π be an admissible irreducible representation of the Hecke algebra $\mathcal{H}(G, K)$, contained in $\mathcal{A}_0(G, \omega)$, as in [GJ] pp. 146-149 and Corollary 1.2.8, i.e. an irreducible component π of the representation of $G(\mathbb{A})$ on $L^2_0(G(F)\backslash G(\mathbb{A}), \omega)$. In other words, π is an irreducible cuspidal representation of $G(\mathbb{A})$ with central character ω.

Let $S(\pi)$ be the set of pairs (Φ, f) where Φ is a gaussian Schwartz-Bruhat function on $M(\mathbb{A})$ and f is an admissible coefficient of π. Denote by $V_\pi \subset L^2_\delta(R)$

the closure of the vector space spanned by all the functions $E(\Phi, f)$ when (Φ, f) runs over all pairs in $S(\pi)$. We then define

$$(16) \qquad \mathcal{H}_\pi = L_\delta^2(R)/V_\pi \,.$$

For any $r \in R$ and $\chi \in L_\delta^2(R)$ let $W(r)(\chi) \in L_\delta^2(R)$ be defined by the formula

$$(W(r)(\chi))(t) = \chi(r^{-1}t) \,.$$

This action of R on $L_\delta^2(R)$ leaves V_π invariant. Indeed, let $z \in Z(\mathbb{A})$ be such that $|\nu(z)| = r$. By changing the variable x to $z\,x$ in the definition (2), we find that $E(\Phi, f)(r^{-1}t)$ is equal to $E(\Phi_z, f)(t)$, where

$$\Phi_z(x) = \Phi(z\,x) \,.$$

When Φ is gaussian, the same is clearly true for Φ_z. This proves our claim. So we get a representation

$$W : R \to \mathrm{Aut}\,(\mathcal{H}_\pi) \,.$$

By the arguments of [C] III, (22)-(26), when F is a number field there exists an infinitesimal generator D_π of this action of $\mathbb{R}_+^*$ on $\mathcal{H}_\pi$:

$$(17) \qquad D_\pi\,\xi = \lim_{\varepsilon \to 0} \frac{1}{\varepsilon}\,(W(e^\varepsilon) - 1)\,\xi$$

for any ξ in the domain of D_π. The operator D_π is closed and unbounded. On the other hand, when F has positive characteristic the image by W of the generator Q of $R = Q^{\mathbb{Z}}$ is a unitary operator that can be written

$$(18) \qquad W(Q) = Q^{D_\pi} \,.$$

Let

$$R^\perp = \{u \in R/t^{iu} = 1 \text{ for all } t \in R\}$$

and

$$\widehat{R} = \mathbb{R}/R^\perp \,,$$

hence $\widehat{R} = \mathbb{R}$ when F is a number field, and $\widehat{R} = \mathbb{R}/((2\pi/\log Q)\,\mathbb{Z})$ when F is a function field. In both cases, the spectrum of D_π lies in $i\,\widehat{R}$.

Let $L(s, \pi)$ be the L-function of the cuspidal representation π. It is defined in [GJ] as an infinite Euler product for $\mathrm{Re}(s)$ large enough, and the main result of op.cit., Corollary 13.8, is that $L(s, \pi)$ can be analytically continued to an entire function of $s \in \mathbb{C}$ (which satisfies a functional equation); see also below §2.7.

Theorem 2. i) *The operator D_π has a discrete spectrum $Sp\,(D_\pi)$ contained in $i\,\widehat{R}$.*

ii) *As a set, $Sp\,(D_\pi)$ coincides with the set of zeroes of $L(s, \pi)$ on the critical line: $\rho \in Sp\,(D_\pi)$ if and only if $L\left(\frac{1}{2} + \rho, \pi\right) = 0$.*

iii) *The multiplicity of $\rho \in Sp\,(D_\pi)$ is the largest integer n such that $n < \frac{1+\delta}{2}$ and n is less or equal to the multiplicity of $\frac{1}{2} + \rho$ as a zero of $L(s, \pi)$.*

2.4. To prove Theorem 2 we need the following lemma:

Lemma 1. *A dense subset of V_π consists of all the functions $b * E(\Phi, f)$ defined by*

$$(b * E(\Phi, f))(t) = \int_R b(r) \, E(\Phi, f) \, (r^{-1} t) \, d^\times r \,,$$

where $b \in C_c^\infty(R)$ and $(\Phi, f) \in S(\pi)$.

Proof of Lemma 1. When F has positive characteristic, this is trivial since we can take for b the delta function at one, so we assume that F is a number field. Since V_π is stable by the action of $R = R_+^*$ and closed in $L_\delta^2(R)$, for any $\xi \in V_\pi$ and any Schwartz function $f \in \mathcal{S}(R)$, the function

$$V(f)\xi = f * \xi$$

lies in V_π. In particular $b * E(\Phi, f)$ lies in V_π when $b \in C_c^\infty(R)$.

On the other hand, we know from [C] p. 80, Lemma 4, that there exists a sequence of functions $f_n \in \mathcal{S}(R)$ such that $V(f_n)$ tends strongly to one in $L_\delta^2(R)$ and the norm of $V(f_n)$ is bounded. Choose $b_n \in C_c^\infty(R)$ such that

$$\int_R |b_n - f_n| \, (1 + (\log t)^2)^{\delta/4} \, d^\times t < \frac{1}{n}$$

(that such a function b_n exists follows from [S], §VII, Th. III and Th. VI). From [C], loc.cit., (4), we deduce that $V(b_n)$ converges strongly to one and the norm of $V(b_n)$ is bounded. Therefore $b_n * E(\Phi, f)$ converges to $E(\Phi, f)$, and the linear span of the functions $b * E(\Phi, f)$, $b \in C_c^\infty(R)$, $(\Phi, f) \in S(\pi)$, is dense in V_π. q.e.d.

2.5. The proof of Theorem 2 now proceeds as the argument of Connes in [C], pp. 82-87, Appendix, Proof of Theorem III.1, using the results of Godement-Jacquet [GJ], proof of Corollary 13.8.

For any function χ on R which is rapidly decreasing at 0 and $+\infty$, we let

$$(19) \qquad\qquad M(\chi, s) = \int_R \chi(t) \, t^s \, d^\times t$$

be its Mellin transform, $s \in \mathbb{C}$. We consider also its Fourier transform $\widehat{\chi}(u)$, $u \in \widehat{R}$, which is characterized by the formula

$$(20) \qquad\qquad \chi(t) = \int_{\widehat{R}} \widehat{\chi}(u) \, t^{iu} \, du \,,$$

where du is Lebesgue measure on $\widehat{R} = \mathbb{R}/R^\perp$. Using Parseval's formula, these definitions extend to the vector spaces $\mathcal{S}'(R)$ and $\mathcal{S}'(\widehat{R})$ of tempered distributions ([S], Chapter VII).

From Theorem 1 we know that, for any $(\Phi, f) \in S(\pi)$, the function $E(\Phi, f)$ has rapid decay at 0 and $+\infty$. Since $d^\times t$ is the quotient of the Haar measures on $G(\mathbb{A})$ and G_1, the Mellin transform

$$M(E(\Phi, f), s) = \int_R t^{n/2} \left(\int_{G_t} \Phi(x) \, f(x) \, dx \right) t^s \, d^\times t$$

can be written

(21)
$$M(E(\Phi, f), s) = Z\left(\Phi, f, s + \frac{n}{2}\right),$$

where

$$Z(\Phi, f, s) = \int_{G(\mathbb{A})} \Phi(g)\, f(g)\, |\nu(g)|^s\, dg$$

is the integral considered in [GJ], § 12 and 13.

Now let $b \in C_c^\infty(R)$. Since the Mellin transform turns convolutions into products, we deduce from (21) that

(22)
$$M(b * E(\Phi, f), s) = M(b, s)\, Z\left(\Phi, f, s + \frac{n}{2}\right).$$

Since $L^2_{-\delta}(R)$ is contained in $\mathcal{S}'(R)$, any $\psi \in L^2_{-\delta}(R)$ is the Fourier transform

(23)
$$\psi(t) = \int_{\widehat{R}} \widehat{\psi}(u)\, t^{iu}\, du$$

of a tempered distribution $\widehat{\psi}$ on the real line ([S], §VII, Th. XIII). By the Parseval formula (i.e. the definition of $\widehat{\psi}$ in [S] loc.cit.), given any $\chi \in \mathcal{S}(R)$ we have

$$\int_R \chi(t)\, \psi(t)\, d^\times t = \int_{R \times \widehat{R}} \chi(t)\, \widehat{\psi}(u)\, t^{iu}\, d^\times t\, du = \int_{\widehat{R}} M(f, iu)\, \widehat{\psi}(u)\, du.$$

When $(\Phi, f) \in S(\pi)$ and $b \in C_c^\infty(R)$, Theorem 1 shows that $\chi = b * E(\Phi, f)$ lies in $\mathcal{S}(R)$, and we can apply the formula above. From (22) we get

(24)
$$\int_R (b * E(\Phi, f))(t)\, \psi(t)\, d^\times t = \int_{\widehat{R}} M(b, iu)\, Z\left(\Phi, f, iu + \frac{n}{2}\right) \widehat{\psi}(u)\, du.$$

2.6. Assume that $\psi \in L^2_{-\delta}(R)$ lies in the subspace $V_\pi^\perp$ orthogonal to V_π. By Lemma 1, this is equivalent to the assertion that

$$\int_R (b * E(\Phi, f))(t)\, \psi(t)\, d^\times t = 0$$

for any $b \in C_c^\infty(R)$ and $(\Phi, f) \in S(\pi)$, i.e., using (24),

(25)
$$\int_{\widehat{R}} M(b, iu)\, Z\left(\Phi, f, iu + \frac{n}{2}\right) \widehat{\psi}(u)\, du = 0.$$

Since $Z\left(\Phi, f, iu + \frac{n}{2}\right)$ is a bounded function of $u \in R$ ([GJ], Proposition 13.9), its product by $\widehat{\psi}(u)$ a tempered distribution $T(u) \in \mathcal{S}'(\widehat{R})$. Let $\widehat{T}(t) \in \mathcal{S}'(R)$ its (inverse) Fourier transform. The function $M(b, iu)$ is (a constant multiple of) the Fourier transform of b. Therefore, by the Parseval formula, (25) is equivalent to

$$\int_R b(t)\, \widehat{T}(t)\, dt = 0.$$

Since $b(t)$ runs over all functions in $C_c^\infty(R)$, this means that $\widehat{T}(t) = 0$, hence $T = 0$. In other words, for all $(\Phi, f) \in S(\pi)$,

$$(26) \qquad\qquad Z\left(\Phi, f, iu + \frac{n}{2}\right) \widehat{\psi}(u) = 0$$

as a distribution on $\widehat{R}$. Conversely, when $\psi \in L^2_{-\delta}(R)$, the condition (26) implies that ψ is orthogonal to V_π.

2.7. It is shown in [GJ] Th. 13.8, that $Z(\Phi, f, s + \frac{n-1}{2})$ is the product of $L(s, \pi)$ by an entire function and that there exists a finite family (Φ_α, f_α), $\alpha \in A$ in $S(\pi)$ such that

$$\sum_\alpha Z(\Phi_\alpha, f_\alpha, s + \frac{n-1}{2}) = L(s, \pi).$$

Therefore (26) is equivalent to the identity of distributions

$$(27) \qquad\qquad L\left(\frac{1}{2} + iu, \pi\right) \widehat{\psi}(u) = 0.$$

The solutions of this equation are discussed in [C] pp. 86-87. It is shown that a basis of solutions of (27) consists of the derivatives of the Dirac distribution at the imaginary parts of the critical zeroes of $L(s, \pi)$. Therefore $\psi(t)$ lies in the orthogonal of V_π iff it is a linear combination of functions of the form

$$\eta_{\rho,a}(t) = t^\rho \log(t)^a$$

where $L\left(\frac{1}{2} + \rho, \pi\right) = 0$ and a is a natural integer which is less than the order of the zero $\frac{1}{2} + \rho$ and less than $(\delta - 1)/2$ (in order that ψ lies in $L^2_\delta(R)$). When ρ is fixed, the action of D_π on the finite dimensional vector space spanned by the functions $\eta_{\rho,a}$ is given by a triangular matrix with diagonal entries equal to ρ. This proves Theorem 2.

References

[C] A. Connes, Trace formula in noncommutative geometry and the zeroes of the Riemann zeta function, *Selecta Math.*, New ser. **5**, 29-106 (1999).

[GJ] R. Godement, H. Jacquet, Zeta functions of simple algebras, Springer Lecture Notes **260** (1972).

[M] C.J. Moreno, Advanced analytic number theory, Part I: Ramification theoretic methods, Contemporary Mathematics 15 (1983).

[S] L. Schwartz, Théorie des distributions, II, (1959) Hermann.

[So] C. Soulé, Sur les zéros des fonctions L automorphes, *C.R. Acad. Sci. Paris*, bf 328, I, 955-958 (1999).

CNRS, Institut des Hautes Études Scientifiques, 35, Route de Chartres, 91440, Bures-sur-Yvette, France

Contemporary Mathematics
Volume **290**, 2001

Artin L-Functions of Graph Coverings

H. M. Stark and A. A. Terras

ABSTRACT. The L-functions considered here are analogues of Artin L-functions of Galois extensions of number fields and Selberg zeta functions of Riemannian manifolds. They are connected to finite unramified normal graph coverings of a graph X. They provide factorizations of the Ihara zeta function of X. Here we survey the properties of these L-functions.

1. Introduction.

In the tree of L-functions there are many branches - including the three we discuss here - that of number theory, that of spectral geometry of Riemannian manifolds and that of graph theory. The number theoretic branch includes the Riemann, Dirichlet, Dedekind zeta function as well as the Artin L-function. The spectral geometry branch includes the Selberg zeta function. The graph theory branch includes the zeta function of Ihara and the corresponding L-functions. Most of our climb will take place on this last branch but let us first take a short look at the other two branches. We will not consider the zeta functions of function fields but those of graph theory are similar in that they are rational functions. See Katz and Sarnak [9] for a discussion of the zeta functions of function fields motivated by quantum chaos. In particular, we have another column for Table 2 in Katz and Sarnak. See Table 1 which follows. Recently L-functions have played a large role in quantum chaos - the investigation of the statistics of energy levels of various non-classical physical systems. See Katz and Sarnak [9] as well as Sarnak's talks at the Spring, 1999 meetings at MSRI available in streaming video at the MSRI website

(http://msri.org/publications/ln/msri/1999/random/sarnak/2/index.html).

Recall that the **Riemann zeta function** is defined for a complex number s with $Res > 1$ by

$$\zeta(s) = \sum_{n \geq 1} n^{-s} = \prod_{p=prime} (1 - p^{-s})^{-1}.$$

Riemann showed that it is possible to extend zeta to a function meromorphic on the whole complex plane with a pole at $s = 1$. Moreover he obtained a functional equation $\Lambda(s) = \pi^{-s}\Gamma(s)\zeta(s/2) = \Lambda(1 - s)$. Using Hadamard factorization of $\Lambda(s)$

1991 *Mathematics Subject Classification.* Primary 11M41.
Key words and phrases. Ihara zeta function, Artin L-function.

one finds that there is a duality between the primes and the non-real zeros of $\zeta(s)$. This led to a proof of the prime number theorem:

$$\#\{p = \text{prime} \mid p \leq x\} \sim \frac{x}{\log x}, \text{ as } x \to \infty.$$

A good reference is Davenport [2]. The Riemann hypothesis says that the non-real zeros of zeta lie on the line $Res = 1/2$. Computer studies have verified this hypothesis for ridiculously many zeros. Moreover the statistics of the high zeros have been investigated by Odlyzko (see [5]) and found to be like the distribution of the eigenvalues of a random Hermitian matrix (or GUE which stands for Gaussian Unitary Ensemble). Sarnak [15] lists the open problems related to this and other zeta functions of number theory. Katz and Sarnak [9] conjecture that the low zeros have level spacings that follow a statistical law coming from a symplectic symmetry. They say: "We infer that in the proposed spectral interpretation of the zeroes of the Riemann Zeta function, the operator should preserve a symplectic form!"

The Riemann zeta function has a generalization corresponding to an algebraic number field $F = \mathbb{Q}(\theta)$, where θ is the root of a polynomial with rational coefficients. The ring O_F of algebraic integers in F consists of roots of monic polynomials with integer coefficients. The **Dedekind zeta function** of F is defined for $Res > 1$ by

$$\zeta_F(s) = \sum_{\mathfrak{a} = ideal \ in \ O_F} N\mathfrak{a}^{-s} = \prod_{\mathfrak{p} = prime} (1 - N\mathfrak{p}^{-s})^{-1}.$$

Here $N\mathfrak{a} = $Norm of $\mathfrak{a} = \#(O_F/\mathfrak{a})$. As with the Riemann zeta function, there is an analytic continuation to the whole complex plane with a pole at $s = 1$ whose residue contains important constants of F such as the class number, discriminant and regulator. And there is a generalization of the prime number theorem known as the prime ideal theorem. See Stark [16] for an elementary introduction. See Lang [10] for a more complete treatment.

The Selberg zeta function is associated to a compact Riemannian manifold $\Gamma\backslash H$, where H is the Poincaré upper half plane with arc length ds defined by $ds^2 = y^{-2}(dx^2 + dy^2)$. Geodesics in H are straight lines and circles orthogonal to the real axis. Then the primes in the Riemann zeta function correspond to primitive closed geodesics $\{\gamma_0\}$ in $\Gamma\backslash H$ of length $\nu(\gamma_0)$ (or equivalently hyperbolic conjugacy classes in Γ). Here "primitive" means that the curve is only traversed once. The **Selberg zeta function** is defined by

$$Z(s) = \prod_{\{\gamma_0\}} \prod_{j \geq 0} (1 - e^{-(s+j)\nu(\gamma_0)}).$$

The outer product is over primitive closed geodesics in Γ. The main tool needed to analyze this function is the Selberg trace formula. Again there is an analogue of the prime number theorem and a duality between the spectrum of the Laplacian on $\Gamma\backslash H$ and the length spectrum which consists of lengths of the prime geodesics. References include Jürgen Elstrodt [4], Dennis Hejhal [7], Audrey Terras [21], Alexei B. Venkov [24], Marie-France Vignéras [27].

Artin L-functions are connected to representations ρ of the Galois group $G = Gal(K/F)$ of a finite extension K of an algebraic number field F. If s is a complex number with $Res > 1$, the **Artin L-function** is

$$L(s, \rho) = \prod_{\mathfrak{p}=prime} \det(I - \rho([K/F, \mathfrak{P}])N\mathfrak{p}^{-s})^{-1}.$$

Here $[K/F, \mathfrak{P}]$ denotes the Frobenius automorphism for a prime $\mathfrak{P}$ of K dividing the prime $\mathfrak{p}$ of F. The expression is messier if $\mathfrak{p}$ is ramified. See Stark [16] or Lang [10] for more information.

Again there is an analogue of the prime number theorem called the **Chebotarev density theorem** saying that given a conjugacy class C in G, the set of primes

$$\{\mathfrak{p} \text{ of } F \mid C = \text{conjugacy class of Frobenius of } \mathfrak{P} \text{ in K dividing } \mathfrak{p}\}$$

has density $\frac{|C|}{|G|}$.

The analytic continuation and functional equation of the Artin L-function are more difficult. The famous conjecture of Artin is still open which says that if the representation ρ is irreducible and non-trivial then $L(s, \rho)$ is entire in s. Conjectures of Stark relate values of derivatives of L and units in K. One of the most useful facts about these Artin L-functions is that they provide a factorization of the Dedekind zeta function

$$\zeta_K(s) = \prod_{\rho \in \hat{G}} L(s, \rho)^{d_\rho}.$$

Here the product over ρ runs over $\hat{G}$ a complete set of inequivalent irreducible unitary representations of G.

There are hybrid Artin-Selberg L-functions studied by Venkov [24], Venkov and Zograf [26]. See Efrat [3] for an application to the factorization of Selberg's zeta function for the modular group and the continued fraction dynamical system.

2. Zeta and L-Functions of Graphs - the 1 Variable Case.

Our subject is the graph theoretic analogue of the Artin L-function. References are Bass [1], Ihara [8], Hashimoto [6], Stark and Terras [18], [19], Sunada [20], Venkov and Nikitin [25]. Consider a connected finite (not necessarily regular) graph with vertex set X and undirected edge set E. For an example look at the tetrahedron or the cube illustrated in Figure 1. We orient the edges of X and label them $e_1, e_2, ..., e_{|E|}, e_{|E|+1} = e_1^{-1}, ..., e_{2|E|} = e_{|E|}^{-1}$. Here the inverse of an edge is the edge taken with the opposite orientation. A **prime** $[C]$ in X is an equivalence class of tailless backtrackless primitive paths in X. Here write $C = a_1 a_2 \cdots a_s$, where a_j is an oriented edge of X. The **length** $\nu(C) = s$. **Backtrackless** means that $a_{i+1} \neq a_i^{-1}$, for all i. **Tailless** means that $a_s \neq a_1^{-1}$. The **equivalence class** of C is $[C] = \{a_1 a_2 \cdots a_s, a_s a_1 a_2 \cdots a_{s-1}, ..., a_2 \cdots a_s a_1\}$; i.e., the same path with all possible starting points. We call the equivalence class **primitive** if $C \neq D^m$, for all integers $m \geq 2$, and all paths D in X. Examples of primes in the tetrahedron and cube are given in Figure 2. On the top right is the prime $[P] = \{abc, bca, cab\}$ in the tetrahedron. The length of this prime is $\nu(P) = 3$.

DEFINITION 1. *The **Ihara zeta function** of X is defined for $u \in \mathbb{C}$ with $|u|$ sufficiently small by*

$$\zeta_X(u) = \prod_{\substack{[C] \ prime \\ in \ X}} (1 - u^{\nu(C)})^{-1}.$$

THEOREM 1. *(Ihara). If A denotes the adjacency matrix of X and Q the diagonal matrix with jth entry $q_j =$ (degree of the jth vertex - 1), then*

$$\zeta_X(u)^{-1} = (1 - u^2)^{r-1} \det(I - Au + Qu^2).$$

Here r denotes the rank of the fundamental group of X. That is $r = |E| - |V| + 1$.

For regular graphs, when $Q = qI$ is a scalar matrix, you can prove this theorem using the Selberg trace formula for the $(q + 1)$-regular tree T. See Terras [22] or Venkov and Nikitin [25]. Here $X = T/\Gamma$, where Γ denotes the fundamental group of X. In the general case there are many proofs. See Stark and Terras [18], [19] for some elementary ones.

Note that since the Ihara zeta function is the reciprocal of a polynomial, it has no zeros. Thus when discussing the Riemann hypothesis we consider only poles. When X is a finite connected $(q + 1)$-regular graph, there are many analogues of the facts about the other zeta functions. For any unramified graph covering (not necessarily normal, or even involving regular graphs) it is easy to show that the reciprocal of the zeta function below divides that above (see Stark and Terras [18]). The analogue of this for Dedekind zeta functions of extensions of number fields is still unproved. There are functional equations. Special values give graph theoretic constants such as the number of spanning trees. See the references mentioned at the beginning of this section for more details. There is an analogue of the Chebotarev density theorem (see Hashimoto [6]).

For example, when X is a finite connected $(q + 1)$-regular graph, we say that $\zeta_X(q^{-s})$ **satisfies the Riemann hypothesis** iff

(2.1)   *for* $0 < \operatorname{Re} s < 1$, $\quad \zeta_X(q^{-s})^{-1} = 0 \iff \operatorname{Re} s = \frac{1}{2}.$

It is easy to see that (2.1) is equivalent to saying that X is a **Ramanujan graph** in the sense of Lubotzky, Phillips, and Sarnak [13]. This means that when λ is an eigenvalue of the adjacency matrix of X such that $|\lambda| \neq q+1$, then $|\lambda| \leq 2\sqrt{q}$. Such graphs are optimal expanders and (when X is non-bipartite) the standard random walk on X converges extremely rapidly to uniform. See Lubotzky [11] or Terras [21] for more information. The statistics of the zeros of the Ihara zeta function of a regular graph can be viewed as the statistics of the eigenvalues of the adjacency matrix. This has been investigated for various families of Cayley graphs such as the finite upper half plane graphs. See Terras [22].

Lubotzky [12] has defined what it means for a finite irregular graph Y covered by an infinite graph X to be X-Ramanujan. It would be useful to reinterpret this in terms of the Ihara zeta function of the graph.

In Table 1 below it is assumed that our graphs are finite, connected and regular. Here "GUE" means that the spacing between pairs of zeros/poles is that of the eigenvalues of a random Hermitian matrix. See Katz and Sarnak [9], Table 2, where the table has as its third column the function field case for which all phenomena in column 1 hold for almost all curves. One should also make a column for Selberg zeta and *L*-functions of Riemannian manifolds. The table assumes the graph *X* is regular.

Table 1. A New Column for Table 2 in Katz and Sarnak [9].

type of ζ or L-function	number field	graph theoretic
functional equation	yes	yes many
spectral interpretation of 0's/poles	?	yes
RH	expect it	iff graph Ramanujan
level spacing of high zeros/poles GUE	expect it	depends on graph [23]

Next let us consider an **unramified finite covering** graph *Y* of *X*. We assume all graphs have no loops or multiple edges. In this case *Y* covers *X* means that there is a covering map $\pi : Y \longrightarrow X$ such that π is an onto graph map and for each $x \in X$ and each $y \in \pi^{-1}(x)$, the set of points adjacent to *y* in *Y* is mapped by π 1-1, onto the set of points adjacent to *x* in *X*. A *d*-**sheeted normal (unramified) graph covering** *Y* of *X* means that there are *d* graph automorphisms $\sigma : Y \longrightarrow Y$ such that $\pi\sigma(y) = \pi(y)$, for all $y \in Y$. Then $Gal(Y/X)$, the **Galois group** of Y/X, is the set of all these $\sigma's$.

An example is to be found in Figure 1. The cube is a quadratic cover of the tetrahedron. The dotted lines represent a spanning tree in the tetrahedron. The cube has 2 copies of this tree forming the 2 **sheets** of the covering.

In Figure 1 we label the vertices of the cube x' or x'', where $x \in \{1, 2, 3, 4\}$ is the label of the corresponding vertex on K_4 The directed edges are labeled with lower case letters. The corresponding capital letter will denote the inverse edge. The non-trivial element of $Gal(Cube/K_4)$ sends x' to x''.

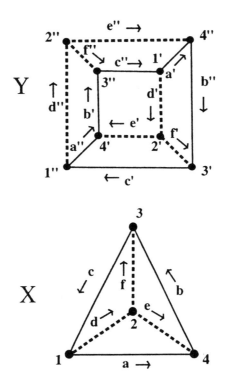

FIGURE 1. The Cube Covers the Tetrahedron (alias K_4).

In general one obtains d-sheeted unramified normal covers Y of X by taking d copies of the spanning tree for X; the vertices of Y are labeled (x, σ), $x \in X$, $\sigma \in Gal(Y/X)$. Then $\tau \in Gal(Y/X)$ acts on vertices of Y via $\tau(x, \sigma) = (x, \tau\sigma)$. And

$$\tau \,(\text{path from } (a, \lambda) \text{ to } (b, \sigma))$$

is the path from $(a, \tau\lambda)$ to $(b, \tau\sigma)$. In [19] we show that there are analogues of all the basic theorems of Galois theory. In particular, there is a 1-1 correspondence between subgroups H of $Gal(Y/X)$ and intermediate covers $\widetilde{X}$ to Y/X. Of course, the concept of intermediate cover needs a careful definition. For example, given H a subgroup of $Gal(Y/X)$, the points of the top graph Y have the form (x, σ), $x \in X$, $\sigma \in Gal(Y/X)$. The points of the intermediate graph $\widetilde{X}$ have the form $(x, H\sigma)$, $x \in X$, $H\sigma \in H\backslash G$. There is an edge between $(x, H\lambda)$ and $(u, H\sigma)$ in $\widetilde{X}$ iff there are h, h' in H such that $(x, h\lambda)$ and $(u, h'\sigma)$ have an edge between them in Y.

Next one finds analogues of the splitting of prime ideals in extensions.

DEFINITION 2. *Let Y/X be a d-sheeted unramified finite normal cover. Suppose that the projection map is $\pi : Y \longrightarrow X$ and that $[D]$ is a prime in Y. Then $\pi(D) = C^f$, where $[C]$ is a prime of X. Then we say $[D]$ is a **prime of Y over the prime** $[C]$ of X. We call $f = f(Y/X, D) = the **residual degree** of D with respect* to Y/X. And we set $g = g(Y/X, D) =$ *the number of primes* $[D]$ *in* Y *over* $[C]$ *in* X. In this paper, $e = e(Y/X, D) = ramification\ index = 1$.

Then, as in number theory, we have the formula

$$efg = d.$$

Example. Look at Figure 2. Consider the prime $[P]$ of length 3 in X (the tetrahedron) with representative $P = abc$. This is covered by a prime $[P']$ of length 6 in the cube.

First recall the labeling of the edges in the cube and tetrahedron given in Figure 1. Next note that in the top of Figure 2, there is only one prime cycle P' in the cube over the prime cycle P in the tetrahedron. For this case we have $f = 2$, since P' is the lift of P^2 and $g = 1$. We always have ramification index $e = 1$. In the bottom of Figure 2, there are two prime cycles R' and R'' in the cube over the prime cycle R in the tetrahedron. In this case $f = 1, g = 2, e = 1$.

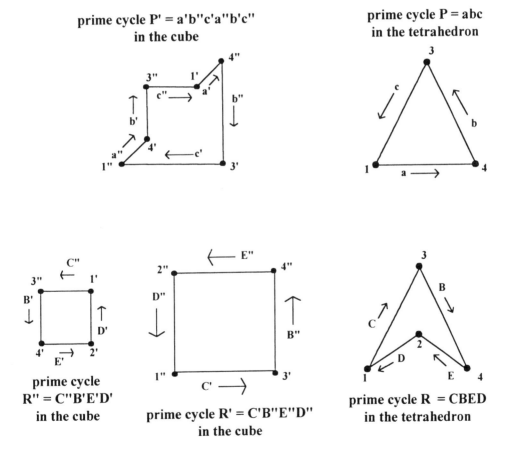

FIGURE 2. Primes in the Cube over Primes in the Tetrahedron.

DEFINITION 3. *Let* C *be a path in* X *that starts at vertex* $a \in X$. *Given* $\lambda \in Gal(Y/X)$ *the path* C *lifts to a unique path* $\widetilde{C}$ *in* Y *of the same length as* C *starting at* (a, λ); *suppose the lift ends at* (a, σ). *If* D *represents a prime of* Y *over*

C, then D is the unique lifting of C^f to Y. The **Frobenius automorphism** of the prime $[D]$ is

(2.2) $$[Y/X, D] = \sigma\lambda^{-1} = Frob(C) \in Gal(Y/X).$$

The **normalized Frobenius** $\sigma(C) = \sigma$ means that we take $\lambda = 1 =$ the identity in (2.2).

Example. Look again at Figure 2. Then the Frobenius automorphism

$$[Cube/K_4, P'] = \text{the non-trivial element of } Gal(Cube/K_4).$$

The Frobenius automorphisms

$$[Cube/K_4, R'] = [Cube/K_4, R''] = \text{identity in } Gal(Cube/K_4).$$

The proof of the following proposition is not hard. See Proposition 2 in [**19**].

PROPOSITION 1. **Some Properties of the Frobenius Automorphism.**
1) The Frobenius is independent of the choice of D in the equivalence class $[D]$.
2) If $\tau \in Gal(Y/X)$, then $[Y/X, \tau \circ D] = \tau[Y/X, D]\tau^{-1}$.

Now suppose that ρ is a representation of the Galois group $Gal(Y/X)$.

DEFINITION 4. *The **Artin L-function** for $u \in \mathbb{C}$ with $|u|$ sufficiently small is*

$$L(u, \rho) = \prod_{\substack{[C] \ prime \\ in \ X}} \det\left(1 - \rho\left([Y/X, D]\right) u^{\nu(C)}\right)^{-1},$$

where we recall (2.2) and the product is over primes $[C]$ of X, with $[D]$ any prime of Y over $[C]$.

THEOREM 2. **Some Properties of the Artin L-Function.**
1)
$$L(u, 1) = \zeta_X(u).$$
2)
$$\zeta_Y(u) = \prod_{\rho \in \hat{G}} L(u, \rho)^{d_\rho}.$$

Here the product is over all inequivalent irreducible unitary representations of the Galois group $G = Gal(Y/X)$.
3)
$$L(u, \rho)^{-1} = (1 - u^2)^{(r-1)d_\rho} \det(I - A_\rho u + Q_\rho u^2),$$
where I denotes the $|X|d_\rho \times |X|d_\rho$ identity matrix and $Q_\rho = Q \otimes I_{d_\rho}$, for Q the diagonal matrix with ath diagonal entry q_a such that $1 + q_a =$ degree of vertex $a \in X$. To define A_ρ, first define for $\sigma \in Gal(Y/X)$, the matrix $A(\sigma)$ to have a, b entry for $a, b \in X$,

$$A(\sigma)_{a,b} = \text{the number of edges in } Y \text{ between } (a, 1) \text{ and } (b, \sigma).$$

Here $1 =$ the identity of $G(Y/X)$. Then set

$$A_\rho = \sum_{\sigma \in Gal(Y/X)} A(\sigma) \otimes \rho(\sigma).$$

PROOF. The proofs of these things can be found in Stark and Terras [**19**].

$\square$

You can find analogues of all the properties of the usual Artin L-functions listed in Lang [10]. For example, one has the usual induction property for the induced representation of $Gal(Y/X)$ from a representation of a subgroup. It is this property which is essential for the proof of the preceding theorem. See Stark and Terras [19].

Example. For our example in Figure 2, the Galois group $G(Y/X) = \{1, \sigma\}$. We write $a' = (a, 1)$ and $a'' = (a, \sigma)$, for $a \in X$. The representations of the Galois group are $\widehat{G} = \{1, \rho\}$, where $\rho(\sigma) = -1$. Then the matrices are

$$A(1) = \begin{pmatrix} 0 & 1 & 0 & 0 \\ 1 & 0 & 1 & 1 \\ 0 & 1 & 0 & 0 \\ 0 & 1 & 0 & 0 \end{pmatrix} \quad and \quad A(\sigma) = \begin{pmatrix} 0 & 0 & 1 & 1 \\ 0 & 0 & 0 & 0 \\ 1 & 0 & 0 & 1 \\ 1 & 0 & 1 & 0 \end{pmatrix}.$$

So

$$A_1 = \begin{pmatrix} 0 & 1 & 1 & 1 \\ 1 & 0 & 1 & 1 \\ 1 & 1 & 0 & 1 \\ 1 & 1 & 1 & 0 \end{pmatrix} = \text{the adjacency matrix of } X$$

and

$$A_\rho = \begin{pmatrix} 0 & 1 & -1 & -1 \\ 1 & 0 & 1 & 1 \\ -1 & 1 & 0 & -1 \\ -1 & 1 & -1 & 0 \end{pmatrix}.$$

It follows that

$$\begin{aligned} \zeta_Y(u) &= L(u, \rho)\zeta_X(u), \\ L(u, \rho)^{-1} &= (1 - u^2)(1 + u)(1 + 2u)(1 - u + 2u^2)^3, \\ \zeta_X(u)^{-1} &= (1 - u^2)^2(1 - u)(1 - 2u)(1 + u + 2u^2)^3. \end{aligned}$$

You can view these formulas as providing a factorization of an 8×8 determinant as a product of two 4×4 determinants.

Irregular examples involving much larger determinants can also be worked out. See the end of Section 3 of Stark and Terras [19].

3. Multivariable Artin L-Functions of Graphs.

3.1. The Multiedge Artin L-Function of a Covering.

Surprisingly it simplifies the theory to introduce more variables than just the one variable u. Let X be our usual connected finite graph with edges labeled as above. For the multiedge L-function we need a $2|E| \times 2|E|$ matrix W called the **multiedge matrix** with ij entry the complex variable w_{ij} if the associated edges e_i and e_j have the ending vertex of e_i equal to the starting vertex of e_j and $e_j \neq e_i^{-1}$. Otherwise the ij entry $w_{ij} = 0$. We also set

(3.1) $$w(e_i, e_j) = w_{ij}.$$

Define W_1 to be the matrix obtained from W by setting all the non-0 variables equal to 1. Then the trace of W_1^n is the number of closed backtrackless tailless paths of length n in X. Compare this with the trace of A^n.

DEFINITION 5. *For the prime $[C]$ of X with $C = a_1 \cdots a_s$, using definition (3.1), define the **multiedge norm** of C to be*

$$N_E(C) = w(a_s, a_1) \prod_{i=1}^{s-1} w(a_i, a_{i+1}).$$

If ρ is a representation of $Gal(Y/X)$, we have the following multiedge Artin L-function.

DEFINITION 6. *The **multiedge Artin L-function** for the multiedge matrix W defined above and the representation ρ of $Gal(Y/X)$ is*

$$L_E(W, \rho) = \prod_{\substack{[C] \ prime \\ in \ X}} \det\left(1 - \rho\left([Y/X, D]\right) N_E(C)\right)^{-1},$$

where we recall (2.2) and the product is over primes $[C]$ of X, with $[D]$ any prime of Y over $[C]$.

Then we obtain the analogues of Theorem 2.

THEOREM 3. *Properties of the Multiedge Artin L-Function.*
1)
$$L_E(W, 1) = \zeta_E(W, X),$$
the multiedge zeta function of X, as considered in [18] and [19].
2)
$$\zeta_E(\widetilde{W}_{spec}, Y) = \prod_{\rho \in \widehat{G}} L_E(W, \rho)^{d_\rho}.$$

Here the product is over all inequivalent irreducible unitary representations of the Galois group $Gal(Y/X)$. The matrix $\widetilde{W}_{spec}$ is the specialized multiedge matrix. To define $\widetilde{W}_{spec}$, recall formula (3.1) and that the W variables of Y are $w_{ij} = w(e'_i, e'_j)$. where e'_j is a directed edge of Y. Then the specialization variables are $\widetilde{w}_{ij} = w(e_i, e_j)$, where $e_j = \pi(e'_j)$ denotes the edge of X projected down from the edge e'_j of Y using the covering map $\pi : Y \longrightarrow X$.
3)
$$L_E(W, \rho) = \det\left(I - W_\rho\right)^{-1},$$

where we define the $2|E|d_\rho \times 2|E|d_\rho$ matrix $W_\rho = (w_{ij}\rho(Frob(e_i)))$, using formula (2.2).

PROOF. See [19] for the proof.   □

One can prove part 3) of Theorem 2 using a determinant identity due to Bass and part 3) of Theorem 3. The induction property of the multiedge Artin L-function is essential for the proof of part 2) of 3 and it is the only tricky one to prove. Along the way one obtains the following proposition.

PROPOSITION 2. *Suppose Y/X is a finite unramified normal cover with Galois group G. If H is the subgroup of G corresponding to an intermediate cover $\widetilde{X}$, let $\chi_1^{\#}$ be the character of G induced from the trivial character of H. The the number of primes $[\widetilde{C}]$ of $\widetilde{X}$ above a prime $[C]$ of X with lengths $\nu(\widetilde{C}) = \nu(C)$ is $\chi_1^{\#}(\sigma(C))$, where $\sigma(C)$ is the normalized Frobenius of Definition 3.*

PROOF. See the proof of the corollary to Lemma 7 in [19].   □

Example. The Multiedge L-function of a Cube covering a Dumbbell.

See Figure 3 for the covering of the cube over the dumbbell. Note that we are allowing loops and multiple edges in this example. Thus our definition of covering needs to be more careful. See Stark and Terras [19]. The multiedge *L*-functions for the representations of the Galois group of Y/X, which is $\mathbb{Z}_4$, require the matrix W which has entries w_{ij}, when edge e_i feeds into edge e_j. For the labeling of the edges of the dumbbell, see Figure 3. We find that the matrix W is:

$$W = \begin{pmatrix} w_{11} & w_{12} & 0 & 0 & 0 & 0 \\ 0 & 0 & w_{23} & 0 & 0 & w_{26} \\ 0 & 0 & w_{33} & 0 & w_{35} & 0 \\ 0 & w_{42} & 0 & w_{44} & 0 & 0 \\ w_{51} & 0 & 0 & w_{54} & 0 & 0 \\ 0 & 0 & 0 & 0 & w_{65} & w_{66} \end{pmatrix}.$$

Next we need to compute $\sigma(e_i)$ for each edge e_i where $\sigma(C)$ denotes the normalized Frobenius automorphism of Definition 3. We will write the Galois group $G(Y/X) = \{\sigma_1, \sigma_2, \sigma_3, \sigma_4\}$, where $(x, \sigma_j) = x^{(j)}$, for $x \in X$. The identification of $G(Y/X)$ with $\mathbb{Z}_4$ sends σ_j to $(j - 1 (mod\ 4))$. Then compute the Galois group elements associated to the edges: $\sigma(e_1) = \sigma_2$, $\sigma(e_2) = \sigma_1$, $\sigma(e_3) = \sigma_2$. The representations of our group are 1-dimensional, given by $\chi_a(\sigma_b) = i^{a(b-1)}$, for $a, b \in \mathbb{Z}_4$.

So we obtain

$$L_E(W, \chi_0, Y/X)^{-1} = \zeta_E(W, X)^{-1}$$

$$= det \begin{pmatrix} w_{11} - 1 & w_{12} & 0 & 0 & 0 & 0 \\ 0 & -1 & w_{23} & 0 & 0 & w_{26} \\ 0 & 0 & w_{33} - 1 & 0 & w_{35} & 0 \\ 0 & w_{42} & 0 & w_{44} - 1 & 0 & 0 \\ w_{51} & 0 & 0 & w_{54} & -1 & 0 \\ 0 & 0 & 0 & 0 & w_{65} & w_{66} - 1 \end{pmatrix};$$

$$L_E(W, \chi_1, Y/X)^{-1} = det(I - W_{\chi_1})$$

$$= det \begin{pmatrix} iw_{11} - 1 & iw_{12} & 0 & 0 & 0 & 0 \\ 0 & -1 & w_{23} & 0 & 0 & w_{26} \\ 0 & 0 & iw_{33} - 1 & 0 & iw_{35} & 0 \\ 0 & -iw_{42} & 0 & -iw_{44} - 1 & 0 & 0 \\ w_{51} & 0 & 0 & w_{54} & -1 & 0 \\ 0 & 0 & 0 & 0 & -iw_{65} & -iw_{66} - 1 \end{pmatrix};$$

$$L_E(W, \chi_2, Y/X)^{-1} = det(I - W_{\chi_2})$$

$$= det \begin{pmatrix} -w_{11} - 1 & -w_{12} & 0 & 0 & 0 & 0 \\ 0 & -1 & w_{23} & 0 & 0 & w_{26} \\ 0 & 0 & -w_{33} - 1 & 0 & -w_{35} & 0 \\ 0 & -w_{42} & 0 & -w_{44} - 1 & 0 & 0 \\ w_{51} & 0 & 0 & w_{54} & -1 & 0 \\ 0 & 0 & 0 & 0 & -w_{65} & -w_{66} - 1 \end{pmatrix};$$

$$L_E(W, \chi_3, Y/X)^{-1} = det(I - W_{\chi_3})$$

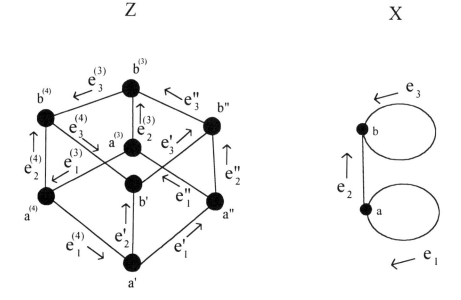

FIGURE 3. The Cube is a $\mathbb{Z}_4$ Covering of the Dumbbell.

$$= det \begin{pmatrix} -iw_{11}-1 & -iw_{12} & 0 & 0 & 0 & 0 \\ 0 & -1 & w_{23} & 0 & 0 & w_{26} \\ 0 & 0 & -iw_{33}-1 & 0 & -iw_{35} & 0 \\ 0 & iw_{42} & 0 & iw_{44}-1 & 0 & 0 \\ w_{51} & 0 & 0 & w_{54} & -1 & 0 \\ 0 & 0 & 0 & 0 & iw_{65} & iw_{66}-1 \end{pmatrix}.$$

Note that, by part 2 of Theorem 3, the product of the preceding four 6×6 determinants has to be a 24×24 determinant:

$$det(I - \widetilde{W}_{spec}) = \prod_{i=0}^{3} det(I - W_{X_i}),$$

where $\widetilde{W}_{spec}$ is given by specializing the multiedge matrix of the cube as follows. For the edge variables of the dumbbell, we write

$$a = w_{11}, b = w_{12}, c = w_{23}, d = w_{26}, e = w_{33}, f = w_{35},$$
$$g = w_{42}, h = w_{44}, j = w_{51}, k = w_{54}, m = w_{65}, n = w_{66}.$$

Using the natural ordering of the edges of Y given by

$e'_1, e''_1, e^{(3)}_1, e^{(4)}_1, e'_2, e''_2, e^{(3)}_2, e^{(4)}_2, e'_3, e''_3, e^{(3)}_3, e^{(4)}_3$, and the inverses in the same order.

the matrix $\widetilde{W}_{spec}$ is the following

$$\left(\begin{smallmatrix}
0 & a & 0 & 0 & 0 & b & 0 & 0 & 0 & 0 & 0 & 0 & 0 & 0 & 0 & 0 & 0 & 0 & 0 & 0 & 0 & 0 & 0 & 0 \\
0 & 0 & a & 0 & 0 & 0 & b & 0 & 0 & 0 & 0 & 0 & 0 & 0 & 0 & 0 & 0 & 0 & 0 & 0 & 0 & 0 & 0 & 0 \\
0 & 0 & 0 & a & 0 & 0 & 0 & b & 0 & 0 & 0 & 0 & 0 & 0 & 0 & 0 & 0 & 0 & 0 & 0 & 0 & 0 & 0 & 0 \\
a & 0 & 0 & 0 & b & 0 & 0 & 0 & 0 & 0 & 0 & 0 & 0 & 0 & 0 & 0 & 0 & 0 & 0 & 0 & 0 & 0 & 0 & 0 \\
0 & 0 & 0 & 0 & 0 & 0 & 0 & 0 & 0 & c & 0 & 0 & 0 & 0 & 0 & 0 & 0 & 0 & 0 & 0 & 0 & 0 & 0 & d \\
0 & 0 & 0 & 0 & 0 & 0 & 0 & 0 & 0 & 0 & c & 0 & 0 & 0 & 0 & 0 & 0 & 0 & 0 & 0 & 0 & d & 0 & 0 \\
0 & 0 & 0 & 0 & 0 & 0 & 0 & 0 & 0 & 0 & 0 & c & 0 & 0 & 0 & 0 & 0 & 0 & 0 & 0 & 0 & 0 & d & 0 \\
0 & 0 & 0 & 0 & 0 & 0 & 0 & 0 & 0 & 0 & 0 & 0 & c & 0 & 0 & 0 & 0 & 0 & 0 & 0 & 0 & 0 & d & 0 \\
0 & 0 & 0 & 0 & 0 & 0 & 0 & 0 & 0 & e & 0 & 0 & 0 & 0 & 0 & 0 & 0 & 0 & f & 0 & 0 & 0 & 0 & 0 \\
0 & 0 & 0 & 0 & 0 & 0 & 0 & 0 & 0 & 0 & e & 0 & 0 & 0 & 0 & 0 & 0 & 0 & 0 & f & 0 & 0 & 0 & 0 \\
0 & 0 & 0 & 0 & 0 & 0 & 0 & 0 & 0 & 0 & 0 & e & 0 & 0 & 0 & 0 & 0 & 0 & 0 & f & 0 & 0 & 0 & 0 \\
0 & 0 & 0 & 0 & 0 & 0 & 0 & 0 & e & 0 & 0 & 0 & 0 & 0 & 0 & 0 & f & 0 & 0 & 0 & 0 & 0 & 0 & 0 \\
0 & 0 & 0 & 0 & g & 0 & 0 & 0 & 0 & 0 & 0 & 0 & 0 & 0 & 0 & 0 & h & 0 & 0 & 0 & 0 & 0 & 0 & 0 \\
0 & 0 & 0 & 0 & 0 & g & 0 & 0 & 0 & 0 & 0 & 0 & 0 & h & 0 & 0 & 0 & 0 & 0 & 0 & 0 & 0 & 0 & 0 \\
0 & 0 & 0 & 0 & 0 & 0 & g & 0 & 0 & 0 & 0 & 0 & 0 & h & 0 & 0 & 0 & 0 & 0 & 0 & 0 & 0 & 0 & 0 \\
0 & 0 & 0 & 0 & 0 & 0 & 0 & g & 0 & 0 & 0 & 0 & 0 & 0 & h & 0 & 0 & 0 & 0 & 0 & 0 & 0 & 0 & 0 \\
j & 0 & 0 & 0 & 0 & 0 & 0 & 0 & 0 & 0 & 0 & 0 & 0 & 0 & 0 & 0 & k & 0 & 0 & 0 & 0 & 0 & 0 & 0 \\
0 & j & 0 & 0 & 0 & 0 & 0 & 0 & 0 & 0 & 0 & 0 & 0 & k & 0 & 0 & 0 & 0 & 0 & 0 & 0 & 0 & 0 & 0 \\
0 & 0 & j & 0 & 0 & 0 & 0 & 0 & 0 & 0 & 0 & 0 & 0 & k & 0 & 0 & 0 & 0 & 0 & 0 & 0 & 0 & 0 & 0 \\
0 & 0 & 0 & j & 0 & 0 & 0 & 0 & 0 & 0 & 0 & 0 & 0 & k & 0 & 0 & 0 & 0 & 0 & 0 & 0 & 0 & 0 & 0 \\
0 & 0 & 0 & 0 & 0 & 0 & 0 & 0 & 0 & 0 & 0 & 0 & 0 & 0 & 0 & 0 & 0 & m & 0 & 0 & 0 & 0 & 0 & n \\
0 & 0 & 0 & 0 & 0 & 0 & 0 & 0 & 0 & 0 & 0 & 0 & 0 & 0 & 0 & 0 & 0 & m & 0 & 0 & n & 0 & 0 & 0 \\
0 & 0 & 0 & 0 & 0 & 0 & 0 & 0 & 0 & 0 & 0 & 0 & 0 & 0 & 0 & 0 & 0 & 0 & m & 0 & 0 & n & 0 & 0 \\
0 & 0 & 0 & 0 & 0 & 0 & 0 & 0 & 0 & 0 & 0 & 0 & 0 & 0 & 0 & 0 & 0 & 0 & m & 0 & 0 & n & 0 & 0
\end{smallmatrix}\right).$$

3.2. The Multipath Artin L-Function of a Covering.

There is one final kind of Artin *L*-function - the multipath *L*-function invented by Stark in [**17**]. Recall that the fundamental group of X can be identified with that generated by edges left out of a spanning tree: $e_1, ..., e_r, e_1^{-1}, ..., e_r^{-1}$. Create the $2r \times 2r$ **multipath matrix** Z with ij entry the complex variable z_{ij} if $e_j \neq e_i^{-1}$ and $z_{ij} = 0$ otherwise. We also define

$$(3.2) \qquad\qquad z(e_i, e_j) = z_{ij}.$$

Note that the multipath matrix is smaller than the multiedge matrix W above but has fewer zeros.

Consider the prime $[C]$ of X with $C = a_1 \cdots a_s$, where $a_j \in \{e_1, ..., e_r, e_1^{-1}, ..., e_r^{-1}\}$ and C is a reduced product in the generators of the fundamental group. Here "reduced" means that $a_{j+1} \neq a_j^{-1}$, for all j, and $a_s \neq a_1^{-1}$.

DEFINITION 7. *Using definition (3.2), the **multipath norm** of the reduced prime* $C = a_1 \cdots a_s$ *is*

$$N_P(C) = z(a_s, a_1) \prod_{i=1}^{s-1} z(a_i, a_{i+1}).$$

Now we can define the multipath *L*-function just as we did the multiedge *L*-function.

DEFINITION 8. *Given a representation ρ of $Gal(Y/X)$, the **multipath Artin L-function** is*

$$L_P(Z,\rho) = \prod_{\substack{[C] \text{ prime} \\ \text{in } X}} \det\left(1 - \rho\left([Y/X, D]\right) N_P(C)\right)^{-1},$$

where we recall $[Y/X, D]$ is defined by (2.2), the multipath matrix Z is defined by (3.2), the multipath norm $N_P(C)$ is in Definition 7, and the product is over primes $[C]$ of X, with $[D]$ any prime of Y over $[C]$.

The multipath Artin L-function has analogous properties to the multiedge L-function. You just have to replace E with P in Theorem 3.

The interesting thing about the multipath L-function is that its variables can be specialized to obtain the multiedge L-function. Suppose that $e_1, ..., e_r$ are the (oriented) edges left out of a spanning tree T of X. The inverse edges are $e_{r+1}, ..., e_{2r}$. Give the edges of T a direction and label them $t_1, ..., t_{|X|-1}$. The inverse edges are $t_{|X|}, ..., t_{2|X|-2}$. If $e_i \neq e_j^{-1}$, write the part of the path between e_i and e_j as the (unique) product $t_{k_1}...t_{k_n}$. Then any prime $[C]$ of X is a product of the generators of the fundamental group of X corresponding to the e_j and thus a product of actual edges e_j and t_k. We now **specialize the multipath matrix Z to $Z(W)$** with entries given by the following polynomials in the W-variables, using (3.2):

$$z_{ij} = w(e_i, t_{k_1}) w(t_{k_n}, e_j) \prod_{\nu=1}^{n-1} w(t_{k_\nu}, t_{k_{\nu+1}}).$$

Via this specialization, we find that

(3.3) $$L_P(Z(W), \rho) = L_E(W, \rho).$$

Example. The Path Zeta Function of the Tetrahedron Specializes to the Edge Zeta Function of the Tetrahedron.

Refer to Figure 1 and label the inverse edges with the corresponding capital letters. List the edges that index the entries of the matrix Z as a, b, c, A, B, C. You will then find that the matrix $Z(W)$ for the tetrahedron is

$$\begin{pmatrix}
w_{aE}w_{ED}w_{Da} & w_{ab} & w_{aE}w_{Ef}w_{fc} & 0 & w_{aE}w_{Ef}w_{fB} & w_{aE}w_{ED}w_{DC} \\
w_{bF}w_{FD}w_{Fa} & w_{bF}w_{Fe}w_{eb} & w_{bc} & w_{bF}w_{Fe}w_{eA} & 0 & w_{bF}w_{FD}w_{DC} \\
w_{ca} & w_{cd}w_{de}w_{eb} & w_{cd}w_{df}w_{fc} & w_{cd}w_{de}w_{eA} & w_{cd}w_{df}w_{fB} & 0 \\
0 & w_{Ad}w_{de}w_{eb} & w_{Ad}w_{df}w_{fc} & w_{Ad}w_{de}w_{eA} & w_{Ad}w_{df}w_{fB} & w_{AC} \\
w_{BE}w_{ED}w_{Da} & 0 & w_{BE}w_{Ef}w_{fc} & w_{BA} & w_{BE}w_{Ef}w_{fB} & w_{BE}w_{ED}w_{DC} \\
w_{CF}w_{FD}w_{Da} & w_{CF}w_{Fe}w_{eb} & 0 & w_{CF}w_{Fe}w_{eA} & w_{CB} & w_{CF}w_{FD}w_{DC}
\end{pmatrix}.$$

As a check specialize all the variables to $u \in \mathbb{C}$ and call the new matrix $Z(u)$. Check that $\det(I - Z(u))$ is the reciprocal of the Ihara zeta function $\zeta_X(u)$.

References

[1] H. Bass, The Ihara-Selberg zeta function of a tree lattice, *Int. J. Math., 3* (1992), 717-797.

[2] H. Davenport, *Multiplicative Number Theory*, Springer-Verlag, N.Y., 1981.

[3] I. Efrat, Dynamics of the continued fraction map and the spectral theory of SL(2,$\mathbb{Z}$), *Inv. Math., 114* (1993), 207-218.

[4] J. Elstrodt, Die Selbergsche Spurformel für Kompakte Riemannsche Flächen, *Jber. d. Dt. Math. Verein., 83* (1981), 45-77.

[5] P.J. Forrester and A. Odlyzko, A nonlinear equation and its application to nearest neighbor spacings for zeros of the zeta function and eigenvalues of random matrices, *Proc. of the Organic Math. Workshop,* http://www.cecm.sfu.ca/~pborwein.

[6] K. Hashimoto, Artin type L-functions and the density theorem for prime cycles on finite graphs, *Intl. J. Math., 3* (1992), 809-826.

[7] D. Hejhal, The Selberg trace formula and the Riemann zeta function, *Duke Math. J. 43* (1976), 441-482.

[8] Y. Ihara, On discrete subgroups of the two by two projective linear group over p-adic fields, *J. Math. Soc. Japan, 19* (1966), 219-235.

[9] N. Katz and P. Sarnak, Zeros of zeta functions and symmetry, *Bull. A.M.S., 36, No. 1* (1999), 1-26.

[10] S. Lang, Algebraic number theory, Addison-Wesley, Reading, MA, 1968.

[11] A. Lubotzky, Discrete Groups, *Expanding Graphs and Invariant Measures,* Birkhäuser, Basel, 1994.

[12] A. Lubotzky, Cayley graphs, eigenvalues, expanders and random walks, in *Surveys in Combinatorics,* 1995, Ed. P. Rowlinson, Cambridge U. Press, Cambridge, 1995, pp. 155-189.

[13] A. Lubotzky, R. Phillips, and P. Sarnak, Ramanujan graphs, *Combinatorica, 8* (1988), 261-277.

[14] P. Sarnak, Arithmetic quantum chaos, *Israel Math. Conf. Proc., 8* (1995), 183-236 (published by the A.M.S.).

[15] P. Sarnak, Some problems in number theory, analysis and mathematical physics, preprint, 1999.

[16] H. M. Stark, Galois theory, algebraic number theory and zeta functions, in *From Number Theory to Physics,* M. Waldschmidt et al (Eds.), Springer-Verlag, Berlin, 1992.

[17] H. M. Stark, Multipath zeta functions of graphs, in *IMA Volumes in Math. and its Applications, Vol. 109, Emerging Applications of Number Theory,* Edited by D. A Hejhal et al. Springer-Verlag, N.Y., 1999, pp. 601-615.

[18] H. M. Stark and A. A. Terras, Zeta functions of finite graphs and coverings, *Adv. in Math., 121* (1996), 124-165.

[19] H. M. Stark and A. A. Terras, Zeta functions of finite graphs and coverings, Part II, *Adv. in Math.,* in press.

[20] T. Sunada, L-functions in geometry and applications, *Lecture Notes in Math., 1201,* Springer-Verlag, N.Y., 1986, 266-284.

[21] A. Terras, *Harmonic Analysis on Symmetric Spaces and Applications. Vol. I,* Springer-Verlag, N.Y., 1985.

[22] A. Terras, *Fourier Analysis on Finite Groups and Applications.* Cambridge U. Press, Cambridge, 1999.

[23] A. Terras, Statistics of graph spectra for some finite matrix groups: finite quantum chaos, to appear in *Proc. Internatl. Workshop on Special Functions,* Hong Kong, 1999.

[24] A.B. Venkov, *Spectral Theory of Automorphic Functions, Proc. Steklov Inst. of Math.* (1982). No. 4, Vol. 153, A.M.S., Providence, R.I., 1982.

[25] A.B. Venkov and A. M. Nikitin, The Selberg trace formula, Ramanujan graphs and some problems of mathematical physics, *Petersburg Math. J., 5,* No.3 (1994), 419-484.

[26] A.B. Venkov and P.G. Zograf, Analogues of Artin's factorization formulas in the spectral theory of automorphic functions associated with induced representations of Fuchsian groups. *Math. U.S.S.R.-Izv., 21* (1983), 435-443.

[27] M.- F. Vignéras, L'équation fonctionelle de la fonction zêta de Selberg de la groupe modulaire $PSL(2,\mathbb{Z})$, *Astérisque, 61* (1979), 235-249.

MATH. DEPT., U.C.S.D., LA JOLLA, CA 92093-0112

Current address: Math. Dept., U.C.S.D., La Jolla, CA 92093-0112
E-mail address: aterras@ucsd.edu
URL: http://www.math.ucsd.edu/~aterras

Selected Titles in This Series

(Continued from the front of this publication)

For a complete list of titles in this series, visit the
AMS Bookstore at **www.ams.org/bookstore/**.